PRINCIPLES AND TECHNIQUES OF HISTOCHEMISTRY

PRINCIPLES AND TECHNIQUES OF
HISTOCHEMISTRY

HENRY TROYER, Ph.D.

Associate Professor of Anatomy,
University of Missouri–Kansas City School of Medicine,
Kansas City, Missouri

LITTLE, BROWN AND COMPANY, BOSTON

QH
613
T76

Man has been truly termed a "microcosm" or a little world in himself, and the structure of his body should be studied not only by those who wish to become doctors, but by those who wish to attain a more intimate knowledge of God, just as close study of the niceties and shades of language in a great poem reveals more of the genius of its author.—GHAZZALI, Alchemy of Happiness

Science and art belong to the whole world, and national barriers vanish before them.—GOETHE

Right from the start I intended to show that histology . . . must be carried out on a chemical basis.—MIESCHER

FOREWORD

It has given me great pleasure to be associated, if only in a small way, with the production of this book.

The technology of histochemistry has made great strides, particularly during the past three decades. Its upward course has been charted by the appearance of many texts, in several different languages; the number of these books clearly reflects the importance of the subject and its utility to workers in different disciplines.

All the more, therefore, I appreciate this opportunity to welcome the latest addition to the list of books on histochemistry, and to write its foreword. It is a pleasant duty, the literary equivalent to breaking the bottle of champagne over the bow of a new vessel, and it provides me with a forum in which to express my belief that this book, written as it is in a frank, forthright, and stimulating manner, will prove a worthy successor to those works that have gone before and a good companion for those that have stood the test of time.

A. G. Everson Pearse

Histochemistry now consists of a large body of literature and offers investigators a wealth of methods for examining biological research problems. I have felt that it would be extremely useful to have a short textbook and laboratory manual that would give a newcomer a somewhat condensed overview of what histochemistry has to offer. This book, then, is intended for young investigators, graduate students, research technicians, and others who wish to acquaint themselves with some of the more common histochemical methods that might be applied to their research. In addition to giving the strengths and limitations of each method, I have always attempted to present the rationale for each step of the method. My aim was to assist the reader in understanding histochemistry rather than to present a series of "cookbook" methods.

I have tried to select methods according to three criteria: the usefulness of the method, the ease with which it can be carried out, and its merit (specifically, the dependability and reproducibility of the method). It was impossible for me to prove each of these methods personally; in many cases I relied on the recommendations of other very capable histochemists about certain methods. It was not difficult to decide which of the older methods should be included because time has a way of preserving the better methods while relegating the less worthy ones to oblivion. However, it was difficult to select the newer methods that should be included.

A book such as this would have been impossible without the very generous help and cooperation of certain people. I am much indebted to some of my former students, particularly Dr. Fred R. Nusbickel, for encouraging me to write it. I also appreciate the encouragement received from Professor A. G. Everson Pearse and Professor P. S. C. Bunning. I thank Dr. Carlin A. Pinkstaff, Dr. Thomas H. Rosenquist, and Dr. Theodore L. McClure for reading parts of the manuscript and for offering valuable and constructive criticism. I am also deeply grateful to Miss Katie Burns, whose skillful typing and editorial assistance greatly lightened my own burden during the final preparation of the manuscript, and to the administration of the Medical College of

Georgia School of Medicine, Augusta, Georgia, and Ahmadu Bello University, Zaria, Nigeria, for providing me time to write this book while I was serving on their faculties.

H. T.

CONTENTS

TABLE OF METHODS

INTRODUCTION TO HISTOCHEMISTRY

Histochemistry can be defined as the chemistry of tissue components and its relation to tissue morphology. The concern for precise chemical identification of tissue components is as great as the concern for accurate localization and distribution of the components.

RELATION OF HISTOCHEMISTRY TO OTHER DISCIPLINES

In order to appreciate the morphological and functional heterogeneity of cells and tissues, the histochemist must have a good background in histology. A sound knowledge of inorganic, organic, and biological chemistry is also essential in order to appreciate the chemical nature of the tissue components being studied and to understand the chemical reactions that form the bases of the histochemical methods. A sound background in chemistry and histology is therefore essential for histochemical studies to be maximally beneficial.

A primary difference between the disciplines of histology and histochemistry is that histology is primarily concerned with the study of microscopic morphology while histochemistry is concerned with the chemistry of cells and tissues related to their morphology. Histology is therefore primarily subserved by empirical staining methods while histochemistry is based on specific chemical reactions.

Biochemical methods may yield highly quantitative information concerning a particular organ or tissue, while histochemistry may yield primarily qualitative information concerning a particular group of cells within the same tissue. For example, a biochemical assay for a particular enzyme in the liver may show only that it was present in a relatively low concentration. A histochemical procedure may actually reveal a high concentration of the enzyme in the Kupffer cells and a total absence of the enzyme in the hepatocytes.

Histochemistry offers the only methods by which chemical components of tissues can be localized regionally within a particular tissue or at a cellular or subcellular level. The practice of histochemistry is based on the premise that there is merit in studying the cellular and subcellular distribution pattern of a given chemical component of tissue. This is based on a keen appreciation for the morphological heterogeneity between cells and the implied functional heterogeneity of the cells.

It is possible with histochemical methods to study cells and tissues that are too small to be studied by other methods. Histochemical methods can be applied to small surgical biopsies or very small tissues, such as the islets of Langerhans, that cannot easily be isolated from the surrounding tissues.

Some biologists tend to consider their newfound histochemical skills to be the ultimate in biological research, but this attitude must be guarded against. Histochemistry offers useful and interesting approaches, but it should not be thought of as having greater absolute value than other research disciplines.

The relationship between histochemistry and other disciplines is capably discussed by Vialli (1966).

ASPECTS OF HISTOCHEMISTRY

Localization and Quantitation

Histochemistry has sometimes been ignored as a modern research method because it is not primarily a quantitative method. However, once a chemical component has been accurately characterized and its distribution related to the morphology, then the urgency for quantitation fades somewhat. Localization and distribution are parameters as important as quantitation. Morphology has at least as important a place in science as quantitation, as has been powerfully emphasized by Elias (1971).

That is not to de-emphasize quantitation: Many histochemists practice some type of quantitation, although sometimes by rather crude methods. There is value in simply being able to say that a histochemical reaction is "strong" in one location as opposed to being "weak" in another. To be able to describe precisely where the reaction is strong or weak lends validity to this kind of quantitation. However, a system that expresses the strengths of reactions in terms of units of measure would obviously have much greater value.

Quantitative histochemical methods are inherently difficult to perform, not only from the standpoint of technical achievement, but also because it is sometimes difficult to find meaningful units in which to express quantitative histochemical results. Despite such difficulties significant progress has been made in recent years; continued efforts will be rewarding and must be encouraged.

Specificity

Some workers prefer to disregard histochemical procedures that are not absolutely specific. This is an unfortunate attitude; there are few histochemical methods that are absolutely specific and always reliable. The words of Adams (1965) seem relevant: "Preoccupation with devising histochemical methods that are 'absolutely specific' under all circumstances is about as realistic as the medieval search for the philosophers' stone." This is to say that care and vigilance must be exercised in interpreting histochemical results rather than abandoning all but foolproof methods.

Integration

Histochemistry offers something of a holistic approach to biology. Structure and function have too frequently been considered to be mutually exclusive disciplines. However, structure and function are indeed related and interdependent. This point has been forcefully made by Weibel (1971):

The order of the elements must be rigidly maintained if the living organism is to function properly, in other words, that the structure and function are strictly interdependent. Modern biology has come to learn that precisely prescribed structure is essential for life at the levels of molecules, of cells and of organs, as well as in the entire organism, and that deviations from these normal structures may lead to disease.

The Scope of Histochemistry

Histochemistry is concerned with the task of identifying chemical components of tissues in situ and relating that knowledge to tissue morphology. Any method that aids toward such an end should be considered to be within the realm of histochemistry. The following are some of the general methodological approaches to histochemistry:

(1) Some dyeing procedures stain certain tissue components selectively and therefore can be considered histochemical methods. An example is alcian blue, which can be made to stain only acid glycosaminoglycans. Many other dyeing procedures (for example, those employing eosin) stain tissue components rather indiscriminately and are therefore useful only as morphological stains.

(2) Some tissue components can be made to react with a nondye reagent, which can then react with another reagent to produce a colored final reaction product. A good example is the colloidal iron reaction with acid glycosaminoglycans. The colloidal form of iron has an affinity for the anionic groups of certain mucous substances. Iron so bound can be demonstrated with the prussian blue reaction.

(3) Tissue components can be modified by a chemical reaction such as oxidation or sulfation after which it will react with a color-producing compound. The periodic acid–Schiff and Feulgen reactions are good examples.

(4) The possibility of visualizing one of the products of an enzyme reaction gives rise to the broad area of enzyme histochemistry.

(5) Tissue components that are antigenic and that can be obtained in purified form can be demonstrated with fluorescein-labeled or enzyme-labeled antibody techniques.

(6) Molecules with radioactive atoms incorporated into them can be detected with autoradiographic techniques. The component is detected by means of a subatomic particle emitted during isotope disintegration that hits a photosensitive emulsion overlying the tissue section. The emulsion is developed, and the silver grains show the presumed sites of the molecule of interest.

(7) Microincineration permits the identification of minerals in minute volumes of tissue. This technique will no doubt give way to the much more sensitive technique of electron probe analysis. (Microincineration is not treated in this book but those seriously interested in this technique should read the article by Kruszynski [1966].)

REFERENCES

Adams, C. W. M. (1965).   Lipid histochemistry. *Adv. Lipid Res.* 7:1–62.

Elias, H. (1971).   Identification of structure by the common-sense approach. *J. Microsc. (Oxf.)* 95:59–68.

Kruszynski, J. (1966).   The microincineration technique and its results (Engl.). In M. Vialli, J. Kruszynski, and M. V. Mayersbach (Eds.), *Handbuch der Histochemie* (W. Graumann and K. Neumann, Series Eds.). Fischer Verlag, Stuttgart. Band I/2, pp. 96–187.

Vialli, M. (1966).   Histochemistry: Its general problems and its relationships
    to the other biological sciences (Engl.). In M. Vialli, J. Kruszynski, and
    M. V. Mayersbach (Eds.), *Handbuch der Histochemie* (W. Graumann and
    K. Neumann, Series Eds.). Fischer Verlag, Stuttgart. Band I/2, pp. 1–95.
Weibel, E. R. (1971).   The value of stereology in analysing structure and
    function of cells and organs. *J. Microsc. (Oxf.)* 95:3–13.

2

TISSUE FIXATION

PURPOSES OF FIXATION

The tissue section to be observed under the microscope should be as similar in appearance to the original living tissue as possible. The way the tissue is handled from the time it is removed from the body until it is ready for viewing under the microscope affects the preservation of its original detail. A piece of tissue that is left in the air following its removal will dry, resulting in gross distortion of its microscopic appearance. If it is placed in water, it will swell; if placed in a strong salt solution, it will shrink. Any of these mistreatments would cause serious distortion of the histology. Handling and processing tissues in an appropriate manner following their removal is therefore of utmost importance. Good discussions on fixation can be found in a paper by Hopwood (1969) and in a monograph edited by Stoward (1973). Dawson (1972) and McDowell and Trump (1976) reviewed fixation with particular reference to pathology.

Fixation is the first, and perhaps the most important, step in preparation of tissue for histochemical study. Fixation guards against four different forces that tend to alter the morphology of tissues and cells.

Prevention of Autolysis

Removing a tissue from its parent organism separates it from its source of oxygen and nutrition, which sets into motion certain degradative chemical processes. Most of the postmortem changes are probably due to the action of certain hydrolytic enzymes. Many hydrolases are always present in tissues under normal conditions but are neatly packed away in small organelles called lysosomes (these hydrolytic enzymes are therefore called lysosomal enzymes). Immediately following cellular death, the lysosomes seem to rupture and spill their enzymes. These enzymes are then free to begin disrupting the integrity of the cells.

Proper fixation prevents autolysis by two mechanisms. The lysosomal enzymes may be inactivated by the fixative, and the tissue components (upon which they would act) may be chemically altered so as to become unsusceptible to the enzymes.

Insolubilization of Tissues

Fixation of tissues renders insoluble certain tissue components that may otherwise leach out during subsequent handling. Lipids are the most difficult component to fix; they are nearly always dissolved during dehydration and clearing (see Chapter 3). No one fixative will insolubilize all tissue components. The component of interest can usually be preserved adequately for histochemical study through proper selection of the fixative. However, if certain tissue components are of no interest to the investigator, their loss may be of little consequence. (In the case of bone, for example, it is highly desirable to remove the mineral salts that are responsible for hardness. A decalcification step is therefore added in processing bone.)

Protection from Putrefaction

Fixation serves to protect the tissues from damage by microorganisms. It is hardly necessary to emphasize that a fresh piece of tissue stored in water or air at ambient temperatures is very susceptible to bacterial or fungal infection. This process is called putrefaction.

Protection from Damage

Fixation helps to protect the tissue from sustaining damage by subsequent preparative operations, such as dehydration, embedding, sectioning, and mounting. Fixation stabilizes the protein skeleton of the cell, giving the cells some structural support to resist deformation or crushing.

ASPECTS OF TIMING

There is an inevitable time lag between the moment a tissue is separated from its source of nutrients and oxygen, and the moment it is first exposed to the fixative. When this time lag is critical, it may be necessary to perfuse the animal, or at least the organ of interest, with the fixative through the blood vascular system. In this way, the tissue is exposed to the fixative at the moment when its nutrients and oxygen supply are cut off. Perfusion may be necessary when tissues with high metabolic rates are to be studied or when a fixative that penetrates slowly is used.

When a block of fresh tissue is placed in a fixative, the tissue at the surfaces of the block will be exposed to the fixative immediately. However, the tissue near the center will not be exposed until the fixative has had time to diffuse through to the center. The size of the tissue block and the rate at which the fixative diffuses through the tissue must both be taken into consideration for proper fixation. Obviously, when a rapidly diffusing fixative is used, larger blocks of tissue are permissible than when a slow-penetrating fixative is used. If the desired histochemical procedure dictates immediate fixation of all areas of the tissue, so that only a very short diffusion time lag can be tolerated, then small blocks of tissue must be used. It is also conceivable that the time during which any portion of the tissue is exposed to the fixative must be limited. In this case also, it would be imperative to limit the block size to reduce the time required for complete diffusion of the fixative. If a large face of tissue is desired, and diffusion of the fixative is a problem, the tissue may be cut as a thin slice of only a few millimeters. This allows rapid diffusion of the fixative into all parts of the tissue, and sections can be cut on the broad face.

The aim in fixing a tissue for enzyme histochemistry is to preserve both enzyme activity and morphology as faithfully as possible. In general, however, the more faithfully a fixative preserves morphology, the more deleterious it is toward enzymatic activity. It is therefore necessary to make a careful choice of the fixative for a particular research project. (This problem is discussed in more detail in Chapter 11.)

SPECIFIC FIXATIVES
It is of great interest for the histochemist to know as accurately as possible what chemical reactions take place between the tissue component and the fixative. Emphasis will therefore be placed on the mechanism of each fixative as far as it is understood.

Formaldehyde
Formaldehyde is an important fixative with widespread use in histology and histochemistry. Because it has been in use much longer than other aldehydes, more is known about it.

Formaldehyde is a rather volatile gas (boiling point, −21°C) whose vapors can be very irritating. It is very soluble in water and it can be obtained commercially as a 37 to 40 percent solution, known as formalin. When formalin is used as a fixative, it is normally diluted with water or buffer in a ratio of one to nine. Its strength can then be expressed in terms of a percentage of formalin or of formaldehyde. A one-to-nine dilution can properly be called 10 percent formalin or 4 percent formaldehyde.

Formaldehyde exists largely as a polymer in a concentrated aqueous solution—that is, 37 percent formaldehyde. Since only the monomeric form is useful as a fixative, full-strength formaldehyde would be almost useless. When formalin is diluted some depolymerization takes place. Ten percent aqueous formalin tends to be rather acidic (pH 2 to 4) and depolymerization is slow, but at a neutral pH depolymerization is rapid. Therefore, a neutral buffered formaldehyde solution is desirable.

The acidity of formaldehyde solutions results from the presence of small amounts of formic acid, an oxidation product of formaldehyde. Methanol prevents the oxidation of formaldehyde, and for this reason commercial formalin contains 10 to 15 percent methanol as a preservative.

The monomer of formaldehyde in an aqueous solution is probably always in the hydrated form—methylene glycol, $CH_2(OH)_2$—and this should be considered the functional form. Because of its relatively low molecular weight, it can penetrate tissues rather rapidly.

In some applications, the methanol preservative of commercial formalin may be undesirable or altogether unacceptable (Hayashi and Freiman, 1966). A methanol-free formaldehyde solution can be made from paraformaldehyde. Paraformaldehyde—the dry, highly polymerized form of formaldehyde—comes as a white powder. It can be brought into solution by adjusting a paraformaldehyde suspension to pH 7.2 to 7.4 and heating it to 60°C. Obviously, lacking the preservative methanol, it will oxidize readily to formic acid. It is preferable to make a fresh solution for each use. If the solution has been stored, its pH should be checked. Decreased pH indicates deterioration.

Formaldehyde functions as a fixative primarily by reacting with proteins. The types of reactions are numerous, but nearly all result in covalent bridges between polypeptides. French and Edsall (1945) summarized the reactions that can occur between formaldehyde and proteins. In general, formaldehyde attacks any side group that contains an active hydrogen. This constitutes an addition reaction, resulting in a hydroxymethyl group in place of the original hydrogen atom:

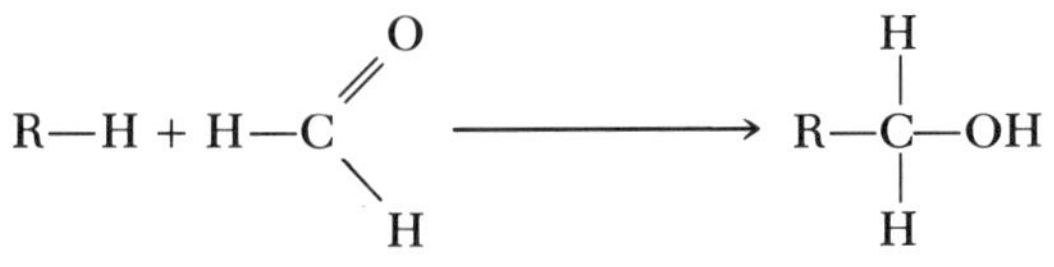

$$R{-}H + H{-}C{\overset{\displaystyle O}{\underset{H}{<}}} \longrightarrow R{-}\underset{\underset{\displaystyle H}{|}}{\overset{\overset{\displaystyle H}{|}}{C}}{-}OH$$

The hydroxymethyl groups so formed can react further with another nearby active hydrogen by a condensation reaction to form a methylene bridge between the polypeptide chains:

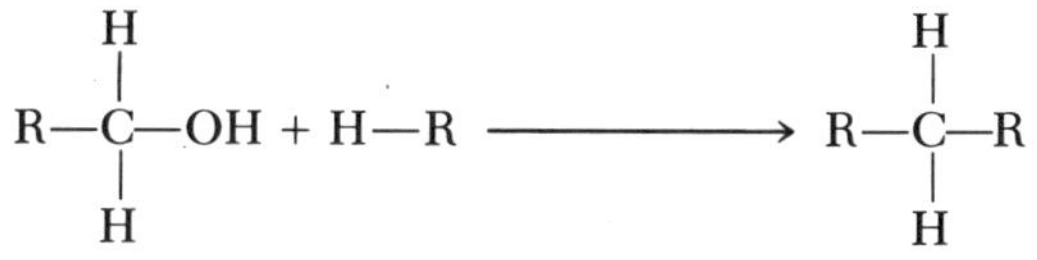

$$R{-}\underset{\underset{\displaystyle H}{|}}{\overset{\overset{\displaystyle H}{|}}{C}}{-}OH + H{-}R \longrightarrow R{-}\underset{\underset{\displaystyle H}{|}}{\overset{\overset{\displaystyle H}{|}}{C}}{-}R$$

The side groups that are attacked by formaldehyde are shown in Table 2-1.

Disulfide linkages may be reduced by formaldehyde to form two sulfhydryl groups. The newly formed sulfhydryl groups can in turn react with formaldehyde to participate in methylene linkages.

In order to achieve maximum fixation with formaldehyde, it is necessary to fix for several days. Long periods of fixation are necessary because the reactions with some of the side groups shown in Table 2-1 are rather slow; other reactions may be somewhat rapid. The only reaction known to be very rapid is the one involving aromatic groups. The condensation reactions are thought to be generally slower than the addition reactions. In many histochemical applications, particularly for enzyme localization, the preferred fixation time is far short of that required for maximum fixation.

An important consideration with formaldehyde is that fixation can be reversed by washing (Flitney, 1965). This reversal occurs somewhat slowly and continues for many days. If storage of tissues is necessary, it is better to store them in the fixative rather than in buffer or saline.

Table 2-1. Protein Side Groups That Will React With Formaldehyde

Protein Group	Formula
Amino (lysine)	$-NH_2$
Amido (glutamine, asparagine)	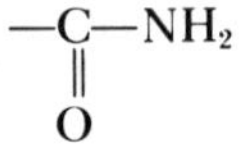 $-\overset{\displaystyle }{\underset{\displaystyle O}{C}}-NH_2$ (C double-bonded to O)
Imidazole (histidine)	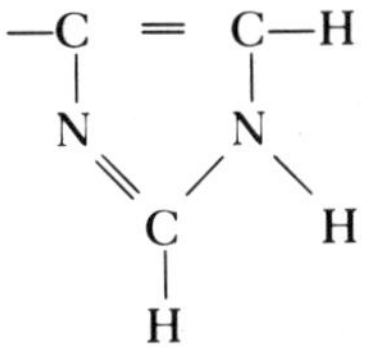 imidazole ring structure
Guanidino (arginine)	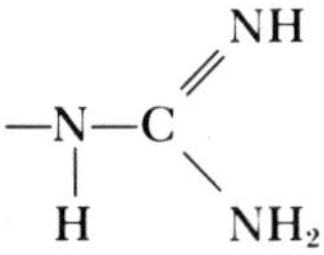 guanidino structure with NH, $-N-C$, H, NH_2
Hydroxyl (serine, threonine)	$-OH$
Carboxyl (aspartic acid, glutamic acid)	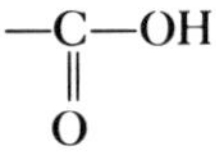 $-\underset{O}{C}-OH$ (C double-bonded to O)
Sulfhydryl (cysteine)	$-SH$
Aromatic groups (phenylalanine, tyrosine, tryptophan)	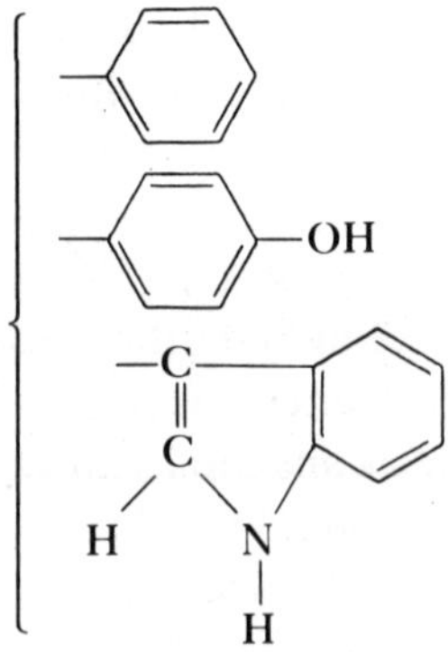 aromatic ring structures
Peptide linkage	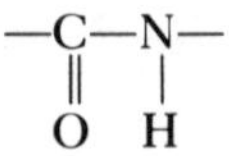 $-\underset{O}{C}-\underset{H}{N}-$ (C double-bonded to O)

Glutaraldehyde

Glutaraldehyde is a five-carbon straight chain dialdehyde with a molecular weight of 100. It is obtainable as a 25 or 50 percent solution. It does not have an unpleasant odor and is not irritating to work with, although it may cause some numbness (fixation) of the hands and fingers after prolonged exposure. Glutaraldehyde was introduced fairly recently as a biological fixative (Sabatini et al., 1963), but it has already been subjected to considerable investigation. It is used very extensively in electron microscopy because of its ability to preserve ultrastructure. It is not widely used in light microscopy because it does not readily penetrate large blocks of tissue.

Glutaraldehyde has been used as a 6 percent solution, and even higher concentrations have been used. However, concentrations higher than 5 percent may cause shrinkage, presumably from osmotic effects (Chambers et al., 1968). For electron microscopy it is typically used as a 3 percent solution in cacodylate buffer, at pH 7.2 to 7.4.

Certain lots of glutaraldehyde may contain certain undesirable impurities. The nature of these impurities has not always been clear, but the tendency to form polymers is well known (Bowes and Cater, 1966). Glutaraldehyde has absorption maxima at 280 and 235 nanometers; the former represents monomeric glutaraldehyde and the latter a contaminant. The concentration of impurities increases with age.

The problems of stability and purity, as well as methods of purification, have been the subject of a number of studies (Fahimi and Drochmans, 1968; Hopwood, 1967; Anderson, 1967; Gillett and Gull, 1972). The purity of glutaraldehyde can be improved by treating it with charcoal, but vacuum distillation (Anderson, 1967) or fractionation on Sephadex G-10 (Hopwood, 1967) is better. On the basis of spectral analysis, the primary contaminant is a glutaraldehyde polymer (Gillett and Gull, 1972). The oxidative product, glutamic acid, is an insignificant contaminant.

Glutaraldehyde fixes tissues in somewhat the same manner as formaldehyde does; that is, it forms crosslinks between polypeptide chains. If the glutaraldehyde monomer is primarily involved (as thought by Hardy et al. [1969]), the reaction may occur in the following manner:

$$2R-NH_2 + \begin{array}{c} O \\ \diagdown \\ C-(CH_2)_3-C \\ \diagup \\ H \end{array} \begin{array}{c} O \\ \diagup \\ \diagdown \\ H \end{array} \longrightarrow R-N-\underset{\underset{H}{|}}{\overset{\overset{OH}{|}}{C}}(CH_2)_3-\underset{\underset{H}{|}}{\overset{\overset{HO}{|}}{C}}-\underset{\underset{H}{|}}{\overset{\overset{H}{|}}{N}}-R$$

There is no agreement, however, that the monomer is the most active form; some authors feel that the polymeric form fixes more efficiently (Robertson and Schultz, 1970).

The side groups that are attacked by glutaraldehyde are approximately the same ones that are attacked by formaldehyde. However, glutaraldehyde forms crosslinks much more rapidly than formaldehyde (Feder and O'Brien, 1968). Flitney (1965) estimated that glutaraldehyde acts 100 times as fast as formaldehyde in fixing albumin in model experiments. Also, a larger number of crosslinks are produced than with most other aldehydes, with the exception of acrolein (Bowes and Cater, 1966). This could probably be explained in terms of greater spatial freedom of glutaraldehyde; that is, as soon as one end of the molecule (or polymer) has reacted with a polypeptide chain, the other end can reach through some distance in space to react with a site on a second polypeptide chain. Formaldehyde, by contrast, is very limited in this respect because of its one-carbon chain length.

The reactions of glutaraldehyde with proteins are listed by Richards and Knowles (1968) and are summarized as follows: (1) The reaction with protein is rapid in an aqueous medium at room temperature. (2) The reaction is apparently irreversible. (3) In some cases the crystalline structure of protein is not disordered. (4) There is a partial loss of lysine. (5) No new Ninhydrin-positive material appears on the standard amino acid chromatogram of the treated protein hydrolysate. (6) A comparison of the hydrogen ion titration curve reveals a shift toward greater acidity during the treatment of proteins.

Richards and Knowles (1968) showed with nuclear magnetic resonance studies that glutaraldehyde existed in aqueous solution primarily as polymers. Their studies indicated that large amounts of α,β-unsaturated aldehydes were present, which apparently formed as a result of aldol condensation. Richards and Knowles believe that only dimers and higher oligomers form crosslinks between proteins by the formation of a double Schiff base. Empirical evidence has shown that unpurified, polymerized glutaraldehyde preserves ultrastructure better than purified glutaraldehyde (Robertson and Schultz, 1970). The involvement of oligomers in protein crosslinking has been disclaimed by Hardy et al. (1969), who felt that only monomeric glutaraldehyde was involved. Gillett and Gull (1972) also felt that purified monomeric glutaraldehyde was a better fixative than the polymeric form. However, they evidently based their judgment on the fact that enzymes were better preserved by the monomeric form, which is reasonable if fewer crosslinks are formed.

The chemistry of glutaraldehyde fixation involves two similar events, both rapid addition reactions, in the formation of each crosslink. This, no doubt, contributes to the rapidity and efficiency of fixation. Formaldehyde, by contrast, depends on two dissimilar events, a rapid addition reaction and a slow condensation reaction.

Glutaraldehyde fixation can be reversed only slightly by prolonged washing. Fixation seems to be permanent, and storage of tissues thus fixed is not critical.

In comparison with formaldehyde, glutaraldehyde penetrates more slowly. Experiments by Chambers et al. (1968) indicated that glutaraldehyde penetrated liver at a maximum rate of 4.5 millimeters at room temperature and 2.5 mm at 4°C in 24 hours. Formaldehyde penetrated 30 to 100 percent further under comparable conditions. Glutaraldehyde, once in contact with tissue, fixes very rapidly compared with formaldehyde. Flitney (1965) has shown that brief exposure of albumin-gelatin films to glutaraldehyde prevents their elution, whereas exposure to formaldehyde, even for prolonged periods, is relatively ineffective. Although the action of glutaraldehyde as a fixative is efficient, the rate of penetration is slow. This points up the need to use small pieces of tissue when one is fixing with glutaraldehyde.

Since glutaraldehyde is a dialdehyde, it may leave a considerable number of free aldehyde groups in the tissue. When the tissue is stained with a procedure that involves the aldehyde groups, erroneous results may be obtained. This is particularly true in use of the periodic acid–Schiff (PAS) or the Feulgen reaction. The free aldehydes, however, can be either destroyed with a mixture of concentrated glacial acetic acid and aniline oil or blocked with phenylhydrazine or borohydride prior to the histochemical procedure.

For a more thorough discussion of glutaraldehyde, the reader is referred to the excellent article by Hopwood (1972).

Acrolein

Acrolein is an α,β-unsaturated aldehyde of low molecular weight (mol wt = 56). It has the following structure:

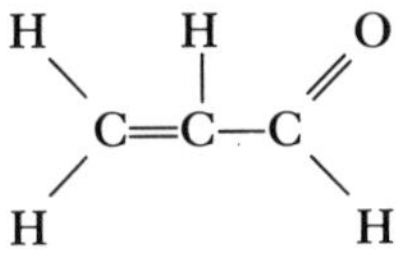

$$
\begin{array}{ccc}
H & H & O \\
\diagdown & | & \diagup\diagup \\
& C{=}C{-}C & \\
\diagup & & \diagdown \\
H & & H
\end{array}
$$

Like formaldehyde, acrolein is a gas at ambient conditions. Its fumes are toxic and extremely irritating, and it should therefore be used in a hood. It is very reactive, penetrates tissues rapidly, and causes very little shrinkage (Luft, 1959). As a fixative, it is generally considered to be superior to formaldehyde and equal to or better than glutaraldehyde (Hündgen, 1968; Jones, 1972). In a penetration experiment, Saito and Keino (1976) found that under similar conditions 2 percent acrolein fixes liver to a depth of 1 mm, whereas glutaraldehyde fixes to a depth of 0.4 mm. Like other aldehydes, acrolein reacts with tissue proteins to produce rigidity in the proteinaceous framework. Flitney (1965) reported that acrolein was superior to 6 other aldehydes in retaining albumin in sections of gelatin. Its superiority as a fixative probably can be explained by the fact that it produces more crosslinks than other aldehydes (Bowes and Cater, 1965). Acrolein reacts rapidly with unsaturated free fatty acids as well as with proteins; it reacts somewhat more slowly with esterified fatty acids (Jones, 1972). However, it has little power to retain lipids in tissues. Hündgen (1968) studied its effects on phosphatases and found it quite satisfactory in most cases. Flitney (1965) even found it possible to demonstrate succinic dehydrogenase if the fixation time was held to 1 minute.

Acrolein should be considered a bifunctional fixative. Its ethylenic bond can react with sulfhydryl, amino, or imido groups, leaving the aldehyde free. Van Duijn (1961) was able to detect proteins histochemically by treating the tissue sections with Schiff's reagent following acrolein treatment.

It has been suggested that acrolein stands an equal chance with glutaraldehyde to become "the fixative of the future" (Jones, 1972). The popularity of acrolein may be impeded by the inconvenience of having to use it in a hood, but its speed and efficiency are certainly in its favor.

Alcohol and Acetone

In the early days of histochemistry, alcohol and acetone were used extensively as fixatives, but they were later replaced by aldehyde fixatives. With the advent of enzyme histochemistry, alcohol and acetone were found to be suitable for fixing tissues to be used for demonstrating enzymes.

Biochemists use alcohol and acetone to precipitate proteins out of solution. Proteins selectively precipitate at different concentrations of alcohol or acetone on the bases of the physiochemical properties of the proteins, such as their dielectric constants, size, conformation, and number of exposed hydrophobic or hydrophilic groups. Proteins thus precipitated may be redissolved in water or buffer. Some proteins with biological activity may, however, lose some of their activity in this process. It has been observed that alcohol or acetone precipitation at low temperatures may preserve biological activity better than if the precipitation took place at higher temperatures. These are all considerations apropos the use of alcohol and acetone as histochemical fixatives.

The classic explanation of the mechanism of fixation is that alcohol and acetone simply replace water in the tissue with the solvent. This would amount to dehydration of the tissue, although the mechanism is probably not that simple. Conformation of at least some proteins is undoubtedly altered in some way, which would explain the mildly altered physical properties that are sometimes observed. Certain intimate associations between proteins and lipids would surely be altered since both alcohol and acetone are powerful lipid solvents.

Cell morphology is poorly preserved with either alcohol or acetone. Cytoplasm will shrink considerably and may pile up against one side of the cell opposite the side at which the fixative enters. Nuclei may also be misshapen. Alcohol is therefore usually used in combination with other agents, such as in Rossman's, Gendre's, and Clark's fixatives. (See Table 2-2 for the composition of fixatives.) Glycogen is sometimes fixed with 80% alcohol, although freeze-substitution has largely supplanted use of alcohol for this purpose. Cold alcohol or acetone is useful in preserving most hydrolytic enzymes. Acetone is useful for preserving esterases, because alcohol inactivates some esterases.

Table 2-2. Fixatives Useful in Histochemistry

Components of Fixative	Quantity
Acetone	
100% acetone at 0–4C	
Ethanol	
80% ethanol at 0–4C	
Formalin	
Formaldehyde, 37% to 40%	10 ml
Distilled water	90 ml
Neutral buffered formalin	
Formaldehyde, 37% to 40%	10 ml
Phosphate buffer, 0.2M, pH 7.0	90 ml
Formol-saline	
Formaldehyde, 37% to 40%	10 ml
Sodium chloride	0.9 gm
Distilled water	90 ml
Formol-calcium	
Formaldehyde, 37% to 40%	10 ml
Calcium chloride	1.1 gm
Distilled water	90 ml
Formol-cetylpyridinium chloride (CPC)	
Formaldehyde, 37% to 40%	10 ml
Cetylpyridinium chloride	0.5 gm
Distilled water	90 ml
Formol-sucrose	
Formaldehyde, 37% to 40%	10 ml
Sucrose	7.5 gm
Phosphate buffer, 0.2M, pH 7.4	90 ml
Alcohol-formol-acetic acid	
Formaldehyde, 37% to 40%	10 ml
Ethanol, absolute	85 ml
Acetic acid	5 ml
Rossman's fixative	
100% ethanol saturated with picric acid	90 ml
Formaldehyde, 37% to 40%	10 ml
Gendre's fixative	
95% ethanol saturated with picric acid	90 ml
Formaldehyde, 37% to 40%	10 ml
Bouin's fixative	
Water saturated with picric acid	75 ml
Formaldehyde, 37% to 40%	20 ml
Glacial acetic acid	5 ml
Clark's fixative	
100% ethanol	75 ml
Glacial acetic acid	25 ml

Picric Acid

Picric acid, also known as trinitrophenol, is useful as a protein and glycogen fixative. Dry picric acid is very explosive, and picric acid is therefore stored as a moist solid. Care should be exercised to keep it from drying out.

Fixatives containing picric acid, particularly Bouin's fixative, are widely used as histological fixatives. Picric acid reacts with proteins to form picrates. Alcoholic formalin solutions containing picric acid (Rossman's fluid and Gendre's fluid) are excellent glycogen fixatives. Smitherman et al. (1972) have shown that Rossman's fluid was the best fixative for glycogen retention and preservation.

Picric acid makes a solution yellow and imparts its color to tissue blocks as well. The color can be removed by prolonged washing in water or alcohol. However, picric acid enhances staining in some cases and need not be removed entirely from the tissue.

Osmium Tetroxide

Osmium tetroxide, often incorrectly referred to as osmic acid, has been used as a fixative for many years. The importance of this fixative increased tremendously after it was observed that osmium tetroxide preserves and fixes cellular detail especially well for electron microscopy.

Wolman (1957) classified most reactions of osmium tetroxide into three categories:

(1) Reactions that cause black precipitates in aqueous solution. The well-known reaction with unsaturated lipids belongs in this group. Double bonds are oxidized to form osmium compounds as follows:

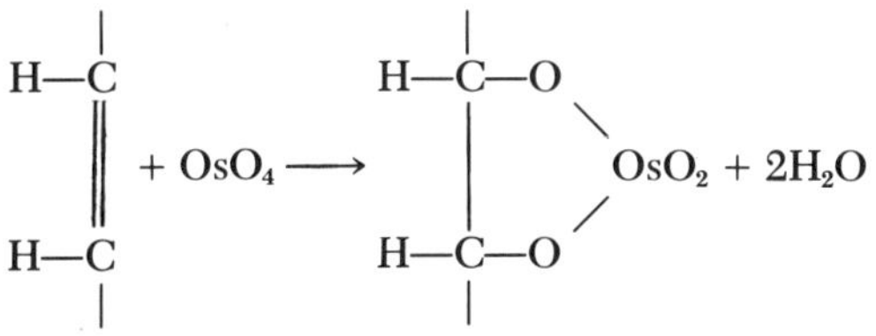

$$
\begin{array}{c}
\overset{|}{\underset{|}{\text{H—C}}} \\
\| \quad + \text{OsO}_4 \longrightarrow \\
\text{H—C}
\end{array}
\quad
\begin{array}{c}
\text{H—C—O} \\
\diagdown \\
\diagup \\
\text{H—C—O}
\end{array}
\text{OsO}_2 + 2\text{H}_2\text{O}
$$

The cyclic ester thus formed is unstable and is easily hydrolyzed.

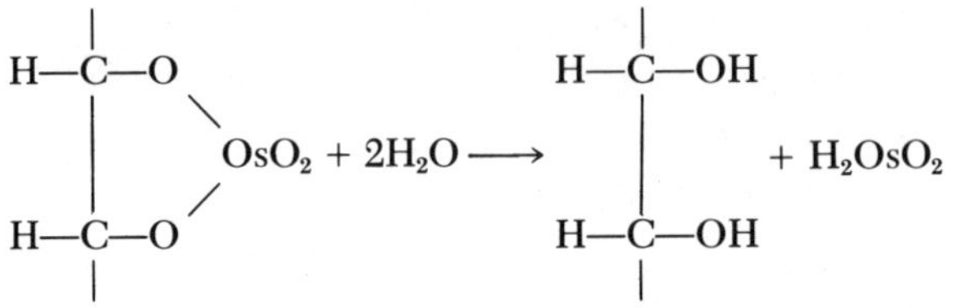

$$
\begin{array}{c}
\text{H—C—O} \\
\hspace{2em}\diagdown \\
\hspace{2em}\text{OsO}_2 + 2\text{H}_2\text{O} \longrightarrow \\
\hspace{2em}\diagup \\
\text{H—C—O}
\end{array}
\qquad
\begin{array}{c}
\text{H—C—OH} \\
\text{H—C—OH}
\end{array}
+ \text{H}_2\text{OsO}_2
$$

The resulting diol can react with another cyclic monoester.

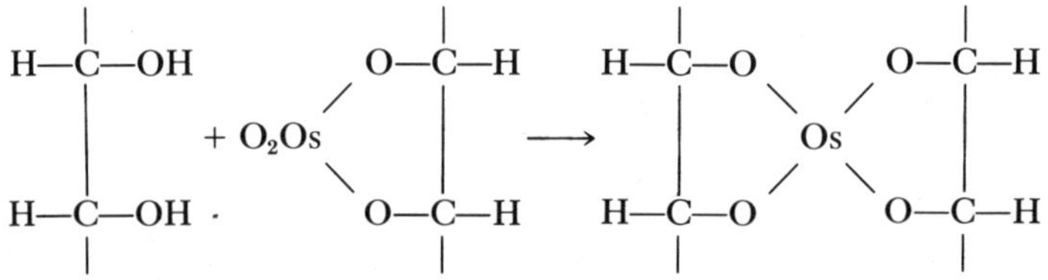

$$
\begin{array}{c}
\text{H—C—OH} \\
\hspace{2em}+ \text{O}_2\text{Os} \\
\text{H—C—OH}\ \cdot
\end{array}
\qquad
\begin{array}{c}
\text{O—C—H} \\
\text{O—C—H}
\end{array}
\longrightarrow
\begin{array}{c}
\text{H—C—O} \\
\hspace{2em}\diagdown\ \diagup \\
\hspace{2em}\text{Os} \\
\hspace{2em}\diagup\ \diagdown \\
\text{H—C—O}
\end{array}
\qquad
\begin{array}{c}
\text{O—C—H} \\
\text{O—C—H}
\end{array}
$$

Sulfhydryl groups of proteins may also be oxidized to form osmium precipitates.

(2) Reactions that cause black precipitates only in nonpolar solvents. The reaction with polyamine sugars, such as chitin and agar, fall in this category.

(3) Reactions that do not leave a black precipitate in aqueous solution. In contrast, the reactions in the first two categories result in precipitates that are visible in light and electron microscopy. The most important reaction in this category is the formation of aldehydes at 1,2-glycols, similar to the reaction caused by periodate or permanganate oxidation. Aldehydes thus generated can be readily demonstrated with Schiff's reagent.

Osmium tetroxide is ineffective in fixing most polysaccharides, proteins, or saturated lipids. Its significance in histochemistry lies in the selective staining of phospholipids in the osmium tetroxide alpha naphthylamine (OTAN) method (Adams, 1959) and in the staining of degenerating myelin with the Marchi methods. (See Adams and Bayliss [1975] for a short discussion.)

Diimidoesters

The use of diimidoesters as fixatives for histochemistry and electron microscopy was recently investigated by Hand and Hassell (1976).

Diimidoesters are nonaldehyde crosslinking agents that have been used by biochemists in studying tertiary protein structure (Wold, 1972; Hunter and Ludwig, 1972). They have the following general structure:

$$
\begin{array}{cc}
\overset{+}{NH_2} & \overset{+}{NH_2} \\
\| & \| \\
R-O-C-R' & -C-O-R''
\end{array}
$$

Unlike aldehydes, diimidoesters do not alter the net charge of proteins, and therefore they faithfully preserve the biological activity of proteins while also preserving morphology. Their reaction with proteins is highly selective compared with that of aldehyde fixatives. At a pH of 7.5, diimidoesters will react with all amino groups, and at a pH of 9.5 they will react only with ϵ amino groups. They can therefore be made to crosslink proteins across ϵ amino groups selectively.

Of the three diimidoesters investigated by Hand and Hassell (1976), the six-carbon, dimethyl suberimidate (DMS) was most effective in rendering rat liver proteins insoluble. Optimal fixation was obtained from a solution containing 16 to 20 mg DMS per ml buffered at pH 9.5 for 2½ hours at room temperature. The Ninhydrin-Schiff reaction for amino groups was diminished somewhat following DMS fixation, but the dihydroxy-dinaphthyl-disulfide reaction for sulfhydryl groups and the Feulgen and PAS reactions were not diminished significantly. Liver cells that were fixed with DMS stained more intensely for glycogen with PAS than those fixed with glutaraldehyde. Hand and Hassell (1976) also found good retention and localization of acid phosphatase, glucose 6-phosphatase, and catalase. The latter two enzymes are readily inactivated by aldehyde fixation. Although DMS requires further investigation, from preliminary data it would appear that it will probably earn itself a respectable place among histochemical fixatives.

One serious drawback with the diimidoesters is their instability, especially on exposure to atmospheric conditions. They should be protected against moisture by storing them in an evacuated desiccator in the cold. The free base is considered more stable than its hydrochloride derivative. Even under optimal storage conditions the free base should be used within a few months of its preparation; the hydrochloride derivative should be used within a few weeks. In aqueous solution the diimidoesters are much less stable; the half-life is probably only a few hours.

REFERENCES

Adams, C. W. M. (1959).   A histochemical method for the simultaneous demonstration of normal and degenerating myelin. *J. Pathol. Bacteriol.* 77:648–650.

Adams, C. W. M., and Bayliss, O. B. (1975).   Lipid Histochemistry. In D. Glick and R. M. Rosenbaum (Eds.), *Techniques of Biochemical and Biophysical Morphology.* Wiley-Interscience, New York. Vol. 2, pp. 99–156.

Anderson, P. J. (1967).   Purification and quantitation of glutaraldehyde and its effect on several enzyme activities in skeletal muscle. *J. Histochem. Cytochem.* 15:652–661.

Bowes, J. H., and Cater, C. W. (1965).   Crosslinking of collagen. *J. Appl. Chem.* 15:296–304.

Bowes, J. H., and Cater, C. W. (1966).   The reaction of glutaraldehyde with proteins and other biological materials. *J. Roy. Microsc. Soc.* 85:193–200.

Chambers, R. W., Bowling, M. C., and Grimley, P. M. (1968).   Glutaraldehyde fixation in routine histopathology. *Arch. Pathol.* 85:18–30.

Dawson, I. M. P. (1972).   Fixation: What should the pathologist do? *Histochem. J.* 4:381–385.

Fahimi, H. D., and Drochmans, P. (1968).   Purification of glutaraldehyde. Its significance for preservation of acid phosphatase activity. *J. Histochem. Cytochem.* 16:199–204.

Feder, N., and O'Brien, T. P. (1968).   Plant microtechnique: Some principles and new methods. *Am. J. Bot.* 55:123–142.

Flitney, F. W. (1965).   The time course of the fixation of albumin by formaldehyde, glutaraldehyde, acrolein and other higher aldehydes. *J. Roy. Microsc. Soc.* 85:353–364.

French, D., and Edsall, J. T. (1945).   The reactions of formaldehyde with amino acids and proteins. *Adv. Protein Chem.* 2:277–335.

Gillett, R., and Gull, K. (1972).   Glutaraldehyde—its purity and stability. *Histochemistry* 30:162–167.

Hand, A. R., and Hassell, J. R. (1976).   Tissue fixation with diimidoesters as an alternative to aldehydes. II. Cytochemical and biochemical studies of rat liver fixed with dimethylsuberimidate. *J. Histochem. Cytochem.* 24:1000–1011.

Hardy, P. M., Nicholls, A. C., and Rydon, H. N. (1969).   The nature of glutaraldehyde in aqueous solution. *Chem. Comm.* Pp. 565–566.

Hayashi, M., and Freiman, D. G. (1966).   An improved method of fixation for formalin-sensitive enzymes with special reference to myosin adenosine triphosphatase. *J. Histochem. Cytochem.* 14:577–581.

Hopwood, D. (1967).   The behaviour of various glutaraldehydes on Sephadex G-10 and some implications for fixation. *Histochemistry* 11:289–295.

Hopwood, D. (1969).   Fixatives and fixation: A review. *Histochem. J.* 1:323–360.

Hopwood, D. (1972).   Theoretical and applied aspects of glutaraldehyde fixation. *Histochem. J.* 4:267–303.

Hündgen, M. (1968).   Der Einfluss verschiedener Aldehyde auf die strukturerhaltung gezüchteter Zellen und auf die Darstellbarkeit von vier Phosphatasen. *Histochemistry* 15:46–61.

Hunter, M. J., and Ludwig, M. L. (1972).   Amidination. *Methods Enzymol.* 25B:585–596.

Jones, D. (1972).   Reactions of aldehydes with unsaturated fatty acids during histological fixation. *Histochem. J.* 4:421–465.

Luft, J. H. (1959).   The use of acrolein as a fixative for light and electron microscopy. *Anat. Rec.* 133:305.

McDowell, E. M., and Trump, B. F. (1976).   Histologic fixatives suitable for diagnostic light and electron microscopy. *Arch. Pathol. Lab. Med.* 100:405–414.

Richards, F. M., and Knowles, J. R. (1968).   Glutaraldehyde as a protein cross-linking reagent. *J. Mol. Biol.* 37:231–233.

Robertson, E. A., and Schultz, R. L. (1970).   The impurities in commercial glutaraldehyde and their effect on the fixation of brain. *J. Ultrastruct. Res.* 30:275–287.

Sabatini, D. D., Bensch, K., and Barrnett, R. J. (1963).   Cytochemistry and electron microscopy. The preservation of cellular ultrastructure and enzymatic activity by aldehyde fixation. *J. Cell Biol.* 17:19–58.

Saito, T., and Keino, H. (1976).   Acrolein as a fixative for enzyme cytochemistry. *J. Histochem. Cytochem.* 24:1258–1269.

Smitherman, M. L., Lazarow, A., and Sorenson, R. L. (1972).   The effect of light microscopic fixatives on the retention of glycogen in protein matrices and the particulate state of native glycogen. *J. Histochem. Cytochem.* 20:463–471.

Stoward, P. J. (1973).   *Fixation in Histochemistry.* Chapman and Hall, London.

Van Duijn, P. (1961).   Acrolein-Schiff, a new staining method for proteins. *J. Histochem. Cytochem.* 9:234–241.

Wold, F. (1972).   Bifunctional reagents. *Methods Enzymol.* 25B:623–651.

Wolman, M. (1957).   The reaction of osmium tetroxide with tissue components. *Exp. Cell Res.* 12:231–240.

3

TISSUE PROCESSING

DEHYDRATION AND EMBEDDING

For most histological and histochemical applications, thin sections about 5 to 10 microns in thickness are required. Monolayers of cultured cells and smears of blood and bone marrow are exceptions. Biological tissues are generally rather soft, making it quite difficult to cut acceptably thin sections directly from the fresh or fixed tissues. Methods must be used to hold the tissues firm, which facilitates cutting thin sections with a sharp knife. Firmness can be achieved either by embedding the tissues in a suitable embedment or by freezing the tissue.

In the early days of microscopy histologists tried to harden tissues artificially with fixatives, in order to be able to cut suitably thin sections for microscopy. Nearly 100 years ago, the method of embedding tissues in paraffin was developed; this represented a great step forward in microscopic technique. Firmness was achieved with a supporting medium (an embedment), rather than by hardening the tissue itself. For many years paraffin served as almost the only embedment. Most of our knowledge from microscopy has been gained from sections cut from paraffin-embedded tissues.

A few other media were developed, such as celloidin, gelatin, and carbowax, but they have found limited applications. For the vast majority of applications in histology and histochemistry, paraffin was adequate as an embedment and there was little incentive to refine embedding and sectioning techniques.

With the advent of electron microscopy, it was at once apparent that the available techniques for producing sections were utterly inadequate. Therefore, near the end of the 1940s there was suddenly a tremendous incentive to develop microtomy methods more suitable for electron microscopy. New synthetic polymers were developed for embedding media, microtomes were improved, the art of making and using glass knives was perfected, and diamond knives began to be available. Following those developments, histologists and histochemists soon learned to take advantage of the new and more sophisticated technology to obtain better-quality histological sections for light microscopy.

Perhaps the most important benefit to histologists and histochemists has been the development and perfection of glycol methacrylate (GMA) as an embedding medium. In 1960 Rosenberg et al. introduced GMA as an embedment for electron microscopy. It was later modified specifically for use in light microscopy by Ashley and Feder (1966), Ruddell (1967a), and Feder and O'Brien (1968). It has also been found useful for autoradiography (North, 1971; Rambourg et al., 1971), in immunohistochemistry (Hoshino and Kobayashi, 1972; Arnold et al., 1974), and in enzyme histochemistry (Feder, 1963; Cope, 1968; Hoshino and Kobayashi, 1971; Ashford et al., 1972; Weber, 1974; Troyer and Nusbickel, 1975). It will no doubt find much more extensive use in the future.

The following attributes must be considered in a good embedding medium: (1) It must be easily converted from a liquid to a solid after it has completely infiltrated a tissue. Solidification may be brought about by crystallization, hydrogen-bonding, or covalent bonding (polymerization). (2) During the liquid phase, it must be able to penetrate cells and organelles without undue difficulty. As pointed out by Baker (1966), the ease with which a liquid embedding medium penetrates a tissue depends not only on the intrinsic qualities of the medium but also on the type of fixative used. The liquid form should not react with cell components, and it should not dissolve or extract them. (3) The medium should not change appreciably in volume during solidification. (4) The medium should have good cutting qualities (homogeneity, hardness, plasticity, and elasticity). (5) The medium either must be easily removed from the tissue section prior to staining or must allow the staining reagents to penetrate without obstruction. (6) The medium should be cheap and readily available.

The technical details of the use of paraffin and GMA will be discussed here, paraffin because of its widespread use and GMA because of its promising future.

PROCESSING TISSUES WITH PARAFFIN
Paraffin is a derivative of crude petroleum. It is a group of variable-length, long-chain hydrocarbons of the methane series. Most paraffins suitable as embedding media melt between 52 and 58°C, which suggests an average chain length of somewhat less than 30 carbons (Baker, 1966).

While in its liquid phase, the embedding medium must be allowed to infiltrate the cells completely. Since most paraffins have a melting point between 52 and 58°C, it must infiltrate the cells while it is hot. Infiltration must be carried out at only a few degrees above the melting point of paraffin.

The critical shrinkage point of collagen is approximately 65°C. Exposure of collagenous tissues to this temperature must be carefully guarded against to avoid excessive shrinkage. Any amount of heat, however, is likely to cause a small amount of shrinkage.

In order to allow paraffin to infiltrate, water must first be completely removed from the cells (that is, the section must be dehydrated) because water and paraffin are completely immiscible. Water can be removed by replacing it with an organic solvent that is miscible with it in all proportions. Ethanol is very suitable and is normally used, but acetone is sometimes used. The organic solvent must replace the water gradually, however, because a great deal of local turbulence is generated at the interface between pure water and pure ethanol. This could cause damage or distortion to cellular components. However, if the tissue is first placed in 50 to 70 percent ethanol, the turbulence is not liable to be so severe as to cause damage or distortion. Another step of 95 percent ethanol (and sometimes also an intervening step of 85 percent ethanol) is inserted before absolute ethanol is used. The number of steps or the gradient differences should be determined by (1) the degree of fixation, (2) the delicacy of the tissue, and (3) the degree of cellular detail to be preserved.

However, ethanol and paraffin are still not miscible. Another organic solvent (generally called the clearing agent or the antimedium) must therefore be inserted between ethanol and paraffin. Xylene (also known as xylol) is most frequently used, but toluene, benzene, and chloroform have also been used. The ring compounds have a hardening effect on collagenous tissues, whereas chloroform does not have that effect, and it is sometimes favored for that reason.

Hot paraffin finally replaces the clearing agent and completely occupies the space formerly occupied by water. Several changes of pure hot paraffin are necessary in order to replace the clearing agent completely. If a trace amount of clearing agent is allowed to remain, it will prevent proper solidification of the paraffin, leading to difficulty in cutting sections.

Recently Paraplast Plus* was introduced on the market. The manufacturer claims that it requires one-third less time to infiltrate than ordinary paraffin. It contains dimethyl sulfoxide (DMSO), a compound with a very high dipole moment that acts to carry the paraffin molecules into the tissue.

The time required for embedding tissues using ethanol dehydration and xylene clearing usually exceeds eight hours. Normally, the tissue is processed overnight with an automatic tissue processing machine. In laboratories where paraffin embedding is done infrequently and where a tissue processing machine is unavailable, it would be desirable to streamline the processing procedure. Using a completely different rationale for dehydration, Prentø (1978) was able to reduce the time required for embedding fixed tissues to less than 3 hours, using far fewer steps. Dimethoxypropane (DMP) served as both the dehydrating and clearing agent. Acidified DMP (as used by Prentø) does not simply replace the water but chemically reacts with water to form methanol and acetone. The reaction is very rapid and endothermic. Therefore, by monitoring the drop in temperature and noting the subsequent equilibrium, it is simple to judge when the reaction is complete. Residual methanol and acetone can be removed with benzene or methyl salicylate.

After the block of tissue has been completely infiltrated with paraffin, it is placed in a mold containing hot paraffin and oriented in the desired manner. The paraffin is then allowed to solidify.

The conventional method for dehydrating, clearing, and embedding is subject to many variations among different laboratories. A typical schedule is given in Table 3-1. Each laboratory will modify the schedule to suit its own needs and preferences. A schedule for dehydrating and clearing with DMP is given in Table 3-2.

Upon solidifying, paraffin shrinks 16.5 percent in volume. Paraplast (the commercial name for purified paraffin) is supposed to shrink less because it consists of polymers whose molecular weight range is limited. It shrinks 14 percent by volume.

No doubt the two most objectionable aspects of paraffin as an embedding medium are the heat required for melting and the shrinkage upon solidification. Thus, it violates the third attribute of a good em-

*Sherwood Medical Industries, St. Louis, Missouri.

Table 3-1. Schedule for Embedding Tissues in Paraffin (Conventional Method)

Step	Reagent
1	70% ethanol
2	95% ethanol
3	95% ethanol
4	Absolute ethanol
5	Absolute ethanol
6	Xylene
7	Xylcnc
8	Molten paraffin
9	Molten paraffin

Note: Each step is 1 hour in duration. Some workers add one or more of these additional steps: a 50% ethanol step at the beginning; an 85% ethanol step between Step 1 and Step 2; an additional absolute ethanol step; an ethanol-xylene step between Step 5 and Step 6; a xylene-paraffin slurry between Step 7 and Step 8; and an additional molten paraffin step.

bedding medium in the list set forth previously. The shrinkage is not even, because the paraffin does not solidify uniformly; it generally caves in at the top of the mold. This phenomenon may be responsible for a certain amount of distortion. The heat shrinkage is probably a less serious problem, because it is principally a matter of guarding against unnecessarily high temperatures. Despite these problems, paraffin has been by far the most widely used embedding medium for many years, and it will probably not be readily replaced by another medium.

Table 3-2. Schedule for Embedding Tissues in Paraffin Using 2,2-Dimethoxypropane (DMP)

Step	Reagent	Duration
1	DMP	15 minutes or until temperature reaches equilibrium
2	DMP	
3	Benzene or methyl salicylate	15 minutes
4	Molten paraffin	45 minutes
5	Molten paraffin	45 minutes
6	Molten paraffin	45 minutes

Note: Sections can be cut from the blocks 1 hour after the end of Step 6.

From a superficial consideration, it may seem reasonable to employ a very long and gradual dehydration and infiltration schedule in order to optimize infiltration and minimize distortion. Other factors must be considered, however. Alcohol is a powerful solvent of certain tissue components, especially lipids. In fact, it is assumed that after a tissue is dehydrated with organic solvents, no significant amount of lipid remains in the tissue. Adipose tissue is recognized by the "holes" that remain, which are normally occupied by lipid. Alcohol also precipitates and insolubilizes certain proteins, and for this reason it is sometimes used as a fixative in its own right. The proteins of lightly fixed tissues may become undesirably coagulated during overexposure to dehydrating alcohols. Excessive exposure to clearing reagents may cause excessive hardness or shrinkage. The heat from hot paraffin will normally cause a small amount of shrinkage and distortion, and prolonged exposure to hot paraffin will exaggerate the effect.

PROCESSING TISSUES WITH GLYCOL METHACRYLATE

Glycol methacrylate (GMA) is an ester of methacrylic acid and glycol (dihydroxyethane) and in the monomeric form has the following structure:

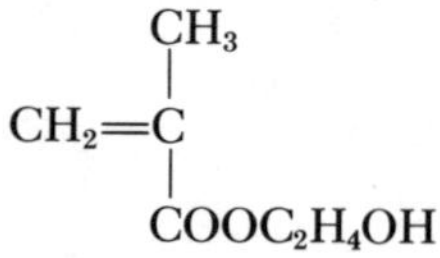

$$CH_2{=}\underset{\underset{COOC_2H_4OH}{|}}{\overset{\overset{CH_3}{|}}{C}}$$

It becomes incorporated into the polymer in the following manner:

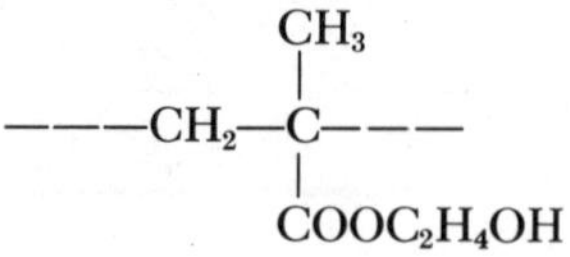

$$----CH_2-\underset{\underset{COOC_2H_4OH}{|}}{\overset{\overset{CH_3}{|}}{C}}---$$

Glycol methacrylate is available from the manufacturer* or from any of a number of chemical supply houses. Various grades of GMA are available and contain different concentrations of inhibitor (to prevent

*Rohm and Haas, Philadelphia, Pennsylvania.

spontaneous polymerization). Only 94 or 96 percent GMA, containing 200 ppm inhibitor, should be used for an embedment of biological material.

A number of authors have contributed toward the development of GMA embedding methods (Ashley and Feder, 1966; Feder and O'Brien, 1968; Ruddell, 1967a, 1967b, 1971; Cole and Sykes, 1974). The method of Ruddell (1967b) or variations of it are generally followed, and an embedding kit is commercially available* for use with that method.

Tissue Preparation
The tissues can be prepared with nearly any histological fixative except those containing osmium tetroxide, manganates, and cupric compounds, all of which may retard or inhibit polymerization (Ruddell, 1971).

The block face must be limited to about 6 by 15 mm if the preferred ⅜-inch glass knives are to be used. The thickness of the tissue should be limited to about 4 or 5 mm in order to allow good penetration of fixative and embedding medium.

Dehydration and Infiltration
Following fixation, the tissues should be washed for 15 to 30 minutes. If glutaraldehyde were used as the fixative, the tissues should be washed for 2 to 4 hours. They are then dehydrated by placing them successively in 70, 95, and 100 percent ethanol for 45 minutes each. Rigorous dehydration, as for paraffin embedding is unwarranted since the GMA monomer is soluble in water. Bennett et al. (1976) have considered it superfluous to use the 100 percent ethanol step. Indeed, dehydration can be accomplished with a graded series of glycol methacrylate solutions, avoiding ethanol altogether. However, this process involves a much longer dehydration schedule (Ashford et al., 1972).

The tissues must be well infiltrated with GMA monomer. Under normal conditions, 4 hours are sufficient, although it is frequently convenient to leave the second change overnight. The tissues should be transferred to a fresh solution after the first 2 hours. The second

*JB-4 Embedding Kit, Cat. No. 0226, Polysciences, Inc., Warrington, Pennsylvania.

solution can be saved and used as the first solution in infiltrating a future batch of tissues.

GMA should never be discarded in a sink because there is a risk that it would polymerize in the drain trap. A waste receptacle that can be discarded should be used.

Embedding

The GMA is activated by adding 1 part of solution B to 25 parts of solution A. Solution B is rather viscous, even at room temperature, and the mixture must be stirred vigorously to ensure thorough mixing. The solution should be mixed in a beaker placed in an ice bath. The polymerization reaction is extremely exothermic. Polymerization of the activated GMA begins when the solution reaches 18 to 20°C. Polymerization of the GMA produces heat, and the temperature of the mixture may rise uncontrollably. The mixture may soon become hot and viscous, and consequently useless as an embedding medium. It is therefore necessary to prevent polymerization from the beginning by keeping it cold.

About 1.5 ml of the activated GMA is transferred by pipette into each plastic molding cup, and a piece of tissue placed into each. All air bubbles must be eliminated; it is particularly important to be wary of tissues with small irregularities that can trap air. For the same reason, tissues that have been infiltrated should not be removed from the GMA mixture for any length of time to prevent drying. Otherwise, air will be trapped in the tiny tunnels and crevices of the tissue surface, which will prevent proper polymerization.

After the GMA and the block of tissue have been placed into the plastic molding cup, the block holder is placed on top of the mold as shown in Figure 3-1. The level of the GMA should be adjusted to about halfway up the block holder. The edges around the block holder must be sealed with hot paraffin. After the paraffin around the edge has solidified, a few drops should be put into the center hole. The paraffin must not be placed into the center hole first because it may displace the GMA to form a lake.

The GMA is now allowed to polymerize at room temperature for several hours. The heat produced by this small volume of plastic will readily dissipate. For larger sizes of tissues, it may be advisable to keep the plastic molding cup on a tray of crushed ice in order to retard polymerization temporarily. This will give the activated GMA more

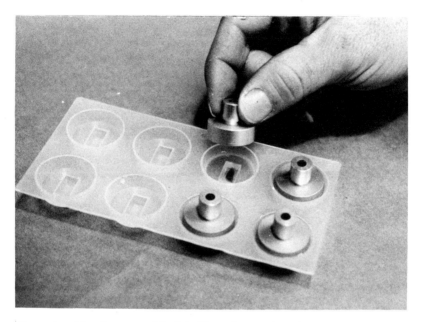

Figure 3-1. The aluminum block holder is being placed on the plastic molding cup, which already contains the activated glycol methacrylate and a piece of tissue.

time to infiltrate the tissue, resulting in a more uniform polymerization throughout the tissue.

Microtomy

Glass slides should be prepared by placing them in an acid-alcohol solution for 5 minutes and washing in running water for another 5 minutes. After drying with a lint-free cloth, the slides are ready to use.

The polymerized GMA blocks are removed from the plastic molds. Small amounts of sticky unpolymerized plastic can be wiped off with a paper towel soaked with 95 percent ethanol. Some of the excess plastic should be trimmed away from around the tissue with a razor blade. A GMA-embedded tissue block is shown in Figure 3-2.

The blocks may be sectioned with an ultramicrotome or with an ordinary rotary microtome, but the Sorvall JB-4 microtome* (shown in Fig. 3-3) is specially built for this purpose. It is designed to use glass, diamond, or steel knives. Glass knives made from plate glass that is 9

*DuPont Company, Sorvall Operations, Newtown, Connecticut.

Figure 3-2. A block of tissue embedded in plastic is shown attached to an aluminum block holder. Some of the excess plastic has been trimmed away from the edges.

Figure 3-3. The motor-driven version of the Sorvall JB-4 microtome is shown in this plastic tissue sectioning setup. It is useful to have a warming tray and a staining jar available for quick staining and evaluation of sections.

mm thick are preferred. Although this glass is thicker than that used for knives in ultramicrotomy, the principle of making them is the same except that water troughs (boats) are omitted.

Bennett et al. (1976) described an ingenious method of making a glass knife in which the cutting edge is obtained along the *width* of a glass strip rather than along the *thickness* of the glass. Bennett and his associates named this the Ralph knife, after its inventor. They reported having made Ralph knives as long as 25 mm. With this kind of knife, tissue blocks with a much larger face can be cut.

Sections of 0.5 to 1 micron in thickness (or even less) can be cut, although sections that are 2 microns thick are easier to produce and handle. The sections are picked up with fine-point forceps as they come off the knives (Fig. 3-4), and are transferred to a clean glass slide

Figure 3-4. Sections of plastic-embedded tissues are being cut with a glass knife. The sections are picked up with a fine-point forceps as they come off the knife.

containing a meniscus of water (Fig. 3-5). The sections are somewhat hydrophobic and momentarily repulse the water so that they may touch the slide surface, as shown in Figure 3-6; this results in numerous small wrinkles in the section. To avoid this problem, a higher meniscus of water must be used. The section will flatten out in a few seconds. It may be advisable to allow the blocks to air-dry for a day or two, so that the sections will be somewhat less prone to wrinkle. The water can then be drained off while the section is held and oriented on the surface of the slide. Excess water can be removed by shaking, by applying a stream of air, or by simply blowing it off. A weak solution of

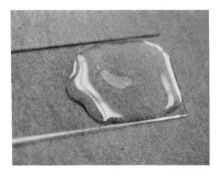

Figure 3-5. A plastic section is floating on a meniscus of water. The section flattens out in only a few seconds. The water is then drained and the slide containing the section is dried.

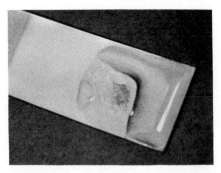

Figure 3-6. A plastic section is repelling the water and is touching the surface of the slide. The section will be unsatisfactory because it will have numerous fine wrinkles. This problem is prevented by using a higher meniscus of water.

ammonium hydroxide has been recommended for floating out the sections in order to accelerate flattening. However, it has been found that this procedure spreads or explodes the sections excessively, and thus it is not recommended.

The slide can be dried within a few minutes on a warming tray. Further drying at 50 to 55°C for 30 minutes or more may be necessary to lessen the chances of losing the section while staining.

Staining

GMA is not removed before staining; it produces a hydrophilic porous molecular meshwork and all stains and reagents can penetrate virtually unhindered. The staining characteristics of the tissue components are virtually unaltered. The GMA monomer is inert toward tissue components (unlike some other plastics), leaving the reactive groups to interact with dye molecules in the normal way. Many of the standard histological and histochemical staining procedures can be used with GMA sections, although staining times may have to be increased in some cases.

More information on GMA methodology can be found in the excellent article by Bennett et al. (1976).

COMPARISON OF PARAFFIN AND GMA TECHNIQUES

Paraffin is removed from the tissue sections and the sections are usually brought to water (rehydrated) before carrying out the histochemical procedure. The cells and tissues have only their own proteinaceous meshwork (which should have been somewhat stabilized by fixation) for their support. The tissues must not be allowed to dry out, since that could cause serious distortion. Because GMA is not removed from the tissues, it lends support to the interstices of the cells and tissues. GMA sections therefore do not have to be rehydrated and dehydrated before and after staining; the dry slides with sections can be plunged directly into the staining solution and dried directly from water after staining.

The problem of paraffin shrinkage has already been pointed out. GMA, however, shrinks only minimally during polymerization.

Paraffin blocks can be sectioned as thin as 4 microns, although it requires a considerable amount of skill to produce sections that thin.

Normally, paraffin blocks are sectioned at a thickness of 5 microns or greater. Most tissues embedded in GMA can be sectioned quite satisfactorily at 1 micron or less. Even tissues containing small amounts of undecalcified bone can be sectioned at 2 microns with no difficulty. The thinness of the glycol methacrylate sections allows greater cellular detail to be demonstrated.

CRYOTOMY

The biological activity of certain tissue components, especially that of certain enzymes and antigens, is readily destroyed during exposure to fixatives. Therefore, the use of fresh-frozen tissue sections offers an alternative to the conventional routine of fixing and embedding of tissues, and it permits histochemical investigation of sensitive, biologically active components of tissue.

Solidity of a piece of tissue can be achieved by freezing it. Ice will give the tissue support, and to a certain extent the ice can be considered the equivalent of an embedding medium.

A superficial consideration of the process of cutting frozen sections may tempt one to consider it a matter of cutting ice. However, considering the brittle nature of ice, this concept is quite inadequate. Thornburg and Mengers (1957) proposed a model, based on solid physical principles, to explain the process of cutting frozen sections. Thus cryotomy now has a good scientific basis. Optimal conditions for cutting good cryostat sections have been, and probably still are, established entirely on empirical trial and error.

The point of departure for Thornburg and Mengers' discussion was the recognition of the physical principle that pressure lowers the freezing point of water. The freezing temperature is lowered 1°C by 10^9 dynes per square centimeter. This is a realistic estimate of the pressure that might appear at the cutting edge of a knife. Of course, this in itself would not be sufficient to melt a zone of tissue at the knife. However, the energy dissipated in overcoming the tissue resistance to the knife will result in a small amount of heat at the leading edge. The two factors, pressure and heat, under the right conditions will cause a small zone of melted tissue just ahead of the cutting edge of the knife. It is the melted zone that is in contact with the sharp knife edge. The melted zone serves as an efficient conductor of the heat, which is dissipated primarily into the knife. The water will flow around the knife edge and refreeze. A thin zone of refrozen tissue will appear on the inside of the section and on the face of the block.

Three deductions were made by Thornburg and Mengers (1957) from the above theoretical considerations. (1) As the temperature of the knife and the block is raised, the micromelting zone is increased. At some point, the width of the zone will include the whole thickness of the section, and the section will pile up on the edge of the knife. (2) As the temperature of the knife is lowered, the heat will be conducted into the knife more readily, and the size of the micromelting zone will be reduced. (3) As the temperature of the block is lowered, the pressure that is required to maintain the micromelting zone will increase (thus increasing the energy to be dissipated). At some point the mechanical strength of the ice will be exceeded by the required pressure, and normal cutting can no longer occur.

The optimal cryostat temperature must finally be determined by trial and error. That temperature will be different for different kinds of tissues, but a theoretical understanding can help one to select the "trials" more rationally.

FREEZE-DRYING AND FREEZE-SUBSTITUTION

Freeze-drying and freeze-substitution are both techniques concerned with removing ice from tissues as an alternative to conventional dehydration. In a sense they are methods of dehydrating tissues while in their frozen state. Following the removal of water, the tissues are embedded in paraffin or in GMA, and they are subsequently processed in the conventional way.

Freeze-drying utilizes a vacuum to facilitate the removal of tissue ice. Water molecules can go directly from the solid phase to the gaseous phase without going through an intermediate liquid phase. The physical process is called sublimation. If the ice (or frozen tissue) is placed in a high vacuum, the rate of drying is greatly increased, rendering the process useful as a method of dehydration. Following complete removal of all the ice, the temperature can be raised, and hot paraffin can then easily infiltrate the dried tissue without exposure to any organic solvents.

Freeze-substitution essentially involves dehydration of the frozen tissue with an apolar organic solvent such as acetone or ethanol. The frozen tissue is exposed to the solvent at a very low temperature for as long as several weeks while the ice is slowly "dissolved" out. Following the successful removal of all the ice, the tissue can be cleared and embedded in paraffin.

Both the freeze-drying and freeze-substitution methods are losing favor with histochemists because they are time-consuming and inconvenient. Furthermore, improved methods of fixation and more refined methods of handling tissues make these methods less necessary. For those who desire to try either of these methods of processing tissue, the monograph by Pearse (1968) should be consulted.

REFERENCES

Arnold, W., Mitrenga, D., and Mayersbach, H. (1974). The preservation of substance and immunological activity of intravenously injected human IgG in the mouse liver. *Acta Histochem.* 49:161–175.

Ashford, A. E., Allaway, W. G., and McCully, M. E. (1972). Low temperature embedding in glycol methacrylate for enzyme histochemistry in plant and animal tissues. *J. Histochem. Cytochem.* 20:986–990.

Ashley, C. A., and Feder, N. (1966). Glycol methacrylate in histopathology. *Arch. Pathol.* 81:391–397.

Baker, J. R. (1966). *Cytological Technique.* Methuen, London.

Bennett, H. S., Wyrick, A. D., Lee, S. W., and McNeil, J. H. (1976). Science and art in preparing tissues embedded in plastic for light microscopy, with special reference to glycol methacrylate, glass knives and simple stains. *Stain Technol.* 51:71–97.

Cole, M. B., and Sykes, S. M. (1974). Glycol methacrylate in light microscopy: A routine method for embedding and sectioning animal tissues. *Stain Technol.* 49:387–400.

Cope, G. H. (1968). Low-temperature embedding in water-miscible methacrylates after treatment with antifreezes. *J. Roy. Microsc. Soc.* 88:235–257.

Feder, N. (1963). Histochemical demonstration of lipids and enzymes in tissue specimens embedded at low temperatures in glycol methacrylate. *J. Cell Biol.* 19:23A.

Feder, N., and O'Brien, T. P. (1968). Plant microtechnique: Some principles and new methods. *Am. J. Bot.* 55:123–142.

Hoshino, M., and Kobayashi, H. (1971). The use of glycol methacrylate as an embedding medium for the histochemical demonstration of acid phosphatase activity. *J. Histochem. Cytochem.* 19:575–577.

Hoshino, M., and Kobayashi, H. (1972). Glycol methacrylate embedding in immunocytochemical methods. *J. Histochem. Cytochem.* 20:743–745.

North, R. J. (1971). Methyl green–pyronin for staining autoradiographs of hydroxyethyl methacrylate–embedded lymphoid tissue. *Stain Technol.* 46:59–62.

Pearse, A. G. E. (1968). *Histochemistry: Theoretical and Applied* (3rd ed.), Vol. 1. Little, Brown, Boston.

Prentø, P. (1978). Rapid dehydration—clearing with 2,2-dimethoxypropane for paraffin embedding. *J. Histochem. Cytochem.* 26:865–867.

Rambourg, A., Bennett, G., Kopriwa, B., and Leblond, C. P. (1971). Détection radioautographique des glycoprotéines de l'épithélium intestinal du rat après injection de fucose-³H. *J. Microsc. (Paris)* 11:163–168.

Rosenberg, M., Bartl, P., and Leško, J. (1960). Water-soluble methacrylate as an embedding medium for the preparation of ultrathin sections. *J. Ultrastruct. Res.* 4:298–303.

Ruddell, C. L. (1967a). Hydroxyethyl methacrylate combined with polyethylene glycol 400 and water; an embedding medium for routine 1–2 micron sectioning. *Stain Technol.* 42:119–123.

Ruddell, C. L. (1967b). Embedding media for 1–2 micron sectioning. 2. Hydroxyethyl methacrylate combined with 2-butoxyethanol. *Stain Technol.* 42:253–255.

Ruddell, C. L. (1971). Embedding media for 1–2 micron sectioning. 3. Hydroxyethyl methacrylate–benzoyl peroxide activated with pyridine. *Stain Technol.* 46:77–83.

Thornburg, W., and Mengers, P. E. (1957). An analysis of frozen section techniques. I. Sectioning of fresh-frozen tissues. *J. Histochem. Cytochem.* 5:47–52.

Troyer, H., and Nusbickel, F. R. (1975). Enzyme histochemistry of undecalcified bone and cartilage embedded in glycol methacrylate. *Acta Histochem.* 53:198–202.

Weber, G. (1974). Glycol methacrylate embedding in enzyme histochemistry: Application to arthropod tissue incubated for demonstration of unspecific esterase and succinic dehydrogenase activity. *Histochemistry* 39:155–161.

4

BUFFERS

It is important that histochemists have a good working knowledge of buffers. The literature of histochemistry is replete with studies that are open to criticism because buffers were used carelessly or irrationally. One method published for demonstrating glucose 6-phosphatase calls for the use of tris buffer at pH 6.7; however, the buffering capacity of tris at this pH is greatly reduced. Admittedly, there are some applications in histochemistry where the pH is not critical and the worker can afford to take shortcuts in buffer preparation. Nevertheless, the histochemist should recognize the fact that a shortcut is being used. In many applications, however, the pH is critical and shortcuts are forbidden.

FUNCTION OF BUFFERS

A plea is being made here for rational use of buffers in histochemistry. Two instances illustrate why it is important to control the pH at which histochemical reactions take place. (1) All enzymes have pH optima; that is, at a certain pH, a given enzyme operates at a maximum velocity. If the pH of the incubation medium is either increased or decreased by only 1 or 2 pH units, the velocity of the enzymes is significantly reduced, often drastically. Oxidoreductases are particularly sensitive to pH changes. (2) Reactive groups of biopolymers are ionizable, and the degree of ionization is a function of the pH. The degree of ionization profoundly influences the reaction of these groups with histochemical reagents. Typically, the reaction proceeds best at a pH where the majority of the groups are ionized.

The function of a buffer is to resist the changes that might result in the pH of a solution when an acid or a base is added. A base added to the solution will contribute OH^- ions, which will associate with some of the H^+ ions and thereby increase the pH. However, the increase in pH will induce the buffer to contribute, or release, additional H^+ ions and therefore minimize the pH change.

A buffer in the acid range is composed of a weak acid and a salt of the same acid. Similarly, a buffer in the basic range is composed of a weak base and its salt. The acid or base will be minimally dissociated, while the salt may be regarded as completely dissociated.

In a $0.1M$ solution of acetic acid, only 1.3 percent of the molecules are dissociated. Therefore, the concentration of both H^+ ions and $C_2H_3O_2^-$ ions will be $0.0013M$ (1.3 percent of $0.1M$ is $0.0013M$), and the concentration of $HC_2H_3O_2$ will be $0.0987M$. From the law of mass action, this can be expressed as a constant:

$$K = \frac{[C_2H_3O_2^-] \times [H^+]}{[HC_2H_3O_2]} = \frac{0.0013 \times 0.0013}{0.0987} = 0.000018M = 1.8 \times 10^{-5}M$$

This is called the dissociation constant (or ionization constant) and can be expressed as its negative logarithm, pK, just as the H^+ ion concentration is generally expressed as its negative logarithm, pH. The pK of acetic acid is 4.75. Table 4-1 shows the pK values of some biologically important compounds.

In biological work the equation is usually expressed in terms of the H^+ ion concentration, which is simply a matter of rearranging the above equation:

$$[H^+] = \frac{K[HC_2H_3O_2]}{[C_2H_3O_2^-]}$$

With two more operations, the equation can be expressed in terms of pH:

(1) Take the logarithm of each side:

$$\log [H^+] = \log K + \log [HC_2H_3O_2] - \log [C_2H_3O_2^-]$$

(2) Multiply each side by -1:

$$-\log [H^+] = -\log K - \log [HC_2H_3O_2] + \log [C_2H_3O_2^-]$$

By definition, $-\log [H^+]$ and $-\log K$ are pH and pK, respectively; therefore

$$pH = pK + \log \frac{[C_2H_3O_2^-]}{[HC_2H_3O_2]}$$

Table 4-1. Dissociation Constants and pK Values of Acids and Bases Useful as Buffers

Compound	Dissociation Constant (K)	pK (25°C)*
Cacodylic acid (1)	2.67×10^{-2}	1.57
Maleic acid (1)	1.42×10^{-2}	1.83
Phosphoric acid (1)	7.52×10^{-3}	2.12
Glycine (1)	4.46×10^{-3}	2.35
Citric acid (1)	7.45×10^{-4}	3.14
Barbituric acid	$9.8 \ \times 10^{-5}$	4.01
Citric acid (2)	1.73×10^{-5}	4.75
Acetic acid	1.76×10^{-5}	4.75
Citric acid (3)	4.02×10^{-6}	5.40
Maleic acid (2)	8.57×10^{-7}	6.07
Cacodylic acid (2)	5.37×10^{-7}	6.27
Carbonic acid (1)	4.30×10^{-7}	6.37
Phosphoric acid (2)	6.23×10^{-8}	7.21
Veronal	$3.7 \ \times 10^{-8}$	7.43
Tris(hydroxymethyl) aminomethane	8.57×10^{-9}	8.07
Boric acid (1)	$7.3 \ \times 10^{-10}$	9.14
Glycine (2)	1.68×10^{-10}	9.87
Carbonic acid (2)	5.61×10^{-11}	10.25
Phosphoric acid (3)	$2.2 \ \times 10^{-13}$	12.67
Boric acid (2)	$1.8 \ \times 10^{-13}$	12.74
Boric acid (3)	$1.6 \ \times 10^{-14}$	13.80

*The values for boric acid are for 20°C.
Compiled from R. C. Weast (Ed.), *Handbook of Chemistry and Physics* (59th edition), Chemical Rubber Co., West Palm Beach, Florida, 1978–1979. Values for tris were obtained from R. G. Bates and H. B. Hetzer, *J. Phys. Chem.* 65:667, 1961.

This relationship is known as the Henderson-Hasselbalch equation. It can be used to calculate the pH of 0.1*M* acetic acid:

$$pH = 4.75 + \log \frac{0.0013}{0.0987}$$

$$= 4.75 + \log 0.0013 - \log 0.0987$$

$$= 4.75 - 2.89 + 1.01 = 2.87$$

This can be readily verified experimentally.

Buffering Capacity

The pH of a 0.1M acetic acid solution can be raised by adding sodium acetate. If the proportion of salt to acid is increased by increments and the pH is measured and plotted, as in Figure 4-1, a sigmoid curve is obtained. It will be noted that at the point where there are equal parts of salt and acid, the pH shift per increment is minimal. This point occurs at a pH of 4.75, which corresponds to the pK of acetic acid. This also follows from a consideration of the Henderson-Hasselbalch equation; when the numerator (concentration of salt) equals the denominator (concentration of acid), the ratio is 1 and the log of 1 is 0. Hence, at this point the pH equals the pK.

The pH of the acetic acid solution could also have been raised with sodium hydroxide. However, the addition of sodium hydroxide would have essentially been the same as making sodium acetate in the solution.

The buffering action is seen when the concentrations of acetic acid and sodium acetate are approximately equal. It will be recalled that the purpose of a buffer is to resist changes in H^+ ion concentration. Figure 4-1 shows that the pH undergoes minimal changes in the central portion of the sigmoid curve; this reflects the buffering capacity of the solution.

It has been shown that the pH of a solution containing 0.1M acetic acid and 0.1M sodium acetate has a pH of 4.75. A 0.000018M solution

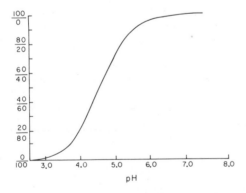

Figure 4-1. Buffer curve of an acetic acid–sodium acetate solution. The vertical axis represents the proportion of sodium acetate to acetic acid. In the middle region of the curve, the pH is least affected by a change in the ratio of acid to salt.

of HCl also has a pH of 4.75. Consider what happens when 0.000018 moles of sodium hydroxide is added to each solution. This amount of sodium hydroxide is sufficient to neutralize the HCl solution, hence raising the pH from 4.75 to 7. HCl is a strong acid and has no buffering capacity. For the acetic acid–sodium acetate solution, the effect of the sodium hydroxide can be calculated:

$$pH = pK + \frac{[salt]}{[acid]}$$

$$= 4.75 + \log \frac{0.1 + 0.000018}{0.1 - 0.000018}$$

$$= 4.75$$

The change in the pH is too slight to be detected. The acetate buffer resisted a change in pH, while the 0.000018M HCl solution had no power to do so.

Buffering capacity is an expression of the efficiency with which a buffer resists a change in pH. Consider what happens when 0.01 moles of HCl is added to a liter of 0.1M acetate buffer (0.05M acetic acid and 0.05M sodium acetate). Before the HCl is added, the pH is 4.75; after adding the HCl the pH will drop:

$$pH = pK + \log \frac{0.05 - 0.01}{0.05 + 0.01}$$

$$= 4.75 + \log \frac{0.04}{0.06}$$

$$= 4.57$$

The pH has dropped 0.18 unit. If the concentration of the buffer is made five times as strong—i.e., with the use of 0.25M acetic acid and 0.25M sodium acetate (0.5M acetate buffer)—then the effect of adding 0.01 mole of HCl will be much less:

$$pH = pK + \log \frac{0.25 - 0.01}{0.25 + 0.01}$$

$$= 4.75 + \log \frac{0.24}{0.26}$$

$$= 4.72$$

The pH of this solution has dropped only 0.03 unit. It is therefore evident that the buffering capacity increases as the molarity of the buffer is increased.

Suppose the acetate buffer is used at a pH that is considerably different from the pK of acetate. The ratio of salt to acid will have to be changed. For the purpose of comparison, however, the final concentration of buffer should be the same as in the last example. Therefore, 0.45M acetic acid and 0.05M sodium acetate will be used, yielding a 0.5M acetate buffer. The pH before the HCl is added is:

$$pH = pK + \log \frac{0.05}{0.45}$$

$$= 4.75 - 0.95$$

$$= 3.80$$

After adding 0.01 mole of HCl, the pH drops:

$$pH = pK + \log \frac{0.05 - 0.01}{0.45 + 0.01}$$

$$= 4.75 - 1.06$$

$$= 3.69$$

The pH has dropped 0.11 unit, as compared with 0.03 unit in the preceding example. It is apparent, therefore, that the buffering capacity of a solution is decreased when it is used at a pH that is significantly different from its pK.

Nevertheless it is frequently desirable to use a buffer at a pH that is somewhat different from its pK. How far can the pH be shifted away from the pK and still be useful? As a rule of thumb, a buffer may be used within 1 pH unit of its pK and still be reasonably effective. Therefore, acetate probably should not be used below pH 3.75 or above pH 5.75. Tris buffer should be used between pH 7.1 and pH 9.1, since its pK is 8.07. Buffering capacity starts dropping off rapidly beyond these somewhat arbitrary limits.

SELECTION OF A BUFFER

The most important factor in the selection of a buffer is the buffering capacity, which includes the molar strength and the proximity of the pH and the pK of the solution. Other factors include the concentration of buffer ions, the chemical reactivity, and the temperature coefficient of the buffer.

Concentration of Buffer Ions

Buffers are commonly used between 0.2M and 0.05M strengths. The buffer must have great enough strength to maintain the proper pH. However, the concentration of buffer ions may also influence chemical reaction. For instance, a high concentration of buffer ions may reduce the staining reaction of certain dyes by competing with the dye for the binding sites. Also certain concentrations of some ions may help to solubilize or insolubilize proteins or carbohydrates of tissues.

Chemical Reactivity

The buffer must be selected not only on the basis of its pK relative to the pH desired for the incubation medium, but also on the basis of its chemical reactivity. Because calcium phosphate will precipitate, phosphate buffers cannot be used when calcium ions are present or when it is anticipated that they will be evolved as a reaction product. Certain buffer ions inhibit certain enzymes; for instance, acetate ions inhibit N-acetyl-β-glucosaminidase and citrate ions inhibit β-glucuronidase.

Temperature Coefficient

Temperature influences the pH of a buffer to a surprising degree, and the effect is frequently overlooked. The temperature coefficient of a buffer is the change in pH caused by raising the temperature of the buffer by 1°C. If a 0.05M tris buffer is adjusted to pH 7.680 at 25°C and the temperature is then raised to 37°C, the pH will drop to 7.361. The temperature coefficient of tris is approximately -0.015 pH unit per degree C, which is one of the highest temperature coefficients. Phosphate buffer, on the other hand, has a temperature coefficient of only -0.0015 pH unit per degree C.

ACETATE BUFFER (0.1 MOLAR)

STOCK SOLUTIONS:
Solution A. 0.1N acetic acid (5.7 ml glacial acetic acid per 1000 ml)
Solution B. 0.1M sodium acetate, $CH_3COONa \cdot 3H_2O$ (13.61 gm per
1000 ml)

WORKING SOLUTION:
The following amounts are used to make 200 ml of a working solution
at the desired pH.

pH	Solution A (ml)	Solution B (ml)
3.6	185	15
3.8	176	24
4.0	164	36
4.2	147	53
4.4	216	74
4.6	102	98
4.8	80	120
5.0	59	141
5.2	42	158
5.4	29	171
5.6	19	181

CITRATE BUFFER (0.1 MOLAR)

STOCK SOLUTIONS:
Solution A. 0.1M sodium citrate, $Na_3C_6H_5O_7 \cdot 2H_2O$ (29.41 gm per
1000 ml)
Solution B. 0.1M citric acid, $C_6H_8O_7 \cdot H_2O$ (21.01 gm per 1000 ml)

WORKING SOLUTION:
The following amounts are used to make 50 ml of a working solution at
the desired pH.

pH	Solution A (ml)	Solution B (ml)
3.0	3.5	46.5
3.2	6.3	43.7
3.4	10.0	40.0

pH	Solution A (ml)	Solution B (ml)
3.6	13.0	37.0
3.8	15.0	35.0
4.0	17.0	33.0
4.2	18.5	31.5
4.4	22.0	28.0
4.6	24.5	25.5
4.8	27.0	23.0
5.0	29.5	20.5
5.2	32.0	18.0
5.4	34.0	16.0
5.6	36.3	13.7
5.8	38.2	11.8
6.0	41.5	9.5
6.2	42.8	7.2

PHOSPHATE-CITRATE (McILVAINE'S) BUFFER

STOCK SOLUTIONS:

Solution A. 0.2M disodium phosphate, Na_2HPO_4 (28.39 gm per 1000 ml)

Solution B. 0.1M citric acid, $C_6H_8O_7$ (19.21 gm per 1000 ml)

WORKING SOLUTION:

The following amounts are used to make 200 ml of a working solution at the desired pH.

pH	Solution A (ml)	Solution B (ml)
2.4	12.4	187.6
2.6	21.8	178.2
2.8	31.7	168.3
3.0	41.1	158.9
3.2	49.4	150.6
3.4	57.0	143.0
3.6	64.4	135.6
3.8	71.0	129.0
4.0	77.1	122.9
4.2	82.8	117.2
4.4	88.2	111.8
4.6	93.5	106.5
4.8	98.6	101.4
5.0	103.0	97.0

	Solution A	Solution B
pH	(ml)	(ml)
5.2	107.2	92.8
5.4	111.5	88.5
5.6	116.0	84.0
5.8	120.9	79.1
6.0	126.3	73.7
6.2	132.2	67.8
6.4	138.5	61.5
6.6	145.5	54.5
6.8	154.5	45.5
7.0	164.7	35.3
7.2	173.9	26.1
7.4	181.7	18.3
7.6	187.3	12.7
7.8	191.5	8.5
8.0	194.5	5.5

PHOSPHATE BUFFER (0.1 MOLAR)

STOCK SOLUTIONS:

Solution A. 0.2M monosodium phosphate, NaH_2PO_4 (27.58 gm per 1000 ml)

Solution B. 0.2M disodium phosphate, Na_2HPO_4 (28.39 gm per 1000 ml)

WORKING SOLUTION:

To make 200 ml of a working solution at the desired pH, the following amounts are used and the resulting mixture diluted with distilled water to make 200 ml.

	Solution A	Solution B
pH	(ml)	(ml)
5.7	93.5	6.5
5.8	92.0	8.0
5.9	90.0	10.0
6.0	87.7	12.3
6.1	85.0	15.0
6.2	81.5	18.5
6.3	77.5	22.5
6.4	73.5	26.5
6.5	68.5	31.5
6.6	62.5	37.5
6.7	56.5	43.5

pH	Solution A (ml)	Solution B (ml)
6.8	51.0	49.0
6.9	45.0	55.0
7.0	39.0	61.0
7.1	33.0	67.0
7.2	28.0	72.0
7.3	23.0	77.0
7.4	19.0	81.9
7.5	16.0	84.0
7.6	13.0	87.0
7.7	10.5	90.5
7.8	8.5	91.5
7.9	7.0	93.0
8.0	5.3	94.7

TRIS-MALEATE BUFFER

STOCK SOLUTIONS:

Solution A. 0.2M tris-maleate solution (24.2 gm tris(hydroxymethyl)-aminomethane + 23.2 gm maleic acid or 10.6 gm maleic anhydrate in 1000 ml)

Solution B. 0.2N NaOH

WORKING SOLUTION:

To make 100 ml of a working solution at the desired pH, the following amount of Solution B is added to 25 ml of Solution A and the volume is brought to 100 ml with distilled water.

pH	Solution B (ml)
5.2	3.5
5.4	5.4
5.6	7.8
5.8	10.3
6.0	13.0
6.2	15.8
6.4	18.5
6.6	21.3
6.8	22.5
7.0	24.0
7.2	25.5
7.4	27.0
7.6	29.0

pH	Solution B (ml)
7.8	31.8
8.0	34.5
8.2	37.5
8.4	40.5
8.6	43.3

TRIS BUFFER (0.05 MOLAR)

STOCK SOLUTIONS:

Solution A. 0.2M tris (24.2 gm tris(hydroxymethyl)aminomethane per 1000 ml)

Solution B. 0.2M HCl (16.8 ml concentrated HCl per 1000 ml)

WORKING SOLUTION:

To make a working solution at the desired pH, the following amount of Solution B is added to 25 ml of Solution A and the volume is brought to 100 ml with distilled water.

pH	Solution B (ml)
7.2	22.1
7.4	20.7
7.6	19.2
7.8	16.3
8.0	13.4
8.2	11.0
8.4	8.3
8.6	6.1
8.8	4.1
9.0	2.5

BARBITAL-ACETATE BUFFER

STOCK SOLUTIONS:

Solution A. 19.43 gm sodium acetate ($NaC_2H_3O_2 \cdot 3H_2O$) + 29.45 gm barbital sodium per 1000 ml)

Solution B. 0.1N HCl

WORKING SOLUTION:

To make 250 ml of a working solution at the desired pH, 50 ml of Solution A is added to the following amount of Solution B and 20 ml 8.5 percent NaCl; the resulting solution is diluted with distilled water to make 250 ml.

pH	Solution B (ml)
3.8	133.5
4.0	125.5
4.2	117.5
4.4	109.5
4.6	101.5
4.8	94.5
5.0	88.0
5.2	82.5
5.4	78.0
5.6	75.0
5.8	72.5
6.0	71.0
6.2	69.5
6.4	68.5
6.6	66.5
6.8	64.0
7.0	60.5
7.2	56.5
7.4	50.5
7.6	43.0
7.8	34.5
8.0	26.0
8.2	18.5
8.4	13.0
8.6	9.0
8.8	6.0
9.0	3.0

BARBITAL (MICHAELIS) BUFFER (0.05 MOLAR)

STOCK SOLUTIONS:

Solution A. 0.2N sodium barbital (41.2 gm per 1000 ml)
Solution B. 0.2N HCl (16.8 ml concentrated HCl per 1000 ml)

WORKING SOLUTION:

To make 200 ml of a working solution at the desired pH, 50 ml of Solution A is mixed with the following amount of Solution B; the result is diluted with distilled water to 200 ml.

pH	Solution B (ml)
6.8	45.0
7.0	43.0
7.2	39.0
7.4	32.5
7.6	27.5
7.8	22.5
8.0	17.5
8.2	12.7
8.4	9.0
8.6	6.0
8.8	4.0
9.0	2.5
9.2	1.5

BORATE BUFFER (0.05 MOLAR)

STOCK SOLUTIONS:

Solution A. 0.1M boric acid, H_3BO_3, in 0.1M KCl (6.18 gm boric acid + 7.46 gm KCl per 1000 ml)

Solution B. 0.1N sodium hydroxide

WORKING SOLUTION:

To make 100 ml of a working solution at the desired pH, 50 ml of Solution A is mixed with the following amount of Solution B and the result is diluted with distilled water to 100 ml.

pH	Solution B (ml)
7.8	2.6
8.0	4.0
8.2	5.9
8.4	8.5
8.6	12.0
8.8	16.3
9.0	21.3
9.2	26.7

pH	Solution B (ml)
9.4	32.0
9.6	36.8
9.8	40.8
10.0	43.9

GLYCINE BUFFER

STOCK SOLUTIONS:
Solution A. 0.1M glycine in 0.1M sodium chloride (7.51 gm glycine + 5.85 gm NaCl per 1000 ml)
Solution B. 0.1N sodium hydroxide (4.0 gm per 1000 ml)

WORKING SOLUTION:
To make 100 ml of a working solution at the desired pH, the following amounts are used.

pH	Solution A (ml)	Solution B (ml)
8.45	95	5
8.79	90	10
9.22	80	20
9.56	70	30
9.98	60	40
10.32	55	45
10.90	51	49
11.14	50	50
11.39	49	51
11.92	45	55
12.21	40	60
12.48	30	70
12.66	20	80
12.77	10	90

5

PROTEINS

Proteins are extremely prevalent throughout animal tissues and have a wide variety of functions. Collagen and other so-called structural proteins play an important role in the maintenance of morphology. Many proteins are involved in carrying out various aspects of metabolism as enzymes, hormones, antibodies, carrier proteins, contractural proteins, and proteins involved in maintaining osmotic balance. The spectacular variety of functions carried out by proteins underscores their importance.

CLASSIFICATION OF PROTEINS

Proteins can be classified according to their composition and physical properties. The general classes of proteins are (1) fibrous proteins, (2) globular proteins, and (3) conjugated proteins.

Fibrous Proteins

Fibrous proteins can also be called structural proteins because they are concerned primarily with the maintenance of morphology, on both the gross and microscopic levels. These proteins include collagen, elastin, and keratin. They are largely insoluble in water and highly resistant to most animal proteases. The notable exceptions are animal collagenase, which can partially digest insoluble collagen, and elastase, which can digest elastin.

Collagen is by far the most abundant fibrous protein; it makes up about 30 percent of the total protein of the human body, or about 6 percent of the dry weight. About one-third of the total number of amino acid residues are glycine. The proline and alanine contents are also high. Some of the proline residues are hydroxylated, which makes collagen almost unique among proteins. The amino acids tryptophan, cysteine, and methionine are totally absent.

A certain amount of collagen can be extracted with cold salt solutions, especially from embryonic tissue or rapidly growing connective tissue. This fraction is known as soluble collagen. Insoluble collagen can be dissolved with hot water or with strong acids or alkali. Collagen so extracted will solidify in a chilled aqueous solution, and it is known as gelatin. It is useful in the scientific laboratory but is perhaps better known as a food product.

Elastins are less prevalent but form an important constituent in arterial walls, tendons, and elastic cartilage. Their amino acid composition is somewhat similar to that of collagen. Amino acid analysis reveals the presence of a unique compound called desmosine, which represents a crosslink point where four lysine residues from four different polypeptide chains were interconnected. Elastin cannot be converted into gelatin. It is worthy of note that collagen and elastin are both of mesodermal origin.

Keratin is quite unrelated to elastin or collagen and is of ectodermal origin. It occurs in hair, wool, nails, hooves, and epidermis and serves a wide variety of functions. Cysteine content is high: 14 percent in human hair and 15 percent in wool.

Globular Proteins

The globular proteins are soluble in dilute solutions of salts, acids, and bases. In solution, they tend to take a spheroidal or ellipsoidal shape. This group consists primarily of albumins, globulins, histones, and protamines.

Albumins (for example, egg albumin and serum albumin) contain a large number of polar amino acid residues and are therefore highly soluble in water. Serum albumin functions as a carrier protein and helps to maintain proper osmotic balance between blood and other body compartments.

Globulins (for example, serum gamma globulins, myoglobin, and hemoglobin) are sparingly soluble in water but can be dissolved in salt solutions. For a given globulin, a particular concentration of salt will maximally dissolve the protein. Higher or lower concentrations of salt will precipitate the globulin. The isolation of globulins takes advantage of this phenomenon, because a particular salt concentration will precipitate one species of globulin but keep others largely in solution.

Histones and protamines are both normally associated with nuclear DNA. Histones are obtained from mammalian sources, primarily from lymphoid tissues; protamines are isolated from fish sperm. They are very basic proteins because of their high content of arginine and lysine. It is thought that they form salt linkages (ionic bonds) with nucleic acids.

Conjugated Proteins

Conjugated proteins are those that contain a nonprotein moiety. This moiety may consist of nucleic acids (nucleoproteins), carbohydrates (proteinpolysaccharides and glycoproteins), lipid material (lipoproteins and proteolipids), or chromoproteins (hemoglobins and cytochromes).

Proteinpolysaccharides (the term *proteinpolysaccharide* is preferred over the term *mucopolysaccharide*) are those proteins containing more than 4 percent carbohydrate. One protein molecule serves for the covalent attachment of several units of chondroitin sulfate, keratan sulfate, dermatan sulfate, or heparin. No doubt many mucous materials from the glands of the digestive tract will fall into this category when biochemically characterized. (Proteinpolysaccharides will be discussed more thoroughly in Chapter 6.)

Glycoproteins are sugar-containing proteins with less than 4 percent carbohydrate. Some hormones, the blood group substances, and some globulins and albumins belong in this category. Glycoproteins are almost invariably periodic acid–Schiff-positive.

Lipoproteins include serum lipoprotein and the proteins with which lecithin and cholesterol are associated. The central nervous system is rich in lipoproteins.

HISTOCHEMICAL DETECTION OF PROTEINS

It is relatively easy to demonstrate all proteins of a tissue section, but it is virtually impossible to demonstrate one species of protein at the exclusion of all others. The best that one can expect to do is to indicate relative concentrations of certain amino acids as they are incorporated into the primary structure of proteins. For example, bovine insulin contains 12 percent cysteine and no tryptophan, whereas bovine glucagon contains 3.4 percent tryptophan and no cysteine. One should therefore be able to demonstrate beta cells of the pancreatic islets with a method specific for $-SH$ and $-SS-$ groups, and the alpha cells with a method specific for tryptophan.

Most histochemical methods useful for demonstrating proteins depend on chemical reactions with specific groups along the polypeptide chain. It should be borne in mind that aldehyde fixatives also react with many of these groups, making aldehyde fixation frequently unsuitable in protein histochemistry. Cryostat sections or freeze-dried sections are usually preferred. Fixation, if desired, can be accomplished with acid-alcohol solution or with one of the standard nonaldehyde histological fixatives.

It is well known that certain groups, especially nonpolar groups, are often hidden in the tertiary structure of a protein molecule and are therefore inaccessible to the histochemical reagents. Acid-alcohol solution denatures and swells proteins enough to make some of these groups more readily available. This problem is discussed by Danielli (1953), who concluded that the problem is not serious until quantitative cytochemical measurements are attempted.

Many of the methods will actually demonstrate several amino acid groups, in which case it will be necessary to block all the unwanted groups selectively before the chromogenic reaction is carried out.

GENERAL METHODS FOR DEMONSTRATING PROTEINS

There are many methods available for demonstrating proteins in general but only three have been selected for consideration here.

Mercury–Bromophenol Blue Method

The virtue of this method lies in its simplicity and in the ease with which it can be performed. It was developed as a method for detecting protein on electrophoresis paper strips (Durrum, 1950), was later adapted for use with histological sections (Mazia et al., 1953), and was further modified by Bonhag (1955). The mechanism of action is not understood but probably depends on a simple protein-dye affinity. It does not appear that mercury or any other mordant is essential; however, Menzies (1961) found that if mercury is omitted, treatment with a blueing or developing agent such as Scott's tap-water substitute is required. Menzies left the tissues in the staining solution for 4 hours in order to show muscle striations. However, both the method of Mazia et al. and Bonhag's modification call for only 15 minutes, which seems quite adequate.

Acrolein-Schiff Method

Acrolein is an unsaturated aldehyde that is useful as a fixative. It turns out that the double bond end of the molecule is more reactive than the aldehyde end, leaving a free aldehyde group attached to the protein.

Van Duijn (1961) took advantage of the reaction in developing a histochemical reaction for protein. The proteins could be visualized by simply treating the sections with Schiff's reagent following the acrolein fixation.

Acrolein reacts with $-SH$, $-NH_2$, $=NH$, and imidazole (histidine) groups. Possibly it may also react with tryptophan, proline, and tyrosine. Strangely, van Duijn could get no reaction with isolated histones. Isolated DNA gave a mild reaction, but van Duijn thought that this was probably due to contaminants. However, Fraschini and Marinozzi (1977) showed that acrolein does react with DNA as well as with basic nuclear proteins.

"Stains-All" Method

The two preceding methods stain proteins rather specifically but they have no power to discriminate among different kinds of proteins. To that degree, their usefulness is limited. However, the stains-all method is capable of distinguishing among glycoproteins, phosphoproteins, nucleoproteins, and other proteins. This method uses a cationic carbocyanine dye, which biochemists have used to visualize and differentiate electrophoretically separated macromolecules. The dye 1-ethyl-2-[3-(1-ethylnaphtho[1,2d]-thiazolin-2-ylidene)-2-methyl-propenyl]-naphtho[1,2d]-thiazolium bromide (hence the desire to call it by a trivial name!) has been adapted for histochemical purposes by Green and Pastewka (1974a and 1974b). The method stains not only simple and conjugated proteins but glycosaminoglycans and nucleic acids as well. It differentiates among these components by staining them different colors. In a very weak Michaelis buffer at pH 4.5, nuclei stain purple, mast cell granules stain red-purple, glycoproteins (from goblet cells and salivary glands) stain blue-green, cartilage matrix stains purple, phosphoproteins stain blue, and most other proteins stain red.

The stain is light-sensitive in solution and must be used in the dark, but otherwise the staining procedure is very straightforward. Its usefulness in research and pathology should be explored vigorously.

METHODS FOR INDIVIDUAL AMINO ACID RESIDUES OF PROTEIN

DNFB Method

Dinitrofluorobenzene (DNFB) was introduced by Sanger (1945) as a reagent for detecting terminal amino groups of proteins. It is also known to react with ε-amino groups of lysine, sulfhydryl groups, histidine, and tyrosine. The reaction with $-NH_2$ and $-SH$ groups produces a small amount of yellow color, but with histidine and tyrosine no color is produced (Danielli, 1953; Porter, 1950). The amino groups of glycosaminoglycans evidently make no contribution to the color, since treatment with hyaluronidase causes no diminution of color. The color of the reaction with $-NH_2$ and $-SH$ groups is too weak to be very useful. However, Zerlotti and Engel (1972) used this method (without the customary subsequent diazotization-coupling step) and increased the visibility of the color with a wedge filter, used at 410-nm transmission. They showed that $-SH$ groups make only a minimal contribution to the color. The method was therefore nearly specific for α-amino and ε-amino groups, visualizing mainly proteins rich in lysine and hydroxylysine.

It is customary to reduce the nitro groups on the attached DNFB to amino groups. These groups are subsequently diazotized and allowed to react with a naphthol compound such as H-acid (8-amino-1-naphthol-3,6-disulfonic acid) to produce a strong color. This procedure now also visualizes tyrosine and possibly histidine, making the reaction rather nonspecific for proteins in general.

It is possible to block $-NH_2$ and $-SH$ groups selectively prior to DNFB treatment, making the method specific for tyrosine. As blocking agents, iodoacetamide has been used for the $-SH$ groups and nitrous acid for the $-NH_2$ groups. Alternatively, naphthoquinone can be used for blocking both groups at once (Danielli, 1949).

A number of reducing reagents can be used for reducing the nitro groups of the incorporated DNFB. The reagent most commonly used by organic chemists for reducing nitro groups is $SnCl_2$. This reagent can be used for the same purpose in histochemistry, although chromous chloride and sodium hydrosulfite (sodium thionite) have been used as well. However, it has been suggested that the preceding compounds are too weak and that the reduction is more complete with titanous chloride.

The subsequent diazotization step is a very common procedure in organic chemistry and in histochemistry; it is accomplished by treating the tissue sections with an acidic solution of sodium nitrite on an ice bath. α-Naphthol or β-naphthol can be used for coupling, but H-acid or S-acid (8-amino-1-naphthol-5-sulfonic acid) is more suitable because these compounds are much more soluble in aqueous solution.

It is envisioned that the reaction occurs as follows:

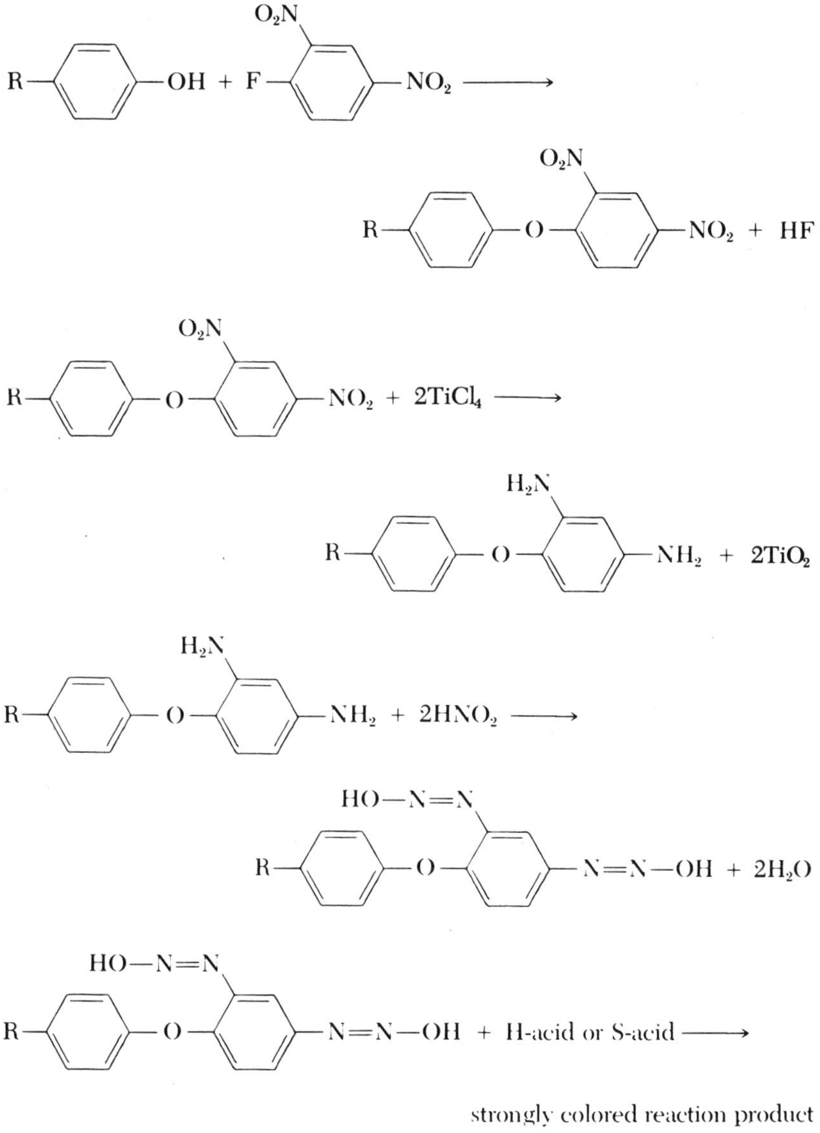

strongly colored reaction product

Methods for Tyrosine

Millon reaction. The classic method for detecting tyrosine biochemically or histochemically is by the Millon reaction. The Millon reagent essentially consists of a mercury salt dissolved in dilute mineral acid with a nitrite salt added. It is thought that phenol of the tyrosine residue is first converted to a nitrosophenol, the $-NO$ group being substituted ortho to the hydroxyl group. The mercury ion then chelates between the hydroxyl and the $-NO$ group, forming essentially a second ring. The complex is red.

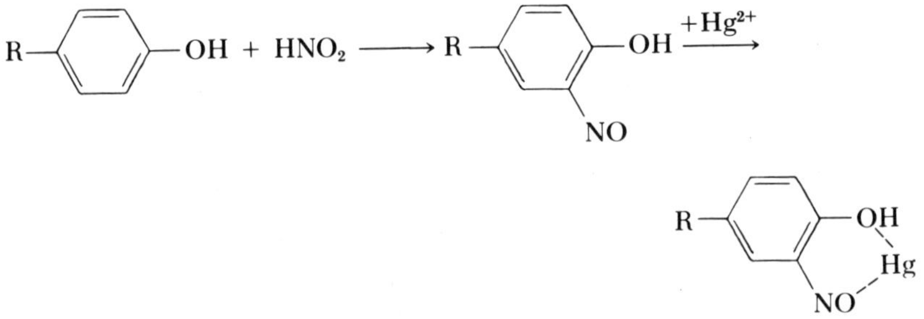

The original Millon method was modified for histochemical purposes by Bensley and Gersh (1933). The method of choice, however, is a further modification by Baker (1956).

Diazotization-coupling method. Another method for demonstrating tyrosine that merits discussion here is based on diazotization of the tyrosine and subsequent coupling with S-acid. The reaction, as originally described by Lillie (1957), is useful only as a general stain for proteins because the reaction also occurs with tryptophan and histidine. However, a later modification of the method (Glenner and Lillie, 1959) has rendered it specific for tyrosine. Nitrite reacts with the tyrosine residue to form a nitroso substitution ortho to the hydroxyl group, just as in the first part of the Millon reaction. Next, tautomerization takes place, followed by a further reaction with nitrite to form a diazonium compound. Finally, the tissue sections are exposed to a slightly basic solution of S-acid. The reaction results in a pinkish red reaction product and is thought to occur as follows:

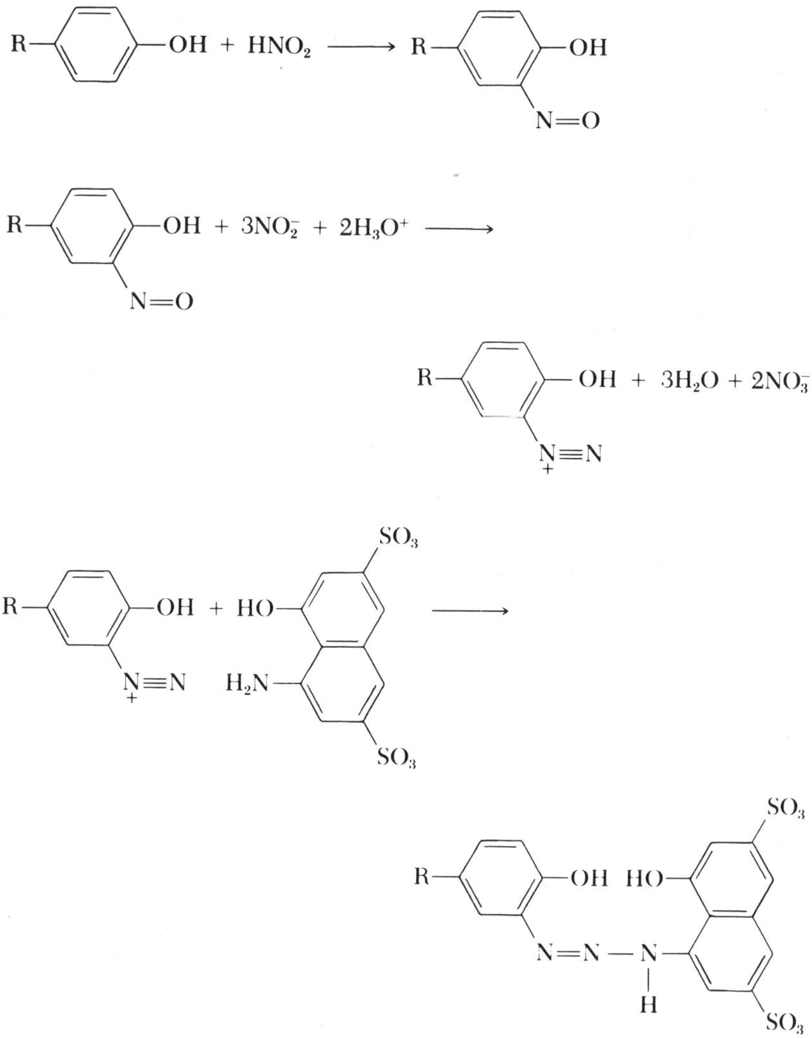

$$R-\!\!\bigcirc\!\!-OH + HNO_2 \longrightarrow R-\!\!\bigcirc\!\!-OH$$
$$\qquad\qquad\qquad\qquad\qquad\qquad N=O$$

$$R-\!\!\bigcirc\!\!-OH + 3NO_2^- + 2H_3O^+ \longrightarrow$$
$$N=O$$

$$R-\!\!\bigcirc\!\!-OH + 3H_2O + 2NO_3^-$$
$$\overset{+}{N}\equiv N$$

$$R-\!\!\bigcirc\!\!-OH + HO-\!\!\bigcirc\!\!\bigcirc\!\!\overset{SO_3}{\underset{SO_3}{}} \longrightarrow$$
$$\overset{+}{N}\equiv N \quad H_2N-$$

$$R-\!\!\bigcirc\!\!-OH \quad HO-\!\!\bigcirc\!\!\bigcirc\!\!\overset{SO_3}{\underset{SO_3}{}}$$
$$N=N-\underset{\underset{H}{|}}{N}-$$

Diazotyrosine is very sensitive to light, and it is therefore necessary to perform the diazotization and coupling steps in the dark.

Glenner and Lillie (1959) tested the suitability of many other compounds for the coupling reaction. They found that only S-acid, α-naphthol, resorcinol, phloroglucinol, pyrogallol, and naphthoresorcinol were suitable.

The diazotization-coupling method is very specific, gives good color, and is very reliable. It rivals Baker's modification of the Millon method as the method of choice for demonstrating tyrosine-rich proteins.

Methods for Tryptophan

DMAB reaction. The most commonly used method for demonstrating tryptophan is the dimethylaminobenzaldehyde (DMAB) reaction perfected by Adams (1957). Procedures utilizing DMAB for this purpose had been developed earlier, but they were less than satisfactory. In Adams' procedure the tissue sections are treated with DMAB dissolved in concentrated HCl. However, an intense blue pigment is developed only after oxidizing the tissue sections in a solution of sodium nitrite, also dissolved in concentrated HCl. The first part of the reaction is thought to occur as follows:

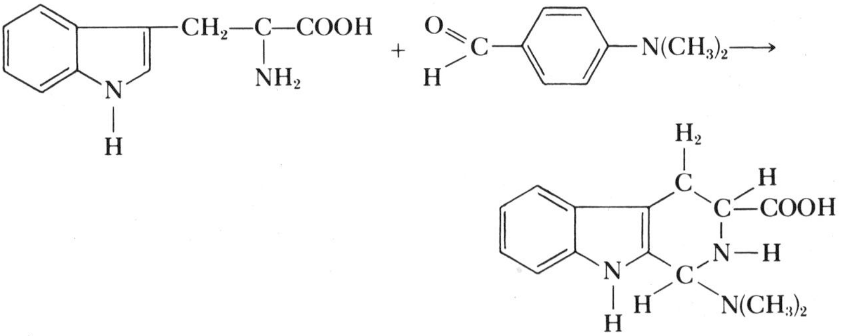

The mechanism by which the reaction product is converted to a blue pigment is not understood.

The blue pigment is seen at sites of proteins containing substantial amounts of tryptophan. Table 5-1 shows that chymotrypsinogen, pepsin, and fibrin contain moderate amounts of tryptophan; it is not surprising therefore that the zymogen granules of the exocrine pancreas, the chief cells of the stomach, and thrombi show strong reaction products. In addition, Paneth cell granules, muscle fibers, neurokeratin, the inner hair root sheath, and many other tissues also give a moderate-to-strong color. Many tissue components do not react; notable among these are collagen, reticulin, parietal cells, goblet cells, lipofuscin, thyroid follicular cells, the neurohypophysis, and the Betz cells.

A blue reaction product is also obtained with other 3-indole derivatives, including tryptamine, serotonin, and 3-indoleacetic acid. The reaction can be completely blocked by prior oxidation of the tissue sections with 90 percent performic acid. Other amino acids also react with DMAB but cannot be oxidized to a blue pigment with nitrite.

Variation using S-acid. A postcoupling variation of this method has been presented by Glenner and Lillie (1957). The tissue sections are treated with diazotized S-acid rather than with nitrite following the DMAB reaction. Although the method is slightly more involved, it may have some merit since it shows the alpha cells of the pancreatic islets more distinctly than Adams' method.

Methods for Sulfhydryl and Disulfide Groups

Of the many methods that have been developed for demonstrating sulfhydryl and disulfide groups, three will be discussed: the ferric ferricyanide method of Chèvremont and Fréderic (1943), the mercury orange method of Bennett (1951), and the DDD (dihydroxy-dinaphthyl-disulfide) reaction of Barnett and Seligman (1952a and 1952b).

It should first be noted that formaldehyde can react with sulfhydryl groups so that they cannot be demonstrated. However, light formalin fixation (for 3 or 4 hours) is generally satisfactory and may be preferred in order to maintain good histological detail. Alternatively, frozen sections may be used.

Ferric ferricyanide method. The ferric ferricyanide method of Chèvremont and Fréderic (1943) is based on the formation of ferric ferrocyanide (Prussian blue) upon sulfhydryl reduction of ferricyanide. First ferric ferricyanide is formed in the reaction medium:

$$2K_3Fe(CN)_6 + Fe_2(SO_4)_3 \longrightarrow 2Fe[Fe(CN)_6] + 3K_2SO_4$$
potassium ferricyanide   ferric ferricyanide

Table 5-1. Amino Acid Composition of Proteins

Protein Source	Ala	Arg	Asn	Asp	Cys	Gln
Expressed as residues per molecule of						
Trypsinogen, bovine	14	2	17	9	12	12
Chymotrypsinogen, bovine	22	4	14	9	10	10
Glucagon, pig	2	1	2	3	0	3
Insulin	1	1	3	0	6	4
Oxytocin, mammalian	0	0	1	0	2	1
Vasopressin, mammalian	0	1	1	0	2	1
Corticotropin, human	3	3	0	2	0	1
Melanocyte-stimulating hormone, bovine	0	1	0	0	0	0
Growth hormone, human	7	10	6	14	4	6
Secretin	1	4	0	2	0	2
Histone IV, bovine	7	14	2	3	0	2
Expressed as moles per 1,000,000 gm protein						
Pepsin, bovine	50.7	5.6	—	125	12.1	—
Fibrin	45	48	—	89.5	16	—
Hair	0	57.4	—	58.6	83–124	—
Wool	45	58	—	55.6	112	—

Source: Data for bovine pepsin from H. Neurath (Ed.), *The Proteins* (2nd ed.), Vol. 1, Academic Press, New York, 1963. Data for fibrin, hair, and wool from R. J. Block and D. Bolling, *The Amino Acid Composition of Proteins and Foods* (2nd ed.), Thomas, Springfield, Mass., 1951. Data for other protein sources from M. O. Dayhoff (Ed.), *Atlas of Protein Sequence and Structure*, Vol. 4, National Biomedical Research Foundation, Silver Spring, Md., 1969.

Ala = alanine; Arg = arginine; Asn = asparagine; Asp = aspartic acid; Cys = cysteine; Gln = glutamine; Glu = glutamic acid; Gly = glycine; His = histidine; Ile = isoleucine; Leu = leucine; Lys = lysine; Met = methionine; Phe = phenylalanine; Pro = proline; Ser = serine; Thr = threonine; Trp = tryptophan; Tyr = tyrosine; Val = valine.

Next, the ferric ferricyanide reacts with sulfhydryl groups of the tissue section:

$$8Fe[Fe(CN)_6] + 6R—S—H \longrightarrow 3R—S—S—R$$
$$+ 2Fe_4[Fe(CN)_6]_3 + 2H_3Fe(CN)_6$$

ferric ferrocyanide (Prussian blue)

All strongly reducing groups in tissue sections will produce the Prussian blue pigment (Adams, 1956). In addition to sulfhydryl groups, these include phenols, indoles, pyrroles, aromatic amines, as-

Table 5-1 (continued)

Glu	Gly	His	Ile	Leu	Lys	Met	Phe	Pro	Ser	Thr	Trp	Tyr	Val
2	25	3	15	14	15	2	3	8	34	10	4	10	18
5	23	2	10	19	14	2	6	9	28	23	8	4	23
0	1	1	0	2	1	1	2	0	4	3	1	2	1
5	3	1	3	3	1	0	3	1	3	3	0	4	4
0	1	0	1	1	0	0	0	1	0	0	0	1	0
0	1	0	0	0	0	0	1	1	0	0	0	1	0
4	3	1	0	1	4	1	3	4	3	0	1	2	3
1	1	1	0	0	1	1	1	1	2	0	1	1	1
20	8	3	8	25	9	3	13	8	18	10	1	8	7
1	2	1	0	6	0	0	1	0	4	2	0	0	1
4	17	2	6	8	11	1	2	1	2	7	0	4	9
77.1	108.0	3.1	76.6	79.6	2.9	13.9	40.8	42.6	125.7	79.8	17.2	51.9	60.6
102	72	18.7	47.0	59.6	62.3	17.4	27.3	46.1	122	61.4	19.1	33.1	47.9
100	60	6.5	37.8	67.2	20.5	6.7	18.2	35–70	72.3	64.7	5.9	18.2	48.7
95.2	90.7	6.5	35.3	68.9	21.2	4.7	24.2	69.6	89.5	54.6	7.4	30.4	46.1

corbic acid, and uric acid. Therefore, parallel control sections must be carried through in which the sulfhydryl groups have been specifically blocked with mercuric chloride. Only sites that are positive in the test and negative in the control can be taken as sites of sulfhydryl groups. Despite a number of newer methods, the Chèvremont-Fréderic method remains valid and useful for demonstrating sulfhydryls.

Disulfide groups can be demonstrated by a modification of the method. Sulfhydryl groups are first blocked and the disulfide groups are then reduced to sulfhydryl groups with thioglycolate. The ferric ferricyanide procedure will then demonstrate sites of disulfide groups.

It has been observed that a green pigment is sometimes produced. This was investigated by Adams (1956), who found that weakly reducing tissue components, such as unsaturated oils, cause the green pigment.

Mercury orange method. The mercury orange method of Bennett (1951) is based on an older biochemical method in which *p*-chloromercuribenzoate was used to form covalent mercaptan linkages with sulfhydryl groups. Bennett (1951) used a related compound, 1-(4-chloromercuriphenylazo)-2-naphthol—also called mercury orange —in his histochemical method. The method has high specificity for tissue sulfhydryl groups. It can be blocked by a variety of sulfhydryl-blocking reagents, including iodoacetate and maleimide compounds as well as other mercaptide-forming reagents.

Bennett (1951) recommended using freeze-dried tissues or tissues fixed in TCA (trichloroacetic acid), but Mescon and Flesch (1952) found that the color developed in formalin-fixed tissues did not differ appreciably from that of freeze-dried tissues.

Mercury orange is not soluble in aqueous solutions. It must be dissolved in ethanol, propanol, butanol, or dimethylformamide, which has been the source of some difficulty with the method. A water-soluble compound, 4-(*p*-dimethylaminobenzene-azo)-phenylmercuric acetate, has been used with good results by Engel and Zerlotti (1964). The method has the disadvantage that the color of the reaction product is light yellow. Engel and Zerlotti had to view the sections with a filter (430 nm) to obtain a good contrast.

The DDD reaction. The DDD (dihydroxy-dinaphthyl-disulfide) method developed by Barrnett and Seligman (1952a and 1952b) seems to be the method of choice for demonstrating sulfhydryl groups. At pH 8.5, DDD reacts with sulfhydryl groups to form a colorless compound. The excess reagent and the reaction by-product are removed from the tissue section by washing thoroughly with alcohol and ether. The colorless reaction product is then converted to an intensely colored compound by coupling it with tetrazotized *o*-dianisidine:

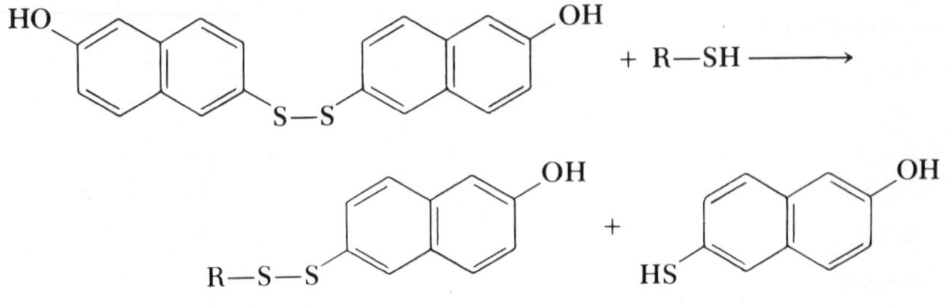

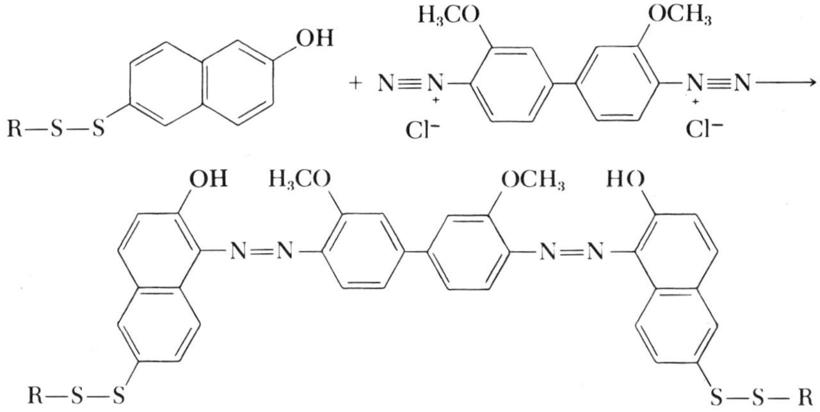

The specificity of this method results from the ability of the disulfides of the reagent to oxidize only sulfhydryl groups of the tissue section. Specific sulfhydryl oxidation with iodine completely prevents the DDD reaction, as does blocking by N-ethylmaleimide and iodoacetate. Development of the colored product with o-dianisidine can be prevented by treating the sections with glutathione after the DDD reaction. Despite this rather convincing evidence of specificity, Gabler and Scheuner (1966) claimed that carboxyl groups in the tissue also react with the DDD reagent. These authors observed that esterification of carboxyl groups diminishes the DDD reaction. Their interpretation assumes, however, that sulfhydryl groups are completely unaffected by the esterification procedure, which may be an unjustified assumption. The specificity of the DDD reaction also depends on the complete removal of all excess DDD reagent and the reaction by-product, 6-thio-2-naphthol. Otherwise a diffuse pink color will develop throughout the tissue.

A blue color was interpreted by Barnett and Seligman (1952a and 1952b) to represent dicoupling (shown above) and to indicate sites of high concentrations of sulfhydryl groups. Red or pink colors were thought to represent monocoupling and to indicate sites of low concentrations of sulfhydryl groups. An intermediate reddish blue color is thought to represent mixed monocoupling and dicoupling and to indicate sites of intermediate concentrations of sulfhydryl groups.

Reactions for Histidine

Method of Lillie and Donaldson. It has long been known that p-diazobenzenesulfonic acid reacts with tyrosine, tryptophan, and histidine and that it could be used as a general stain for proteins. The reaction was rendered specific for histidine by selective nitration blocking of tyrosine and tryptophan with the xanthoproteic reaction (Brunswik, 1923). This reaction, which requires that tissue sections be treated with strong (40 percent) nitric acid, is so destructive to tissue that it is rarely used. Lillie and Donaldson (1972) modified the method by diluting the nitric acid to 20 percent with glacial acetic acid. This modification causes considerably less tissue damage, while the specificity for histidine remains fairly high.

DHT method. Diazo-1-H-tetrazole (DHT) has been used by biochemists for assaying histidine and tyrosine residues in proteins (Horinishi et al., 1964; Takenaka et al., 1969), and the method has been adapted for histochemical purposes (Staple, 1970; Fiedler and Hahn von Dorsche, 1970; Hahn von Dorsche and Fiedler, 1970). DHT has the following structure:

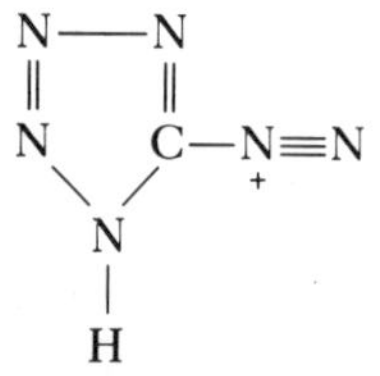

$$
\begin{array}{ccc}
N\!\!-\!\!N & & \\
\parallel \quad \parallel & & \\
N \quad \; C\!-\!\!N\!\!\equiv\!\!N \\
\;\diagdown \diagup \; {\scriptstyle +} \\
\quad N \\
\quad \mid \\
\quad H
\end{array}
$$

DHT is not stable and should be made up fresh each time by diazotization of its 5-amino analog. Other amino acids also react with this reagent, but only tyrosine and histidine form colored reaction products. However, the color obtained with tyrosine is different from and weaker than that obtained with histidine. Furthermore, the reaction with tyrosine can be blocked (with methylation, benzoylation, or iodination) to make it specific for histidine (Hahn von Dorsche and Fiedler, 1970).

The color produced with DHT is somewhat weak and unstable in light. Hahn von Dorsche and Fiedler (1970) complexed the reaction product with a heavy metal salt (copper or lead), which increased the contrast and stabilized the reaction product against light.

Method for Arginine

Only one type of reaction exists for demonstrating arginine. It was originally developed by Sakaguchi (1925), and many modifications have since been made. The method is based on the reaction of α-naphthol with guanidino groups of arginine, which can then be converted to a colored product by treating the tissue sections with hypochlorite or hypobromite. McLeish et al. (1957) introduced the use of dichloronaphthol as a substitute for α-naphthol, resulting in a somewhat greater color intensity.

The only component in mammalian tissues that gives a positive reaction is arginine. However, in lower vertebrates and invertebrates, other rare amino acids can also give positive reactions.

The procedure itself is not difficult to perform except that the tissue sections are difficult to hold on the slide in the alkaline reaction medium. Also, the color is difficult to maintain in the mounted sections. Incorporation of various amines into the mounting medium has improved color preservation, although it is never completely satisfactory.

As a fixative, 10 percent neutral buffered formalin or ethanol-acetic acid may be used. However, Lewitsky's fluid (10 percent formalin with 1 percent chromic acid) is preferred by some workers because it is thought to unmask some of the guanidino groups without destroying them.

MERCURY–BROMOPHENOL BLUE METHOD FOR PROTEINS
(After Mazia et al., 1953; Bonhag, 1955)

TISSUES:
Any fixative except osmium tetroxide may be used. Use paraffin sections.

PREPARATION OF MERCURY–BROMOPHENOL BLUE SOLUTION:
Dissolve 0.4 gm $HgCl_2$ in 40 ml of 2 percent acetic acid, and add 20 mg bromophenol blue.

PROCEDURE:
(1) Deparaffinize sections and bring to water.
(2) Stain with mercury–bromophenol blue solution for 15 minutes.

(3) Wash in 3 changes of 0.5 percent acetic acid for 5 minutes each.
(4) Place directly into *tert*-butyl alcohol for 1 minute.
(5) Place into two more changes of *tert*-butyl alcohol for a total of 3 hours or overnight.
(6) Clear in xylene and mount.

RESULTS:
All proteins are stained a grayish blue color.

ACROLEIN-SCHIFF METHOD FOR PROTEINS
(van Duijn, 1961)

TISSUES:
Fix tissues either with Carnoy's fixative or with 10 percent formalin. Use paraffin sections.

STAINING PROCEDURE:
(1) Deparaffinize sections and bring to 95 percent ethanol.
(2) Incubate in 5 percent acrolein in 95 percent ethanol for 15 to 60 minutes.
(3) Pass through 3 changes of 95 percent ethanol for 5 minutes each.
(4) Transfer to water.
(5) Treat with Schiff's reagent (see page 106) for 10 to 20 minutes.
(6) Dehydrate through graded alcohols, clear in xylene, and mount in synthetic medium.

RESULTS:
The red-to-magenta color indicates the distribution of proteins.

STAINS-ALL PROCEDURE FOR PROTEINS
(Green and Pastewka, 1974a and 1974b)

TISSUES:
Fix tissues with any formalin fixative. Use paraffin sections.

PREPARATION OF STAINS-ALL STOCK SOLUTION:
Dissolve 0.1 gm of stains-all dye* in 100 ml of N,N-dimethyl formamide. Store in the dark or in a brown bottle.

PREPARATION OF STAINS-ALL WORKING SOLUTION:
Prepare Michaelis buffer, pH 3.6, and dilute 1:20. Add 2 ml of stains-all stock solution to 38 ml of the diluted Michaelis buffer.

PROCEDURE:
(1) Deparaffinize sections and bring to water.
(2) Stain in working solution for 1 hour in the dark.
(3) Rinse in tap water.
(4) Dehydrate in *tert*-butyl alcohol *or* air-dry and desiccate overnight in the dark.
(5) Mount with a synthetic resin.

RESULTS:
Glycoproteins stain blue-green, phosphoproteins stain blue, and most other proteins stain red. In addition, nuclei and cartilage matrix stain purple and mast cell granules stain red-purple.

DNFB METHOD FOR PROTEINS
(After Danielli, 1953)

TISSUES:
Fix tissues with alcohol, acetone, or Carnoy's fixative and embed in paraffin. Freeze-dried tissues or fresh-frozen cryostat sections may also be used.

PREPARATION OF DNFB SOLUTION:
Add 1 gm of sodium bicarbonate to 100 ml of 70 percent ethanol to make a saturated solution, and then add 0.5 ml DNFB.

---

*1-ethyl-2-[3-(1-ethylnaphtho [1,2d] thiazolin-2-ylidene)-2-methylpropenyl[-naphtho [1,2d] thiazolium bromide

PREPARATION OF NITROUS ACID:
Dissolve 0.5 gm sodium nitrite in 100 ml 0.1N HCl and keep it at 0 to 4°C. Prepare fresh each time.

PREPARATION OF H-ACID:
Dissolve 2 gm H-acid (8-amino-1-naphthol-3,6-disulfonic acid) in veronal acetate buffer, pH 9.4.

PROCEDURE:
(1) Deparaffinize sections and bring to 70 percent ethanol.
(2) Place in DFNB solution for 1 hour.
(3) Wash three times in 70 percent ethanol.
(4) Place in 5 percent sodium hydrosulfite for 30 minutes at 45°C.
(5) Wash in two changes of water for 3 minutes each.
(6) Place in nitrous acid for 5 minutes at 0°C.
(7) Wash in water.
(8) Place in H-acid solution for 15 minutes at 0° to 4°C.
(9) Wash in tap water.
(10) Dehydrate through graded alcohols, clear in xylene, and mount with synthetic mountant.

RESULTS:
Proteins with amino groups, sulfhydryl groups, histidine, and tyrosine will be colored reddish purple.

MILLON REACTION
(Modified from Baker, 1956)

TISSUES:
Fix tissues in formol-saline or in Heidenhaim's fixative overnight. Use paraffin sections. (Baker recommended embedding in celloidin.)

PREPARATION OF MERCURIC SULFATE STOCK SOLUTION:
Add 10 ml of concentrated sulfuric acid to 90 ml distilled water. Add 10 gm mercuric sulfate and heat until it is dissolved. Cool the solution and add distilled water to make 200 ml. The solution is stable and can be stored at room temperature.

PREPARATION OF INCUBATION MEDIUM:
Add 3 ml of a 0.25 percent sodium nitrite solution to 30 ml of the mercuric sulfate stock solution.

PROCEDURE:
(1) Deparaffinize sections and bring to water.
(2) Place sections in the incubation medium, which has been brought to the boiling point. Continue to apply heat to keep it boiling. Remove sections after 10 to 30 seconds.
(3) Wash the sections in three changes of distilled water for 2 minutes each.
(4) Dehydrate through graded alcohols, clear in xylene, and mount with synthetic medium.

RESULTS:
Proteins containing tyrosine are stained red or orange.

DIAZOTIZATION-COUPLING METHOD FOR TYROSINE
(Glenner and Lillie, 1959)

TISSUES:
Use paraffin sections of formalin-fixed tissues.

PREPARATION OF DIAZOTIZING SOLUTION:
Dissolve 6.9 gm $NaNO_2$ in $2N$ acetic acid.

PREPARATION OF COUPLING SOLUTION:

S-acid*	0.5 gm
KOH	0.5 gm
NH_4 sulfamate	0.5 gm
Ethanol, 70 percent	50 ml

PROCEDURE:
(1) Deparaffinize sections and bring to water.
(2) Place slides in diazotizing solution for 24 hours at 3°C in the dark.

*8-amino-1-naphthol-5-sulfonic acid

(3) Wash four times in distilled water at 3°C in the dark.
(4) Place slides in coupling solution for 1 hour at 3°C in the dark.
(5) Wash three times in 0.1N HCl for 5 minutes each.
(6) Rinse in running water for 10 minutes.
(7) Dehydrate through graded alcohols, clear with xylene, and mount with Permount.

RESULTS:

Tyrosine-containing proteins stain deep purple.

DMAB-NITRITE METHOD FOR TRYPTOPHAN
(Adams, 1957)

TISSUES:

It is preferable to use 1 percent trichloroacetic acid in 80 percent ethanol, but 10 percent formol-saline is also satisfactory if fixation is limited to 12 hours. Use paraffin sections.

PROCEDURE:

(1) Deparaffinize sections and bring to absolute ethanol and allow then to just dry in air. Alternatively, sections may be taken from absolute ethanol and dipped directly into 0.25 percent celloidin and allowed to just dry.
(2) Immerse in a 5 percent solution of p-dimethylamino-benzaldehyde (DMAB) in concentrated HCl for 1 minute.
(3) Transfer immediately to a 1 percent solution of sodium nitrite in concentrated HCl for 1 minute.
(4) Rinse in absolute ethanol.
(5) Wash in running tap water for 30 seconds.
(6) Dehydrate, clear, and mount.

RESULTS:

Proteins with tryptophan form a strong blue pigment. Tryptamine, serotonin, and 3-indoleacetic acid also form the blue pigment.

FERRIC FERRICYANIDE METHOD FOR SULFHYDRYL GROUPS
(Chèvremont and Fréderic, 1943)

TISSUES:

Fix tissues for 3 to 4 hours with a formalin fixative. Use paraffin sections or frozen sections.

INCUBATION MEDIUM:

Ferric sulfate ($Fe_2(SO_4)_3$), 1 percent solution	90 ml
Potassium ferricyanide ($K_3Fe(CN)_6$), 0.1 percent solution	30 ml

Adjust the pH to 2.4. The ferric sulfate solution is stable but the potassium ferricyanide solution must be made fresh each time.

PROCEDURE:
(1) Deparaffinize sections and bring to water.
(2) Place successively in three changes of the incubation medium for a total of 20 to 25 minutes (10 to 20 minutes for frozen sections).
(3) Wash in distilled water.
(4) Differentiate with a brief rinse in 60 percent ethanol containing 2 percent NaOH. (This step is optional.)
(5) Dehydrate, clear, and mount with synthetic medium in the usual way. (Mount frozen sections with glycerin jelly.)

RESULTS:

Sulfhydryl groups are stained blue.

MERCURY ORANGE METHOD FOR SULFHYDRYL GROUPS
(Bennett, 1951; Mescon and Flesch, 1952)

TISSUES:

Use paraffin sections of freeze-dried or formalin-fixed tissues, or use fresh-frozen sections.

PREPARATION OF MERCURY ORANGE SOLUTION:
Prepare a saturated alcoholic solution of mercury orange by dissolving
3 mg of the dye in 100 ml absolute ethanol; stir for 30 to 60 minutes at
room temperature. Filter the solution and dilute with 25 ml distilled
water.

PROCEDURE:
(1) Deparaffinize sections and bring to absolute ethanol.
(2) Place in the mercury orange solution for 1 to 3 hours.
(3) Dehydrate through 95 and 100 percent ethanol, clear in xylene,
    and mount with synthetic mountant.

RESULTS:
Proteins with sulfhydryl groups are stained orange.

DDD METHOD FOR SULFHYDRYL GROUPS
(Barnett and Seligman, 1952b)

TISSUES:
Fixation with 1 percent trichloroacetic acid in 80 percent ethanol is
preferred, but Carnoy's fixative, Bouin's fixative, and formalin are also
satisfactory. Paraffin sections are used.

PREPARATION OF THE DDD REAGENT:
Dissolve 25 mg of dihydroxy-dinaphthol-disulfide (DDD) in 15 ml of
absolute ethanol. Dilute with 35 ml of $0.1M$ veronal-acetate buffer, pH
8.5.

PREPARATION OF FAST BLUE B SALT SOLUTION:
Dissolve 50 mg fast blue B salt in 50 ml of $0.1M$ phosphate buffer, pH
7.4. Use at room temperature. This solution must be made immediately before use.

PROCEDURE:
(1) Bring sections to water.
(2) Incubate in DDD reagent for 1 hour at 50°C.
(3) Cool to room temperature.
(4) Rinse briefly in distilled water.

(5) Wash in two changes of distilled water, acidified to pH 4 to 4.5. (This converts the sodium salts of the naphthyls to the free naphthols.)

(6) Wash in 70, 80, 95, and 100 percent ethanol and then in two changes of absolute ether, 5 minutes in each of the six changes. (This step removes the reaction by-product.)

(7) Rinse in distilled water.

(8) Treat with fast blue B salt solution for 2 minutes.

(9) Wash in running tap water.

(10) Dehydrate, clear, and mount.

RESULTS:

Sites of high concentrations of $-SH$ groups stain blue, while sites of low concentrations stain red.

METHOD FOR HISTIDINE

(Lillie and Donaldson, 1972)

TISSUES:

Fix tissues with neutral buffered formalin or with chloroform-methanol solution. Use paraffin sections.

PREPARATION OF NITRATING SOLUTION:

Carefully mix 20 ml glacial acetic acid with 16 ml nitric acid and cool, preferably in an ice bath. Slowly add 2 ml acetic anhydride.

PREPARATION OF COUPLING SOLUTION:

Treat 195 mg sulfanilic acid with 2 ml concentrated HCl. Add to 25 ml hot (60° to 80°C) distilled water to dissolve the acid completely. Cool the solution to 3°C. Add 150 to 200 mg $NaNO_2$ dissolved in 1 ml ice water and diazotize for 20 minutes at 3°C. Following diazotization, the excess $NaNO_2$ may be destroyed by adding 100 mg urea. Finally, add 12 to 15 ml of 16 percent Na_2CO_3, precooled.

PROCEDURE:

(1) Treat with nitrating solution for 4 hours at 3°C.

(2) Wash in 2 changes of distilled water for 5 minutes.

(3) Treat with coupling solution for 3 minutes at 3°C.

(4) Wash in 3 changes of 0.1N HCl for at least 5 minutes each.

(5) Stain with 0.02 percent azure A in 1 percent HCl for 20 minutes.

(6) Rinse in two 10-second changes of distilled water.

(7) Dehydrate in several changes of fresh acetone.

(8) Clear in xylene.

(9) Mount with xylene cellulose tricaprate.

RESULTS:

Sites containing histidine are stained various shades of green and yellow.

DHT METHOD FOR HISTIDINE

(Staple, 1970; Hahn von Dorsche and Fiedler, 1970)

TISSUES:

Fix tissues with Carnoy's fixative or with alcoholic formalin. Use paraffin sections.

PREPARATION OF TYROSINE-BLOCKING REAGENT:

Mix 30 ml of Gram's iodine with 2 ml 3 percent ammonia. The pH of the solution should be about 10.

PREPARATION OF GRAM'S IODINE:

Iodine crystals	1 gm
Potassium iodide	2 gm
Distilled water	300 ml

PREPARATION OF DHT:

Dissolve 1 gm 5-amino-1-H-tetrazole in 23 ml 1.6N HCl. Cool the solution on an ice bath. Then slowly add 0.7 gm sodium nitrite dissolved in 10 ml distilled water. Adjust the pH to 8 with $5M$ potassium hydroxide and then to pH 8.8 by adding $0.05M$ potassium hydroxide one drop at a time. Dilute the solution with an equal volume of $0.67M$ carbonate buffer, pH 8.8. (Caution: Remember that DHT is not stable and should be prepared fresh each time. Do not try to prepare a higher concentration of DHT because the precipitate that may form is dangerously explosive.)

PROCEDURE:

(1) Deparaffinize the section and bring to water.
(2) Treat with tyrosine-blocking reagent for 5 minutes.
(3) Wash in water for 5 minutes.
(4) Incubate in DHT solution for 1 hour at room temperature.
(5) Wash in running water for 5 minutes.
(6) Rinse in distilled water.
(7) Treat with a solution of either copper acetate or lead acetate (0.1 gm per 100 ml distilled water) for 5 minutes.
(8) Wash in water for 5 minutes.
(9) Dehydrate through graded alcohols, clear in xylene, and mount with a synthetic resin.

RESULTS:

Histidine-containing proteins stain red (when complexed with copper acetate) or grayish yellow (when complexed with lead acetate). (If the metal complexing step is omitted, the color is yellow.) If the blocking step is omitted, both histidine- and tyrosine-containing proteins stain.

METHOD FOR ARGININE
(After McLeish et al., 1957)

TISSUES:

McLeish and co-workers used this method for smears. Fix the smears for 15 minutes in Lewitsky's fixative (equal volumes of 10 percent formalin and 1 percent chromic acid). Paraffin sections may also be used of tissues fixed with various fixatives including formalin and Carnoy's and Bouin's fixatives.

PREPARATION OF 2,4-DICHLORO-1-NAPHTHOL SOLUTION:

NaOH, 1 percent	30 ml
2,4-dichloro-1-naphthol, 1 percent dissolved in 70 percent ethanol	0.6 ml
Sodium hypochlorite, 1 percent	1.2 ml

Prepare immediately before use.

PROCEDURE:

(1) Deparaffinize sections and bring to water, or, if smears are used, pass them quickly through a similar graded alcohol series.
(2) Place in 2,4-dichloro-1-naphthol solution for 6 minutes.
(3) Rinse quickly in 5 percent urea.
(4) Place in 1 percent NaOH for 5 minutes.
(5) Mount in a mixture of 9 volumes of glycerol and 1 volume of 10 percent NaOH.

RESULTS:

Areas of tissue containing arginine (or arginine-rich proteins) are colored red. The color fades slowly over a few days in this alkaline mounting medium. If a nonalkaline mounting medium is used, the color is much less stable, quickly fading to orange and then yellow.

REFERENCES

Adams, C. W. M. (1956). A stricter interpretation of the ferric ferricyanide reaction with particular reference to the demonstration of protein-bound sulfhydryl and disulfide groups. *J. Histochem. Cytochem.* 4:23–35.

Adams, C. W. M. (1957). A *p*-dimethylaminobenzaldehyde-nitrite method for the histochemical demonstration of tryptophane and related compounds. *J. Clin. Pathol.* 10:56–62.

Baker, J. R. (1956). The histochemical recognition of phenols, especially tyrosine. *Q. J. Microsc. Sci.* 97:161–164.

Barnett, R. J., and Seligman, A. M. (1952a). Demonstration of protein-bound sulfhydryl and disulfide groups by two new histochemical methods. *J. Natl. Cancer Inst.* 13:215–216.

Barnett, R. J., and Seligman, A. M. (1952b). Histochemical demonstration of protein-bound sulfhydryl groups. *Science* 116:323–327.

Bennett, H. S. (1951). The demonstration of thiol groups in certain tissues by means of a new colored sulfhydryl reagent. *Anat. Rec.* 110:231–247.

Bensley, R. R., and Gersh, I. (1933). Studies of cell structure by the freezing-drying method. II. The nature of the mitochondria in the hepatic cell of Amblystoma. *Anat. Rec.* 57:217–237.

Bonhag, P. F. (1955). Histochemical studies of the ovarian nurse tissues and öocytes of the milkweed bug, *Oncopeltus fasciatus* (Dallas). I. Cytology, nucleic acids and carbohydrates. *J. Morphol.* 96:381–439.

Brunswik, H. (1923). Über den eindeutigen makro- und mikrochemischen Nachweis des Histidins am Eiweisskomplex. *Hoppe Seylers Z. Physiol. Chem.* 127:268.

Chèvremont, M., and Fréderic, J. (1943). Une nouvelle méthode histochimique de mise en évidence des substances à fonction sulfhydrile. Application à l'epiderme, au poil et à la levure. *Arch. Biol.* 54:589–605.

Danielli, J. F. (1949).   Studies on the cytochemistry of proteins. *Cold Spring Harbor Symp. Quart. Biol.* 14:32–39.

Danielli, J. F. (1953).   *Cytochemistry.* Wiley, New York.

Durrum, E. L. (1950).   A microelectrophoretic and microionophoretic technique. *J. Am. Chem. Soc.* 72:2943–2948.

Engel, M. B., and Zerlotti, E. (1964).   The histochemical visualization of protein-bound sulfhydryl groups with an azomercurial reagent. *J. Histochem. Cytochem.* 12:156–163.

Fiedler, H., and Hahn von Dorsche, H. (1970).   5-Diazonium-1H-tetrazol— ein neues Reagens zum histotopochemischen Nachweis von Histidin. *Acta Histochem.* 35:414–416.

Fraschini, A., and Marinozzi, V. (1977).   Critical analysis of the use of the acrolein-Schiff method as a possible DNA reaction. *Histochemistry* 50:197–206.

Gabler, W., and Scheuner, G. (1966).   Zur Spezifität der Dihydroxy-dinaphthyl-disulfid (DDD) Reaktion. *Acta Histochem.* 23:102–109.

Glenner, G. G., and Lillie, R. D. (1957).   The histochemical demonstration of indole derivatives by the post-coupled *p*-dimethylaminobenzylidene reaction. *J. Histochem. Cytochem.* 5:279–296.

Glenner, G. G., and Lillie, R. D. (1959).   Observation on the diazotization-coupling reaction for the histochemical demonstration of tyrosine: Metal chelation and formazan variants. *J. Histochem. Cytochem.* 7:416–422.

Green, M. R., and Pastewka, J. V. (1974a).   Simultaneous differential staining by a cationic carbocyanine dye of nucleic acids, proteins and conjugated proteins. I. Phosphoproteins. *J. Histochem. Cytochem.* 22:767–773.

Green, M. R., and Pastewka, J. V. (1974b).   Simultaneous differential staining by a cationic carbocyanine dye of nucleic acids, proteins and conjugated proteins. II. Carbohydrate and sulfated carbohydrate-containing proteins. *J. Histochem. Cytochem.* 22:774–781.

Hahn von Dorsche, H., and Fiedler, H. (1970).   Histochemischer Nachweis von Histidin mit Diazonium-1H-tetrazol. *Acta Histochem.* 38:263–269.

Horinishi, H., Hachimori, Y., Kurihara, K., and Shibata, K. (1964).   States of amino acid residues in proteins. III. Histidine residues in insulin, lysozyme, albumin and proteases as determined with a new reagent of diazo-1-H-tetrazole. *Biochim. Biophys. Acta* 86:477–489.

Lillie, R. D. (1957).   Adaptation of the Morel Sisley protein diazotization procedure to the histochemical demonstration of protein bound tyrosine. *J. Histochem. Cytochem.* 5:528–532.

Lillie, R. D., and Donaldson, P. T. (1972).   Histochemical azo coupling of protein histidine: Brunswik's nitration method. *J. Histochem. Cytochem.* 20:929–937.

McLeish, J., Bell, L. G. E., La Cour, L. F., and Chayen, J. (1957).   The quantitative cytochemical estimation of arginine. *Exp. Cell Res.* 12:120–125.

Mazia, D., Brewer, P. A., and Alfert, M. (1953).   The cytochemical staining and measurement of protein with mercuric bromophenol blue. *Biol. Bull.* 104:57–67.

Menzies, D. W. (1961).   Bromophenol blue as a stain for muscle striations. *Stain Technol.* 36:285–287.

Mescon, H., and Flesch, P. (1952).   Modification of Bennett's method for the histochemical demonstration of free sulfhydryl groups in skin. *J. Invest. Dermatol.* 18:261–266.

Porter, R. R. (1950).   The reactivity of the imidazole ring in proteins. *J. Biochem.* 46:304–307.

Sakaguchi, S. (1925).   Über eine neue Farbreaktion von Protein und Arginin. *J. Biochem.* 5:25–31.

Sanger, F. (1945).   The free amino groups of insulin. *Biochem. J.* 39:507–515.

Staple, P. H. (1970).   Observations on the use of diazo-1-H-tetrazole and phenylglyoxal in protein histochemistry as applied to human gingiva. *Histochem. J.* 2:109–121.

Takenaka, A., Suzuki, T., Takenaka, O., Horinishi, H., and Shibata, K. (1969).   States of amino acid residues in proteins. XVIII. A revised way of using diazonium-1-H-tetrazole for reactivity examination of histidine and tyrosine residues. *Biochim. Biophys. Acta* 194:293–300.

van Duijn, P. (1961).   Acrolein-Schiff: A new staining method for proteins. *J. Histochem. Cytochem.* 9:234–241.

Zerlotti, E., and Engel, M. B. (1972).   The reactivity of proteins of some connective tissues and epithelial structures with 2,4-dinitrofluoroben-zene. *J. Histochem. Cytochem.* 10:537–546.

6

CARBOHYDRATES

This chapter deals with the histochemistry of *poly*saccharides. The histochemistry of monosaccharides, disaccharides, and oligosaccharides is virtually nonexistent because it is nearly impossible to fix, immobilize, or otherwise preserve low-molecular-weight saccharides for histochemical demonstration. High-molecular-weight polysaccharides are not actually much easier to fix but they are less likely to diffuse or leach out because of their size.

CLASSIFICATION OF CARBOHYDRATES

The classification of polysaccharides has been very confusing and often entirely unsatisfactory. Histochemists sometimes use terms that are different from those used by biochemists, and, what is worse, the terms are often poorly defined. Furthermore, many polysaccharides exist that have not been isolated, purified, or chemically defined; they can be characterized only on the basis of their histochemical reactions.

The term *mucopolysaccharides* has served nobly in the past to describe any chemical compound that consists of a sugar moiety and a protein moiety. Other terms such as *mucoproteins, mucosaccharides, mucosubstances, mucins,* and *mucoids* are terms with equally obscure meanings. These terms (especially *mucopolysaccharides*) are still used today and are convenient when referring to a broad class of carbohydrate compounds or when it is necessary to express a degree of ignorance.

Attempts have been made to develop systems of classification that are more meaningful in terms of what is known of the chemical composition of the compounds. Jeanloz (1960) made the first and most sweeping proposal for such a system of classification. Pearse (1968) presented a classification system that was derived primarily from Jeanloz (1960) but modified from other authors. Pearse's system has been adopted for this discussion with only one change: the unwieldy term *glucosaminoglucuronoglycans* has been replaced with the more manageable term *acid glycosaminoglycans*. Pearse's (1968) biochemical classification system is presented in the following outline.

I. Glycans (polysaccharides or oligosaccharides)
  A. Homoglycans (one monosaccharide component)
    1. Glycogen
    2. Starch
    3. Cellulose
    4. Dextran
    5. Galactan (Helix)
  B. Homopolyaminosaccharides (Chitin)
  C. Homopolyuronosaccharides
    1. Pectic acid
    2. Alginic acid
  D. Heteroglycans (two or more monosaccharide components)
    1. Glycosaminoglycans
      a. Keratan sulfate
      b. Sialoglycans
    2. Acid glycosaminoglycans
      a. Hyaluronic acid
      b. Chondroitin-4-sulfate (chondroitin sulfate A)
      c. Chondroitin-6-sulfate (chondroitin sulfate C)
      d. Dermatan sulfate (chondroitin sulfate B)
      e. Heparin
II. Polysaccharide-protein complexes
  A. Chondroitin sulfate–protein
  B. Hyaluronic acid–protein
  C. Chitin-protein
III. Glycoproteins and glycopolypeptides
  A. Ovomucoid
  B. Salivary gland mucoid (sialoglycoproteins)
  C. Serum glycoproteins (including immunoglobulins)
  D. Blood group specific substances
IV. Glycolipids
  A. Cerebrosides
  B. Gangliosides
V. Glycolipid-protein complexes
  A. Ox brain mucolipid
  B. Strandin

This system of classification still does not resolve the differences between biochemical description and histochemical characterization, because frequently the histochemical characterization of a compound still does not provide us with the kind of information that allows us to place it in the biochemists' system. A system of histochemical classification of carbohydrates has therefore been proposed by Spicer et al. (1965). This system does justice to our knowledge of the histochemistry of carbohydrates but also follows the biochemical classification previously outlined as closely as possible. An adaptation of the histological classification of Spicer et al. is given in the following outline.

I. Neutral mucosubstances—neutral glycoproteins, immunologically reactive glycoproteins, fucomucins, mannose-rich mucosubstances in epithelia and connective tissues (all periodate reactive); *symbol:* G-mucin or G-mucosubstance

II. Acid mucosubstances, sulfated
   A. Connective tissue mucopolysaccharides (periodate unreactive)
      1. Resistant to testicular hyaluronidase
         a. Alcianophilic in the presence of $1.0M$ (or less) $MgCl_2$—keratan sulfate, heparin; *symbol:* S-mucopolysaccharide A ($1.0M$ $MgCl_2$)
         b. Alcianophilic in the presence of $0.7M$ (or less) $MgCl_2$—dermatan sulfate; *symbol:* S-mucopolysaccharide A ($0.7M$ $MgCl_2$)
      2. Susceptible to testicular hyaluronidase
         a. Alcohol-resistant affinity for 0.02 percent azure A at or above pH 2.0—chondroitin sulfates in cartilage;* *symbol:* S-mucopolysaccharide B2.0T
         b. Alcohol-resistant affinity for 0.02 percent azure A at or above pH 4.0—chondroitin sulfates in vascular tissues;* *symbol:* S-mucopolysaccharide B4.0T
   B. Epithelial sulfomucins (testicular hyaluronidase resistant)
      1. Periodate unreactive
         a. Sulfate esters on *vic*-glycols; *symbol:* GS-mucin
         b. -Sulfate esters not on *vic*-glycols; *symbol:* S-mucin

*These mucopolysaccharides show the same metachromasia in both wet and dehydrated sections, unlike epithelial sulfomucins, which often lose metachromasia after alcohol dehydration.

(1) Alcohol-resistant affinity for 0.02 percent azure A at or above pH 2.0; *symbol:* S-mucin B2.0

(2) Alcohol-resistant affinity for 0.02 percent azure A at or above pH 4.5; *symbol:* S-mucin B4.5

2. Periodate reactive—acid glycoproteins (?)

   a. Alcohol-resistant affinity for 0.02 percent azure A at or above pH 2.0; *symbol:* SG-mucin B2.0

   b. Weak or negligible alcohol-resistant affinity for 0.02 percent azure A at or above pH 4.5; *symbol:* SG-mucin B4.5

III. Acid mucosubstances, nonsulfated

   A. Hexuronic acid–rich mucopolysaccharides—hyaluronic acid, chondroitin; *symbol:* U-mucopolysaccharide

   B. Sialic acid–rich mucosubstances

      1. Connective tissue mucopolysaccharides containing sialic acid (?); *symbol:* C-mucopolysaccharide

      2. Epithelial sialomucins—acid glycoproteins

         a. Highly susceptible to *Vibrio cholerae* sialidase, periodate reactive, and stainable metachromatically with azure A; *symbol:* CG-mucin BS

         b. Slowly digestible with *V. cholerae* sialidase

            (1) Periodate reactive; *symbol:* CG-mucin $S^{\pm}$

            (2) Periodate unreactive; *symbol:* C-mucin $S^{\pm}$

         c. Resistant to *V. cholerae* sialidase

            (1) Rendered metachromatic and susceptible to enzyme by prior saponification; *symbol:* s-mucin (Sap) BS

            (2) Sialidase resistant after saponification

               (a) Periodate reactive; *symbol:* GC-mucin

               (b) Periodate unreactive; *symbol:* C-mucin

HISTOCHEMICAL METHODS FOR CARBOHYDRATES
Periodic Acid–Schiff Reaction

The periodic acid–Schiff (PAS) reaction has been of tremendous importance in carbohydrate histochemistry. It is a two-step reaction; oxidation by periodic acid produces aldehydes in the tissue, and the aldehydes are demonstrated with Schiff's reagent. The method was introduced as a histochemical tool by McManus (1946), who used it to demonstrate "mucins." Lillie (1947) also utilized the method to demonstrate glycogen. Hotchkiss (1948) published a similar method, which he developed independently. However, probably the earliest use and most extensive development of the technique was carried out by Shabadash (1947), who may have employed it as early as 1939. (See the discussion by Aterman and Norkin [1963].)

Periodic acid oxidation. Periodic acid is an oxidizing reagent that is known for its ability to oxidize 1,2-glycols to form dialdehydes. The mechanism of action is not fully understood, although it is thought that a monoester is first formed, and then a cyclic diester is formed as an intermediate compound (Bobbitt, 1956). The latter probably breaks down to form dialdehydes directly. The reaction is thought to take place as follows:

$$
\begin{array}{l}
\text{H}-\overset{|}{\text{C}}-\text{OH} \\
\quad\quad\quad + \text{H}_5\text{IO}_6 \longrightarrow \\
\text{H}-\overset{|}{\text{C}}-\text{OH}
\end{array}
\quad
\begin{array}{l}
\text{H}-\overset{|}{\text{C}}-\text{O}-\text{IO}_3\text{H}_4 \\
\quad\quad\quad\quad\quad\quad + \text{H}_2\text{O} \longrightarrow \\
\text{H}-\overset{|}{\text{C}}-\text{OH}
\end{array}
$$

$$
\begin{array}{l}
\text{H}-\overset{|}{\text{C}}-\text{O} \\
\quad\quad\quad\quad\diagdown \\
\quad\quad\quad\quad\quad\text{IO}_4\text{H}_3 + \text{H}_2\text{O} \longrightarrow \\
\quad\quad\quad\quad\diagup \\
\text{H}-\overset{|}{\text{C}}-\text{O}
\end{array}
\quad
\begin{array}{l}
\text{H}-\overset{|}{\text{C}}{=}\text{O} \\
\quad\quad\quad\quad\quad\quad + \text{H}_2\text{O} + \text{HIO}_3 \\
\text{H}-\overset{|}{\text{C}}{=}\text{O}
\end{array}
$$

The normal conditions for periodate oxidation involve the use of a 1 percent solution for 10 minutes at room temperature. Many variations have been evolved, usually for limited and specialized applications.

Many other oxidizing reagents could theoretically be used as well, and some have actually been substituted for periodic acid. However, other oxidants are often less convenient to use (they are either unstable or difficult to prepare). They also may overoxidize the sections (aldehydes are further oxidized to carboxyl groups). This accounts for the overwhelming choice of periodic acid as the oxidant in this method.

Schiff's reagent. Schiff's reagent is the most widely used reagent for detecting aldehydes in tissues. It is prepared from certain triarylmethane dyes. Basic fuchsin, the most commonly used, is a mixture of varying percentages of three similar triarylmethane dyes: rosanilin, pararosanilin, and new fuchsin. The percentages of the three components help to determine the exact color of the PAS reaction, which ranges from magenta to purple. Basic fuchsin is known to have a considerable amount of impurities; the amount varies from batch to batch, and some batches are simply not suitable for preparing Schiff's reagent. Pararosanilin, obtained as such, is preferred by some workers because it supposedly contains fewer impurities and is thought, therefore, to give more consistent results. Methods have even been given for purifying pararosanilin for use in Schiff's reagent (Yarbo et al., 1954; van Duijn and Riddersma, 1973). One objection that might be raised is that pararosanilin gives the lightest color of the four components of basic fuchsin.

A number of methods have been given for preparing Schiff's reagent. Usually an acidified solution of basic fuchsin is treated with a reducing reagent such as sodium metabisulfite (de Tomasi, 1936), with thionyl chloride (Barger and De Lamater, 1948), or with a stream of sulfur dioxide gas (Itikawa and Ogura, 1954). Lillie (1951) published a method for preparing "cold Schiff," in which sodium metabisulfite is used but heat is not required as in most other methods. This is no doubt the easiest to make and probably now the most widely used. The reduced basic fuchsin, or Schiff's reagent, has also been called leucofuchsin or fuchsin sulfurous acid. A properly prepared solution is clear and colorless (or nearly so), but when the molecules react with aldehydes, they regain their original color.

Following the reaction with Schiff's reagent, the tissue sections are given several rinses in 0.5 percent aqueous sodium metabisulfite as a reducing rinse. If the tissue sections are rinsed in water, the residual Schiff's reagent will readily reoxidize to its original triarylmethane dyes. These dyes have affinities for certain tissue components, which will then be stained substantially the same color as the authentic PAS reaction. The rationale for the reducing rinse, therefore, is to remove the residual Schiff's reagent without allowing it to reoxidize—that is, the reagent should be removed while the molecules are in their reduced state. Some workers have dispensed with the bisulfite rinses, and even Pearse (1968) considered running water adequate. It would be difficult to maintain the same degree of confidence in the reaction using only running water. I therefore consider the bisulfite rinses essential.

Both periodic acid and Schiff's reagent have been prepared as alcoholic solutions in an effort to preserve and demonstrate dextran or other water-soluble carbohydrates (Mowry et al., 1952; Bedi and Horobin, 1976). It appears from both studies that periodate oxidation will not take place unless at least a small amount of water is present in the solution.

Schiff's molecules are largely insoluble in alcohol; therefore it has not been possible to obtain a good alcoholic Schiff's reagent. However, Bedi and Horobin (1976) learned to form a complex of phosphotungstic acid (PTA) and Schiff's reagent that is fully soluble in alcohol. The complex is stored in its crystalline form because it is stable for only a few hours in an alcoholic solution.

Summary of PAS reaction. It should be emphasized that there are two key steps in the PAS procedure: (1) oxidizing the tissue sections with periodic acid (formation of aldehyde groups) and (2) treating the sections with Schiff's reagent (visualizing the aldehyde groups). Between these two steps, it is important to wash out all the residual periodic acid. Following the treatment with Schiff's reagent, it is necessary to remove all unreacted Schiff's reagent while in its reduced form.

Uses of the PAS reaction. The PAS reaction is the second most commonly used histological or histochemical technique. It is frequently used to show certain morphological features such as brush borders, basal lamina, goblet cells, and cartilage, without regard for the chemistry underlying the reaction. When it is used in such a way, it is not truly a histochemical procedure.

As a histochemical technique, the PAS reaction is useful for localizing glycoproteins or any mucous substances containing neutral sugars. Collagen has often been observed to be PAS-positive, which is almost certainly due to covalently bound monosaccharides and disaccharides. The PAS procedure colors glycogen, and the procedure can be used to localize glycogen if a diastase-digested control is used. Amino alcohols are PAS-positive; indeed, the earliest interest in periodic acid was to facilitate an assay for amino alcohols (Malaprade, 1928). Sphingomyelin, alginic acid, and pectinic acid can be thought of as amino alcohols, and they are PAS-positive. The glucosamine portion of chondroitin sulfate would be expected to be PAS-positive on the same basis, but in actual practice it is very resistant to periodate oxidation. It requires 24 hours of oxidation at 30°C (Scott and Dorling, 1969).

Metachromatic Staining Methods

The metachromatic dyes are cationic dyes that have an affinity for certain polyanions. Probably the most popular of these are the thiazan dyes, azure A and toluidin blue, and the azan dye, safranin O. The dye molecules bind the polyanions by means of salt linkages. Salt complexes formed between the dye molecules and chondroitin sulfate have been isolated and studied by Pal and Schubert (1961).

A metachromatic dye is one that stains certain groups of tissue components a color different from that of a weak aqueous solution $(10^{-5}M)$ of the dye. The color of the weak aqueous solution of the dye is called the orthochromatic color. If the concentration of the dye in the solution is increased, the absorption will change to a shorter wave length; the color is then known as the metachromatic color. The dye will stain certain tissue components a metachromatic color, but it may also stain other components of the same tissue section orthochromatically.

Historical aspects. The phenomenon of metachromasia, which has been known and studied for a long time, was reported independently in the same year by Cornil (1875), Heschl (1875), and Jürgens (1875). Ehrlich (1878) applied the term *metachromasia* to the color change of the dye in the tissue. Undoubtedly the term was derived from *metachromatism*, the word that had been used earlier to refer to the color change observed when certain chromogenic salt solutions were heated (Ackroyd, 1876). Bouma (1883) made an earnest attempt at explaining the color change observed when cartilage was stained with safranin. Metachromasia was recognized and carefully defined as a separate and distinct phenomenon (Paneth, 1888). The theory and practice of metachromasia has interested many workers over the years, the result being a vast body of literature on the subject. Several excellent reviews have treated the subject rather thoroughly (Kelly, 1956 and 1958; Schubert and Hamerman, 1956; Bergeron and Singer, 1958). Pearse (1968) also presented a good discussion on the subject, although it is less comprehensive than the earlier ones.

Mechanism of action. It is thought that the mechanism responsible for producing the shift in color is that the dye molecules crowd each other as they attach to the polyanions and become strained in such a way that their π-bonds become distorted. It is observed that polyanions with their charges close together induce metachromasia much more readily than those with greater intercharge distances. This suggests that, in order for a dye molecule to be attached to every negative charge, each dye molecule must interfere with the domain of other dye molecules so as to alter their electron configuration.

Some of the dye is often removed during alcohol dehydration of the tissue sections. It appears that the amount of dye washed out depends on the properties of the component stained as well as on the length of time that the sections are exposed to dehydration. Most workers have learned to dehydrate rapidly to avoid losing an undue quantity of stain.

Another observation about dehydration is that some of the components revert to the orthochromatic color (or partially so) during dehydration. This has been thought of as a problem, but it can be taken advantage of in order to differentiate between different forms of polysaccharides (see the outline on pp. 91–92).

Limitations of metachromatic staining methods. In the orthopedic literature, safranin O is used to assess the content of acid glycosaminoglycans in cartilage (Weiss and Mirow, 1972; Mankin and Lippiello, 1970; Altman et al., 1973; Mankin et al., 1971), because it is thought that the color intensity is approximately proportional to the concentration of this polysaccharide. The justification for doing so is based on a study by Rosenberg (1971), who showed that in the test tube safranin O reacted stoichiometrically with the anionic groups of chondroitin sulfate and keratan sulfate. Workers frequently assume that the same stoichiometric relationship holds true in histological sections. A quantitative histochemical study (Troyer, 1974) showed this assumption to be ill-founded. Cartilage sections that had previously been subjected to hyaluronidase digestion failed to stain with azure A or safranin O even though a significant amount of acid glycosaminoglycans remained in the tissue sections. A section of diseased cartilage showing diminished safranin O staining may grossly exaggerate the loss of acid glycosaminoglycans from the tissue. It is preferable to use other stains, such as alcian blue or high iron diamine, which more accurately reflect the concentration of acid glycos-aminoglycans.

Alcian Blue

Few histochemical staining methods have gained as widespread acceptance and use as rapidly as alcian blue. Alcian blue 8GS, a textile dye, was adopted as a histochemical stain specific for "mucins" by Steedman (1950). The specificity was improved by Mowry (1956) by using a higher concentration of the dye at a lower pH. It was soon noted that new batches of alcian blue were more soluble and less resistant to decolorization (Mowry, 1960). The suppliers then admitted that a new type, known as alcian blue 8GX, was being supplied. This is what is in widespread use today.

Structure of alcian blue. Alcian blue is a copper phthalocyanin dye containing four isothiouronium groups. Its exact structure has been kept a trade secret, to the annoyance of histochemists (Scott, 1972a). After great expenditure of effort, the structure of alcian blue 8GX was elucidated (Scott, 1972b). The basic structure of alcian blue 8GX is as follows:

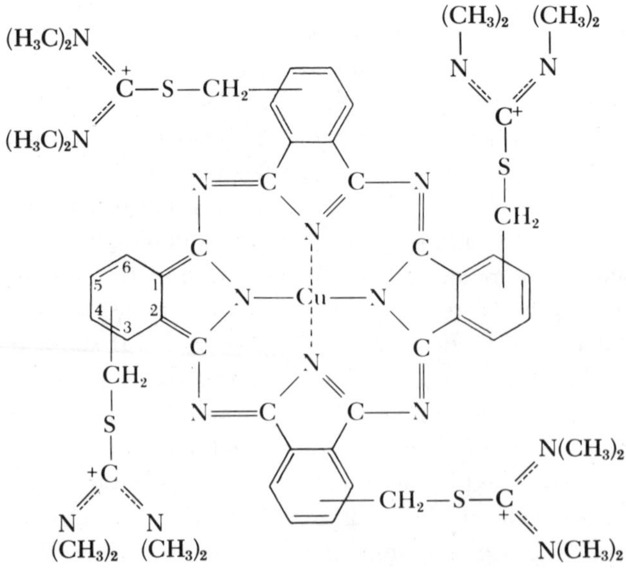

Each of the four isothiouronium groups may be attached at the 4 or 5 position, giving rise to four different isomers of alcian blue 8GX.

Impurities. Alcian blue is frequently contaminated with large quantities of impurities, particularly boric acid. Impurities sometimes constitute as much as 75 percent of the weight. Scott (1972c) presented a relatively simple method of removing boric acid from alcian blue. The percentage of purity in unknown samples of alcian blue 8GX can be estimated from the knowledge that pure samples give an absorption coefficient $E_{1cm}^{1\%}$ of about 425 at a wave length of 623 to 624 nm.

Use of alcian blue. Compared with other mucopolysaccharide stains, alcian blue is unique in three ways: (1) it does not have affinity for nucleic acids; (2) it stains carboxyl groups at a pH substantially below their pK; and (3) it has a high solubility even in the presence of high salt concentrations.

When used in a 3 percent acetic acid solution (pH 2.5 to 2.7), alcian blue stains both sulfated and nonsulfated acid glycosaminoglycans, but at a pH of 1, only sulfated ones stain (Pearse, 1968). Sialoglycoproteins are also stained at pH 2.5 (Spicer et al., 1967).

Jones and Reid (1973a and 1973b) carefully explored the possibility of further distinguishing among mucosubstances. They were able to show that between pH 1.5 and 1.7, alcian blue stains sialidase-resistant sialomucin but not sialidase-sensitive sialomucin. Both sialomucins stain above pH 1.7 but neither stains below pH 1.5. They also found in the canine submandibular gland a sulfomucin that stains *only* below pH 1.5, whereas sulfomucins normally stain anywhere between pH 2.6 and pH 0.5. At pH 1.0, sialomucins can be distinguished from sulfomucins, because at this pH only the sulfomucins stain with alcian blue.

The sensitivity of alcian blue can be increased by restaining after an intermediate alkaline hydrolysis step (Scott, 1972d). The process can be repeated as often as desired. When tissue-bound alcian blue is subjected to weak alkali at room temperature, tetramethylurea is split off its cationic tetramethylisothiouronium groups. Thus the main portion of the alcian blue molecule is freed; however, it nonetheless remains in situ because the process renders it extremely insoluble. The tissue anionic groups are also freed and can now be restained with alcian blue.

Critical electrolyte concentration technique. The critical electrolyte concentration technique offers a method for staining various mucopolysaccharides differentially. The technique has been used with other basic dyes (Bollet, 1967); however, it is most appropriately discussed here because alcian blue is the best dye for this purpose.

Using a less acidic staining solution than is normally used (pH 5.7), Scott et al. (1964) showed that the staining of various polyanions spotted on filter paper could be selectively prevented at specific concentrations of $MgCl_2$. Scott and Dorling (1965) showed that hyaluronic acid staining could be prevented at $0.1M$, ribonucleic acid (RNA) and deoxyribonucleic acid (DNA) at $0.2M$, chondroitin sulfate A at $0.65M$, and heparin at $0.8M$. Keratan sulfate was not prevented from staining even at $1.0M$ $MgCl_2$.

Stockwell and Scott (1965) used this technique to good advantage in studying the distribution of chondroitin sulfate and keratan sulfate in aging human cartilage. In the presence of $0.4M$ $MgCl_2$, both chondroitin sulfate and keratan sulfate are stained with alcian blue, but in the presence of $0.9M$ $MgCl_2$ only high-molecular keratan sulfate is stained. The validity of their findings was further substantiated by selectively digesting away the chondroitin sulfate with testicular hyaluronidase prior to the staining procedure (Scott and Stockwell, 1967).

Nature of alcian blue color. The absorption maximum of alcian blue remains unchanged upon reaction with the chromotrope, unlike that of the metachromatic stains. It is therefore said to be a nonmetachromatic dye. However, if the absorption spectrum of the monomeric dye (as seen in an alcoholic solution, for instance) is taken as the basis of orthochromasia, then alcian blue in an aqueous solution (or tissue stained in alcian blue) is extremely metachromatic (Scott, 1970).

Colloidal Iron

The technique that is generally known as Hale's colloidal iron method provides another way of demonstrating acid glycosaminoglycans. Although the technique was used as early as 1888 by Mörner, Hale (1946) is universally credited with its origin.

The colloidal iron staining reaction is dependent on the affinity of certain tissue polyanions for the colloidal form of ferric ions. After the sections have been treated with an acidic solution of colloidal iron, the unbound ferric ions are carefully washed out of the tissue sections. The bound ferric ions are then visualized by treating the sections with an acidic potassium ferrocyanide solution—the classic Prussian blue reaction.

Early methods commonly employed a relatively high pH (about pH 5); the method was not very specific for acid glycosaminoglycans because it also stains collagen and reticulin (Lillie and Mowry, 1949) and probably other components as well (Davies, 1952). The method, therefore, has been severely criticized, resulting in numerous modifications aimed at increasing its specificity. It has been shown that strongly acidifying the colloidal iron solution (pH 1.8) suppresses the cytoplasmic and nuclear background staining and increases the specificity for carboxyl- and sulfate-containing polyanions (Mowry, 1958). Neuraminic acid is also stained by this method (Haemmerli et al., 1972).

Diamine Methods

A series of methods for staining mucopolysaccharides selectively were developed by Spicer and Jarrel (1961) and Spicer (1961 and 1965) on the basis of the observation that an aqueous solution of N,N-dimethyl-*p*-phenylenediamine could be used to detect periodate-engendered aldehydes as well as acid groups (Spicer and Jarrel, 1961). After standard periodate oxidation, the diamine reacts rapidly with aldehydes to form a brown reaction product, whereas it reacts much more slowly with acid groups to form black reaction products. Tissue components containing both periodate-engendered aldehydes and acid groups give intermediate-colored reaction products.

When the *meta* isomer is added to the *p*-diamine solution, acid polysaccharides stain purple-red. If the diamine mixture is used along with periodate oxidation, the different types of polysaccharides are more readily differentiated. Tissue components containing 1,2-glycols and acid groups in close proximity now stain light yellow-gray or orange-brown following periodate oxidation, as opposed to purple-red in the unoxidized sections. It is thought that the *m*-diamine rapidly condenses with the engendered aldehydes and tends to neutralize the attraction of the acid groups for the colored *p*-diamine complex. The condensation of the colored *p*-diamine complexes requires up to 72 hours for completion.

It was discovered that oxidizing agents, particularly $FeCl_3$, would accelerate the reaction. Again, $FeCl_3$ was found useful in differentiating tissue polyanions by varying the amount of $FeCl_3$ added. A method known as the low iron diamine technique visualizes many components containing sialic acid, carboxyl groups, and sulfate groups. The high iron diamine technique demonstrates sulfated mucopolysaccharides more selectively.

Spicer's claim that new oxidation products are formed from the diamines that are responsible for the stain has been substantiated by Gad and Sylvén (1969), who also showed that polyanions with high-charge densities (e.g., heparin and dextran sulfates) stain dark purple or violet colors, whereas others with lower-charge densities give brownish, reddish, or purplish shades.

The central role that iron plays in making high iron diamine specific for sulfated mucosubstances was shown by Sorvari (1972a and 1972b). Ferric ions were shown to bind to sialic acid groups, carboxyl groups, and nucleic acids; it was suggested that ferric ions compete successfully against the diamine molecules for these groups. This was substantiated with further work by Sorvari and Arvilommi (1973). They were able to separate the high iron diamine staining solution into three fractions. The separate fractions, devoid of iron, stained acidic groups rather indiscriminately, but when the concentration of $FeCl_3$ was restored to the original concentration, the fractions stained sulfomucins very selectively.

Methods for Sialic Acid

Toch and Pearse (1965) were of the opinion that the most satisfactory method for sialic acid was the Ravetto (1964) modification of the Bial reaction. Since then two other noteworthy methods have been developed.

Ravetto's method is based on a biochemical method for detecting sialic acid. Orcinol is used in the presence of a strong acid and copper sulfate, which develops a colored product at the site of sialic acid. The reaction mechanism is not understood.

Sialic acid can be stained with many basic dyes, although not to the exclusion of all other components. Selective enzymatic removal of sialic acid with sialidase (neuraminidase) opens the possibility of an effective control. Rahmann and Katusic (1975) took advantage of the specificity of sialidase and the basophilia of sialic acid to survey the sialic acid-containing compounds of fish brains. Adjacent sections were incubated simultaneously, one in sialidase solution and the other in the buffer solution without the enzyme. Afterward, both sections were stained, with either alcian blue, azure A, or colloidal iron. The difference in coloration between the two sections is therefore presumed to result from sialic acid. The procedure is not very suitable in tissues in which sialic acid makes a minor contribution to the staining intensity.

Sorvari and Lauren (1973) studied the effect of various fixatives on neuraminidase digestibility of the tissue sections. They found that formalin fixation was quite satisfactory. Alcoholic fixatives were somewhat less satisfactory, and methanolic cyanuric chloride completely prevented the enzymatic removal of sialic acid.

It has been observed several times that mild periodate oxidation of sialoglycoproteins results in the selective oxidation of the sialyl residues, while the remainder of the molecule is unmodified (Van Lenten and Ashwell, 1971; Liao et al., 1973). This principle forms the basis of two histochemical methods for sialic acid. Weber et al. (1975) used $5mM$ periodate (about one-tenth the concentration normally used in the PAS reaction) to engender aldehyde groups on sialyl residues. Dansylhydrazine was allowed to react with the resulting aldehydes to form dansylhydrozones, which were then reduced with dimethylamine borane to form stable fluorescent hydrazines.

The method of Roberts (1977) utilizes $0.4mM$ periodate and the standard Schiff's reagent for detecting the aldehydes. The only essential difference between this method and the ordinary PAS reaction is the carefully controlled mild periodate oxidation.

Both Weber et al. (1975) and Roberts (1977) claimed that a minimum of aldehydes were formed on other components, suggesting that their methods are highly selective for sialyl residues.

Differential staining of acetylated and nonacetylated sialic acid. Certain epithelial mucins contain sialic acid residues that are O-acetylated. Such sialic acid residues are not normally PAS-positive but can be rendered so by prior saponification (Culling et al., 1974). In order to specifically demonstrate O-acetylated sialic acid, it was first necessary to block components that are normally PAS-positive; this was accomplished with a periodate-borohydride sequence. Tissue sections were then saponified with alcoholic potassium hydroxide, followed by an ordinary PAS sequence. Since O-acetylated sialic acid occurs almost exclusively in the colon, the method has been found useful in identifying metastases of colonic tumors (Culling et al., 1975), and in diagnosing perianal Paget's disease (Wood and Culling, 1975). A double-staining method was also developed with which it was possible to demonstrate ordinary PAS-positive material and O-acetylated sialic acid simultaneously and differentially (Culling et al., 1976). The former is stained blue with a thionin-derived Schiff's reagent while the latter is stained red with a basic fuchsin-type Schiff's reagent.

STANDARD PAS REACTION

TISSUES:
Fix tissues with any suitable fixative. Use paraffin sections.

PROCEDURE:
(1) Deparaffinize sections and bring to water.
(2) Oxidize with 1 percent aqueous periodic acid for 10 minutes.
(3) Wash in running water for 5 minutes and shake off excess water.
(4) Treat with Schiff's reagent for 10 minutes.
(5) Wash in three changes of 0.5 percent sodium metabisulfite for 2 minutes each.
(6) Wash in water.
(7) Dehydrate through graded alcohols, clear in xylene, and mount with synthetic medium.

RESULTS:
Substances containing *vic*-glycols are stained magenta.

PAS REACTION WITH ABSOLUTE ETHANOL
(Mowry et al., 1952)

TISSUES:
Fix for 48 hours in absolute ethanol at −5°C. Use paraffin sections.

PREPARATION OF ALCOHOLIC SCHIFF'S REAGENT:
Combine 11.5 ml Schiff's reagent (preferably Lillie's) with 0.5 ml concentrated HCl and 28 ml absolute ethanol.

PROCEDURE:
(1) Deparaffinize sections and bring to 70 percent ethanol.
(2) Oxidize in 1 percent periodic acid (dissolved in 70 percent ethanol) for 10 minutes.
(3) Wash in 70 percent ethanol for 5 minutes.
(4) Treat with alcoholic Schiff's reagent for 15 minutes.
(5) Wash in three changes of sodium metabisulfite solution (0.5 gm dissolved in 100 ml absolute ethanol containing 1 ml concentrated HCl) for 2 minutes each.
(6) Dehydrate with absolute ethanol, clear with xylene, and mount with a synthetic medium.

RESULTS:
Substances containing *vic*-glycols are stained magenta.

PAS REACTION WITH ABSOLUTE ETHANOL
(Bedi and Horobin, 1976)

TISSUES:
Bedi and Horobin used a variety of fixatives; however, for substances that are highly water-soluble, alcohol–formol–acetic acid fixative or absolute ethanol should be used. Use paraffin sections.

PROCEDURE:
(1) Deparaffinize sections and bring to 70 percent ethanol.
(2) Oxidize in 1 percent periodic acid in absolute ethanol for 1 hour.
(3) Wash in absolute ethanol for 2 or 3 minutes.
(4) Treat with ethanolic PTA-Schiff reagent (p. 106).
(5) Wash in absolute ethanol for 20 minutes.
(6) Clear in xylene and mount with synthetic medium.

RESULTS:
Substances containing *vic*-glycols are stained magenta.

PREPARATION OF SCHIFF'S REAGENTS

DE TOMASI (1936) SCHIFF'S REAGENT:
Slowly dissolve 1 gm of basic fuchsin in 200 ml of boiling distilled water. Shake it for 5 minutes and allow it to cool to 50°C. Filter and add 20 ml 1N HCl to the filtrate. Allow it to cool to 25°C and add 1 gm sodium (or potassium) metabisulfite. Let the solution stand in the dark for 14 to 24 hours. Add 2 gm of activated charcoal, shake well, and filter. Store it in the dark at 4°C.

DE TOMASI SCHIFF'S REAGENT, LILLIE'S (1951) MODIFICATION:
Boil 100 ml distilled water and remove from the source of heat. Immediately dissolve 1 gm of basic fuchsin (or pararosanilin). Allow it to cool to 60°C and filter. Add 2 gm sodium (or potassium) metabisulfite and 20 ml 1N HCl. Close the vessel tightly and store it for 18 to 24 hours. Add 300 mg activated charcoal, shake well, and filter. Store at 0° to 5°C.

SCHIFF'S REAGENT AS MODIFIED BY ITIKAWA AND OGURA (1954):
Boil 200 ml distilled water and carefully add 1 gm basic fuchsin. Shake until it is dissolved. Cool and filter it into a flask. Bubble SO_2 gas slowly through the solution, shaking it occasionally. When the solution becomes a transparent red color, the process is completed. Close the vessel tightly and leave it in the dark overnight. Add 1 gm of activated charcoal, shake for 1 minute, and filter. Store at 0° to 4°C.

COLD SCHIFF'S REAGENT (LILLIE, 1951):
Dissolve 1 gm basic fuchsin and 1.9 gm sodium metabisulfite in 100 ml 0.15N HCl. Shake the solution at intervals (or place on a mechanical shaker) for 2 hours. The solution should now be clear and yellow to light brown in color. Add 0.5 gm activated charcoal and shake for 1 minute. Filter into a graduated cylinder and wash the residue with a little distilled water to restore the original 100 ml volume.

ETHANOLIC PTA-SCHIFF'S REAGENT (BEDI AND HOROBIN, 1976):
Prepare a fresh batch of Schiff's reagent (Bedi and Horobin suggested making it by de Tomasi's method). The PTA-Schiff complex is made by combining equal volumes of Schiff's reagent and 1 percent aque-

ous phosphotungstic acid (PTA). A white precipitate of PTA-Schiff complex should result. Collect the precipitate by centrifugation and discard the supernatant. Dry the precipitate and store it in a freezer. The PTA-Schiff reagent is prepared by redissolving the precipitate in the same volume of absolute ethanol as was originally used to prepare the PTA-Schiff complex. This reagent must be used immediately, since it is stable for only a few hours.

AZURE A STAINING PROCEDURES
(Spicer et al., 1967)

TISSUES:
Use any formalin fixative. Use paraffin sections.

PREPARATION OF AZURE A STAINING SOLUTIONS:
Dissolve 10 mg azure A in one of the following solutions:

pH 0.5—50 ml 0.5N HCl
pH 1.0—50 ml 0.1N HCl
pH 1.5—30 ml 0.1N HCl + 20 ml 0.1M KH_2PO_4
pH 2.0—20 ml 0.1N HCl + 30 ml 0.1M KH_2PO_4
pH 2.5—48 ml H_2O + 2 ml 0.1M citric acid
pH 3.0—48 ml H_2O + 1.65 ml 0.1M citric acid + 0.35 ml 0.2M Na_2HPO_4
pH 3.5—48 ml H_2O + 1.40 ml 0.1M citric acid + 0.60 ml 0.2M Na_2HPO_4
pH 4.0—48 ml H_2O + 1.25 ml 0.1M citric acid + 0.75 ml 0.2M Na_2HPO_4
pH 4.5—48 ml H_2O + 1.10 ml 0.1M citric acid + 0.90 ml 0.2M Na_2HPO_4
pH 5.0—48 ml H_2O + 1.0 ml 0.1M citric acid + 1.0 ml 0.2M Na_2HPO_4

STAINING PROCEDURE:
(1) Deparaffinize sections and bring to water.
(2) Stain with azure A staining solution for 30 minutes.
(3) Dehydrate through 70, 95, and two 100 percent alcohol solutions for 1 minute each.
(4) Clear in xylene and mount with synthetic resin.

RESULTS:

Only sulfated mucosubstances stain metachromatically at pH 2.0 or below in dehydrated sections. Many sialomucins stain metachromatically above the pK of sialic acid (about pH 3 and above). Hyaluronic acid in some sites stains metachromatically at pH 4 and above. Many epithelia containing secretions that are sulfated or rich in sialic acid may show metachromasia at pH 4.0 to 4.5 before dehydration but revert to orthochromasia or lose their stain entirely during dehydration.

ALCIAN BLUE (pH 2.5)

TISSUES:

Use any formalin fixative. Use paraffin sections.

STAINING PROCEDURE:

(1) Deparaffinize sections and bring to water.
(2) Stain with alcian blue for 1 hour (1 percent alcian blue in 3 percent acetic acid).
(3) Wash in running tap water for 5 minutes.
(4) Dehydrate through graded alcohols, clear in xylene, and mount.

RESULTS:

Weakly acidic sulfated mucosubstances, hyaluronic acid, and sialomucins are colored dark blue. Strongly acidic sulfated mucosubstances are stained weakly or not at all.

ALCIAN BLUE (pH 1.0)

TISSUES:

Use any formalin fixative. Use paraffin sections.

STAINING PROCEDURE:

(1) Deparaffinize sections and bring to water.
(2) Stain with alcian blue for 1 hour (1 percent alcian blue in 0.1N HCl).

(3) Blot dry on filter paper.

(4) Dehydrate through two changes of absolute ethanol, clear in xylene, and mount.

RESULTS:

Only sulfated mucosubstances stain selectively. The most strongly acid substances stain moderately.

ALCIAN BLUE–CRITICAL ELECTROLYTE CONCENTRATION TECHNIQUE

TISSUES:
Use any formalin fixative. Use paraffin sections.

STAINING METHOD:

(1) Deparaffinize sections and bring to water.

(2) Stain for 1 hour in a 0.1 percent solution of alcian blue in $0.05M$ acetate buffer, pH 5.7, containing $0.1M$, $0.2M$, $0.4M$, $0.6M$, $0.8M$, $0.9M$, or $1.1M$ $MgCl_2$.

(3) Wash in water for 5 minutes.

(4) Dehydrate quickly through alcohols, clear in xylene, and mount.

RESULTS:

Hyaluronic acid, sialomucins, and some weakly acid sulfomucins are stained at or above $0.1M$ $MgCl_2$. Most sulfated mucosubstances, including those metachromatically stained with azure A at pH 0.5, stain strongly and selectively at $0.2M$ $MgCl_2$. The various sulfated mucopolysaccharides lose their affinity for alcian blue at different levels of increased $MgCl_2$ concentration, with mast cells, cornea, and some epithelia persisting selectively at $1.0M$ $MgCl_2$. In cartilage, the stain can be made selective for chondroitin sulfate and keratan sulfate with $0.4M$ $MgCl_2$, and for keratan sulfate with $0.9M$ $MgCl_2$.

COLLOIDAL IRON STAINING METHOD
(Mowry, 1963)

TISSUES:
Use any formalin fixative. Use paraffin sections.

PREPARATION OF STOCK COLLOIDAL IRON SOLUTION:
Bring 250 ml distilled water to a boil. Pour in 4.4 ml of a 29 percent ferric chloride solution while keeping the solution boiling. When the solution turns dark red, allow it to cool.

PREPARATION OF COLLOIDAL IRON STAINING SOLUTION:

Distilled water	18 ml
Glacial acetic acid	12 ml
Stock colloidal iron solution	10 ml

PREPARATION OF POTASSIUM FERROCYANIDE SOLUTION:
Mix equal volumes of 2 percent potassium ferrocyanide and $0.25N$ HCl. Prepare fresh each time.

STAINING METHOD:
(1) Deparaffinize sections and bring to water.
(2) Place sections in colloidal iron staining solution for 1 hour.
(3) Rinse in three changes of 30 percent acetic acid for 10 minutes each.
(4) Wash in running distilled water for 5 minutes.
(5) Treat with potassium ferrocyanide solution for 20 minutes.
(6) Wash in running tap water for 5 minutes.
(7) Dehydrate through graded alcohols, clear in xylene, and mount.

RESULTS:
Acidic mucosubstances—including sialomucins, goblet cell mucins, some connective tissue mucins, and mast cell granules—are colored Prussian blue. Staining resembles that of alcian blue (pH 2.5), except that some epithelial sites stain more weakly. Strongly acidic mucins that do not stain with alcian blue similarly do not stain with colloidal iron.

MIXED DIAMINE METHOD
(Spicer, 1965)

TISSUES:
Use paraffin sections of formalin-fixed tissues.

PREPARATION OF STAINING SOLUTION:

Dissolve 30 mg N,N-dimethyl-*m*-phenylenediamine·2HCl and 5 mg N,N-dimethyl-*p*-phenylenediamine·HCl in 50 ml distilled water. Adjust the pH to 3.4 to 4.0 with $0.2M$ Na_2HPO_4.

PROCEDURE:

(1) Deparaffinize two sections and bring to water.
(2) Hydrolyze both sections in $1N$ HCl preheated to 60°C for 10 minutes (Feulgen hydrolysis) to remove nucleic acids.
(3) Wash in water for 5 minutes.
(4) Oxidize one section in 1 percent periodic acid for 10 minutes and wash in running water for 5 minutes.
(5) Stain both sections for 20 to 48 hours.
(6) Dehydrate with two changes of 95 percent ethanol and two changes of absolute ethanol. (Omit rinse in water.)
(7) Clear in xylene and mount.

RESULTS:

Most sulfated and sialic acid–containing mucosubstances are stained purple. Mucosubstances that react with periodic acid stain brown-gray or do not stain at all following periodate oxidation.

LOW IRON DIAMINE METHOD

(Spicer, 1965)

TISSUES:

Use paraffin sections of formalin-fixed tissues.

PREPARATION OF STAINING SOLUTION:

Dissolve 30 mg N,N-dimethyl-*m*-phenylenediamine·$2H_2O$ and 5 mg N,N-dimethyl-*p*-phenylenediamine·HCl simultaneously in 50 ml distilled water. When the reagents are dissolved, pour this solution immediately into a Coplin jar containing 0.5 ml 10 percent $FeCl_3$.

PROCEDURE:

(1) Deparaffinize two slides and bring to water.
(2) Oxidize one section in 1 percent periodic acid for 10 minutes (optional) and wash in running water.

(3) Stain both sections for 18 hours.
(4) Rinse quickly in water.
(5) Stain in 1 percent alcian blue in 3 percent acetic acid for 30 minutes. (This step is optional.)
(6) Rinse quickly in water.
(7) Dehydrate, clear, and mount as in the mixed diamine method.

RESULTS:
Most acid mucosubstances are stained gray-purple to black. Some sialomucins are not stained with diamine but are stained blue with alcian blue. If periodate oxidation is included, neutral mucosubstances are stained with diamine as well.

HIGH IRON DIAMINE METHOD
(Spicer, 1965)

TISSUES:
Use paraffin sections of formalin-fixed tissues.

PREPARATION OF STAINING SOLUTION:
Dissolve 120 mg N,N-dimethyl-*m*-phenylenediamine·2HCl and 20 mg N,N-dimethyl-*p*-phenylenediamine·HCl in 50 ml distilled water. When the reagents are dissolved, pour this solution in a Coplin jar containing 1.4 ml 10 percent $FeCl_3$.

PROCEDURE:
The steps of this procedure are as in the low iron diamine method.

RESULTS:
Sulfated mucosubstances are stained gray-purple to black. Neutral mucosubstances are stained similarly following periodate oxidation. Nonsulfated and sialic acid–containing mucosubstances are not stained with the diamines but are stained blue with alcian blue.

BIAL METHOD FOR SIALIC ACID
(Ravetto, 1964)

TISSUES:
Use cryostat sections of formalin-fixed tissues. Attach the sections on slides and allow them to dry.

PREPARATION OF BIAL REAGENT:

Orcinol	200 mg
$CuSO_4$, $0.1M$	0.25 ml
Concentrated HCl	80 ml

Add distilled water to make 100 ml. Allow the solution to stand for at least 4 hours before using. It can be stored in a refrigerator for at least a week.

PROCEDURE:
(1) Spray the sections with Bial reagent.
(2) Place the sections suspended face down over concentrated HCl that has been preheated to 70°C. Leave in position for 5 to 10 minutes.
(3) Dry the sections in air.
(4) Clear in xylene and mount with synthetic resin.

RESULTS:
Sites of sialic acid are colored various shades of red. The color fades within a day.

MODIFIED PAS METHOD FOR SIALIC ACID
(Roberts, 1977)

TISSUES:
Fix tissue blocks with phosphate-buffered formalin. Use paraffin sections.

PREPARATION OF METABISULFITE SOLUTION:
Prepare $22.5mM$ potassium metabisulfite in $0.05N$ HCl.

PROCEDURE:
(1) Deparaffinize sections and bring to water.
(2) Block all preexisting carbonyl groups by treating the sections with $0.1M$ sodium borohydride in $0.2M$ sodium borate buffer, pH 7.6, for 1 hour.*

*It is doubtful if this precaution is essential. Alternatively, a control slide could be added in which the periodate oxidation is omitted. Any red coloration in the control section can be presumed to result from preexisting carbonyl groups.

(3) Wash in running tap water for 5 minutes.
(4) Oxidize with $0.4mM$ sodium metaperiodate for 30 minutes at 20°C.
(5) Wash in two changes of distilled water for 5 minutes.
(6) Stain in Schiff's reagent for 3 minutes.
(7) Wash in two changes of metabisulfite solution.
(8) Wash in water for 5 minutes.
(9) Dehydrate in graded alcohols, clear in xylene, and mount with synthetic resin.

RESULTS:
Areas of red or magenta color indicate sites of sialic acid.

DIFFERENTIAL STAINING OF ACETYLATED AND NONACETYLATED SIALIC ACID
(Culling et al., 1976)

TISSUES:
Fix tissues with formol-calcium solution for 24 hours. Use paraffin sections.

PREPARATION OF THIONIN-SCHIFF REAGENT:
Dissolve 1 gm thionin in 100 ml distilled water by heating. Cool the solution, add 0.75 percent thionyl chloride and leave it overnight. Add 2 percent charcoal, shake, and filter immediately. Store in a dark brown bottle and keep refrigerated.

PREPARATION OF BOROHYDRIDE SOLUTION:
Dissolve 0.1 percent sodium borohydride in 1 percent disodium phosphate. Must be made up fresh in a fume hood.

PROCEDURE:
(1) Deparaffinize sections and bring to water.
(2) Oxidize in 1 percent periodic acid for 30 minutes.
(3) Wash in running water for 10 minutes.
(4) Place in thionin-Schiff reagent for 30 minutes.
(5) Wash in running tap water for 10 minutes.
(6) Oxidize with 1 percent periodic acid.

(7) Wash for 10 minutes.

(8) Treat with borohydride solution for 30 minutes.

(9) Rinse in 70 percent ethanol.

(10) Treat with 0.5 percent potassium hydroxide in 70 percent ethanol for 30 minutes.

(11) Wash gently in tap water.

(12) Place in fresh 1 percent periodic acid for 10 minutes.

(13) Wash gently in tap water for 10 minutes.

(14) Place in standard Schiff's reagent for 30 minutes.

(15) Wash gently in tap water for 10 minutes.

(16) Dehydrate, clear, and mount with synthetic mountant.

COMMENTS:

Steps 6, 7, and 8 are not generally required. Borohydride treatment is used to ensure that no free aldehyde groups remain before saponification (step 10). Saponification tends to loosen the sections from the slide, and it is therefore necessary to handle the slides very gently following that step.

RESULTS:

PAS-positive material (including nonacetylated sialic acid) stains blue while O-acetylated sialic acids are stained red. Sites with a mixture of both are stained purple.

HYALURONIDASE DIGESTION PROCEDURE

TISSUES:

Use any suitable formalin fixative. Use paraffin sections.

PREPARATION OF HYALURONIDASE SOLUTION:

Dissolve hyaluronidase in phosphate-citrate buffer, pH 5.8, at the rate of 200 NF units per milliliter. A highly purified preparation of hyaluronidase is preferred, to reduce the risk of protease contamination.

PROCEDURE:

(1) Deparaffinize two sections (test and control) and bring to water.

(2) Incubate the control section in buffer and the test section in the hyaluronidase solution, both for 3 to 4 hours at 37°C.

(3) Wash sections in running water for 5 minutes.
(4) Stain with azure A, pH 4.0, or with alcian blue, pH 2.5.

RESULTS:
Areas that are stained in the control section but not in the test section represent hyaluronidase-susceptible substances (hyaluronic acid, chondroitin 4-sulfate, and chondroitin 6-sulfate).

NEURAMINIDASE DIGESTION PROCEDURE
(Sorvari and Lauren, 1973)

TISSUES:
Fix tissues with any formalin fixative. Use paraffin sections.

PREPARATION OF NEURAMINIDASE SOLUTION:
Dissolve 100 units of neuraminidase (from *Vibrio cholerae*) per milliliter of a 0.05M acetate buffer, pH 5.5, containing 0.9 gm per liter of $CaCl_2$.

PROCEDURE:
(1) Deparaffinize two sections and bring to water.
(2) Surround tissue sections with paraffin wells.
(3) Expose the experimental section to about 0.1 ml of the neuraminidase solution and the control to the same buffer but containing no enzyme.
(4) Stain the sections with alcian blue, pH 2.5, or with the azure A, pH 3.5, or by the low iron diamine procedure.

RESULTS:
Sites that stain in the control but fail to stain in the experimental section are presumed to be sites of sialic acid.

REFERENCES
Ackroyd, W. (1876). Metachromatism or colour change. *Chem. News* 34:75–77.
Altman, R. D., Pita, J. C., and Howell, D. S. (1973). Degradation of proteoglycans in human osteoarthritic cartilage. *Arthritis Rheum.* 16:179–185.

Aterman, K., and Norkin, S. (1963). The periodic acid–Schiff reaction. *Nature* 197:1306.

Barger, J. D., and De Lamater, E. D. (1948). The use of thionyl chloride in the preparation of Schiff's reagent. *Science* 108:121–122.

Bedi, K. S., and Horobin, R. W. (1976). An alcohol-soluble Schiff's reagent: A histochemical application of the complex between Schiff's reagent and phosphotungstic acid. *Histochemistry* 48:153–159.

Bergeron, J. A., and Singer, M. (1958). Metachromasy: An experimental and theoretical reevaluation. *J. Biophys. Biochem. Cytol.* 4:433–457.

Bobbitt, J. M. (1956). Periodate oxidation of carbohydrates. *Adv. Carbohyd. Chem.* 11:1–41.

Bollet, A. J. (1967). Connective tissue polysaccharide metabolism and the pathogenesis of osteoarthritis. *Adv. Intern. Med.* 13:33–60.

Bouma, G. (1883). Ueber Knorpeltinction mittels Saffranin. *Centr. Med. Wiss.* 48:865–867.

Cornil, M. V. (1875). Sur la dissociation du violet de méthylaniline et sa séparation en deux couleurs sous l'influence de certains tissus normaux et pathologiques, en particulier par les tissus en dégénérescence amyloide. *C. R. Acad. Sci.* (Paris) 80:1288–1291.

Culling, C. F. A., Reid, P. E., Clay, M. G., and Dunn, W. L. (1974). The histochemical demonstration of O-acetylated sialic acid in gastrointestinal mucïns. Their association with the potassium hydroxide–periodic acid–Schiff effect. *J. Histochem. Cytochem.* 22:826–831.

Culling, C. F. A., Reid, P. E., Burton, J. D., and Dunn, W. L. (1975). A histochemical method of differentiating lower gastrointestinal tract mucin from other mucins in primary or metastatic tumors. *J. Clin. Pathol.* 28:656–658.

Culling, C. F. A., Reid, P. E., and Dunn, W. L. (1976). A new histochemical method for the identification and visualization of both side chain acetylated and nonacetylated sialic acids. *J. Histochem. Cytochem.* 24:1225–1230.

Davies, D. V. (1952). Specificity of staining methods for mucopolysaccharides of the hyaluronic acid type. *Stain Technol.* 27:65–70.

de Tomasi, J. A. (1936). Improving the technic of the Feulgen stain. *Stain Technol.* 11:137–144.

Ehrlich, P. (1878). Contributions to the Theory and Practice of Histological Staining. In *The Collected Papers of Paul Ehrlich.* Translation by F. Himmelweit, M. Marquardt, and H. Dale. Pergamon Press, New York, 1956. Vol. 1, pp. 65–98.

Gad, A., and Sylvén, B. (1969). On the nature of the high iron diamine method for sulfomucins. *J. Histochem. Cytochem.* 9:156–160.

Haemmerli, G., Genter, L., and Straüli, P. (1972). Cytophotometric evaluation of the Hale reaction. *Acta Cytol.* 16:31–36.

Hale, C. W. (1946). Histochemical demonstration of acid polysaccharides in animal tissues. *Nature* 157:802.

Heschl, R. (1875). Eine hübsche à vista-Reaktion auf Amyloid degenerirte Gewebe. *Wien. Med. Wochenschr.* 25:714–715.

Hotchkiss, R. D. (1948). A microchemical reaction resulting in the staining of polysaccharide structures in fixed tissue preparations. *Arch. Biochem.* 16:131–141.

Itikawa, O., and Ogura, Y. (1954). Simplified manufacture and histochemical use of the Schiff reagent. *Stain Technol.* 29:9–11.

Jeanloz, R. W. (1960). The nomenclature of mucopolysaccharides. *Arthritis Rheum.* 3:233–237.

Jones, R., and Reid, J. (1973a). The effect of pH on Alcian Blue staining of epithelial acid glycoproteins. I. Sialomucins and sulphomucins (singly or in simple combinations). *Histochem. J.* 5:9–18.

Jones, R., and Reid, J. (1973b). The effect of pH on Alcian Blue staining of epithelial acid glycoproteins. II. Human bronchial submucosal gland. *Histochem. J.* 5:19–27.

Jürgens, R. (1875). Eine neue Reaktion auf Amyloidkorper. *Virchows Arch. [Pathol. Anat.]* 65:189–196.

Kelly, J. W. (1956). The Metachromatic Reaction. *Protoplasmatologia.* Band II. D/2. Springer-Verlag, Vienna. Edited by L. V. Heilbrunn and F. Weber.

Kelly, J. W. (1958). The use of metachromasy in histology, cytology· and histochemistry. *Acta Histochem.* Suppl. 1:85.

Liao, T.-H., Gallop, P. M., and Blumenfeld, O. O. (1973). Modification of sialyl residues of sialoglycoprotein(s) of the human erythrocyte surface. *J. Biol. Chem.* 248:8247–8253.

Lillie, R. D. (1947). Reticulum staining with Schiff reagent after oxidation by acidified sodium periodate. *J. Lab. Clin. Med.* 32:910–912.

Lillie, R. D. (1951). Simplification of the manufacture of Schiff reagent for use in histochemical procedures. *Stain Technol.* 26:163–165.

Lillie, R. D., and Mowry, R. W. (1949). Histochemical studies on absorption of iron by tissue sections. *Bull. Int. Assoc. Med. Museums* 30:91–98.

Malaprade, L. (1928). Action des polyalcools sur l'acide périodique application analytique. *Bull. Soc. Chim.* 43:683.

Mankin, H. J., Dorfman, H., Lippiello, L., and Zarins, A. (1971). Biochemical and metabolic abnormalities in articular cartilage from osteo-arthritic human hips. II. Correlation of morphology with biochemical and metabolic data. *J. Bone Joint Surg.* [Am.] 53:523–537.

Mankin, H. J., and Lippiello, L. (1970). Biochemical and metabolic abnormalities in articular cartilage from osteo-arthritic human hips. *J. Bone Joint Surg.* [Am.] 52:424–434.

McManus, J. F. A. (1946). Histological demonstration of mucin after periodic acid. *Nature* 158:202.

Mörner, C. T. (1888). Histochemische Beobachtungen über die hyalin Grundsubstanz des Trachealknorpels. *Hoppe Seylers Z. Physiol. Chem.* 12:396–404.

Mowry, R. W. (1956). Alcian Blue technics for the histochemical study of acidic carbohydrates. *J. Histochem. Cytochem.* 4:407.

Mowry, R. W. (1958). Improved procedure for the staining of acidic polysaccharides by Müller's colloidal (hydrous) ferric oxide and its combination with the Feulgen and the periodic acid–Schiff reactions. *Lab. Invest.* 7:566–576.

Mowry, R. W. (1960). Revised method producing improved coloration and acidic polysaccharides with Alcian Blue 8GX supplied currently. *J. Histochem. Cytochem.* 8:323–324.

Mowry, R. W. (1963).   The special value of methods that color both acidic and vicinal hydroxyl groups in the histochemical study of mucins. *Ann. N. Y. Acad. Sci.* 106:402–423.

Mowry, R. W., Longly, J. B., and Millican, R. C. (1952).   Histochemical demonstration of intravenously injected dextran in kidney and liver of the mouse. *J. Lab. Clin. Med.* 39:211–217.

Pal, M. K., and Schubert, M. (1961).   Specific absorption of metachromatic compounds of chondroitin sulfate by insoluble calcium salts. *J. Histochem. Cytochem.* 9:673–680.

Paneth, J. (1888).   Über die secernirenden Zellen des Dunndarm-Epithels. *Arch. Mikrosc. Anat.* 31:113–191.

Pearse, A. G. E. (1968).   *Histochemistry: Theoretical and Applied* (3rd ed.), Vol. 1. Little, Brown, Boston.

Rahmann, H., and Katusic, J. (1975).   Histochemischer Nachweis Sialin-säure-haltige Verbindungen im ZNS von Teleosteern. *Histochemistry* 44:291–302.

Ravetto, C. (1964).   Histochemical identification of sialic (neuraminic) acids. *J. Histochem. Cytochem.* 12:306.

Roberts, G. P. (1977).   Histochemical detection of sialic acid residues using periodate oxidation. *Histochem. J.* 9:97–102.

Rosenberg, L. (1971).   Chemical basis for the histological use of Safranin O in the study of articular cartilage. *J. Bone Joint Surg.* [Am.] 53:69–82.

Schubert, M., and Hamerman, D. (1956).   Metachromasia: Chemical theory and histochemical use. *J. Histochem. Cytochem.* 4:159–189.

Scott, J. E. (1970).   Histochemistry of Alcian Blue. I. Metachromasia of Alcian Blue, Astrablau and other cationic phthalocyanin dyes. *Histochemistry* 21:277–285.

Scott, J. E. (1972a).   Lies, damned lies—and biological stains. *Histochem. J.* 4:387–389.

Scott, J. E. (1972b).   Histochemistry of Alcian Blue. II. The structure of Alcian Blue 8GX. *Histochemistry* 30:215–234.

Scott, J. E. (1972c).   The histochemistry of Alcian Blue. Note on the presence and removal of boric acid as the major diluent in Alcian Blue 8GX. *Histochemistry* 29:129–133.

Scott, J. E. (1972d).   Amplification of staining by Alcian Blue and similar ingrain dyes. *J. Histochem. Cytochem.* 20:750–752.

Scott, J. E., and Dorling, J. (1965).   Differential staining of acid glycosaminoglycans (mucopolysaccharides) by Alcian blue in salt solutions. *Histochemistry* 5:221–233.

Scott, J. E., and Dorling, J. (1969).   Periodate oxidation of acid polysaccharides. III. A PAS method for chondroitin sulphates and other glycosamino-glycurocans. *Histochemistry* 19:295–301.

Scott, J. E., and Stockwell, R. A. (1967).   On the use and abuse of the critical electrolyte concentration approach to the localization of tissue polyanions. *J. Histochem. Cytochem.* 15:111–113.

Scott, J. E., Dorling, J., and Quintarelli, G. (1964).   Differential staining of acid glycosaminoglycans by Alcian Blue in salt solutions. *Biochem. J.* 91:4p–5p.

Shabadash, A. L. (1947).   An improved histochemical test for glycogen and its theoretical basis (in Russian). *Izv. Akad. Nauk. S.S.S.R.* [Biol.] 6:745–760. (Abstract in *Chem. Abstr.* 42:col. 8861, 1948.)

Sorvari, T. E. (1972a).   Histochemical observations on the role of ferric chloride in the high-iron diamine technique for localizing sulphated mucosubstances. *Histochem. J.* 4:193–204.

Sorvari, T. E. (1972b).   Binding of ferric ions to nuclei and other tissue sites in sections stained for sulfated mucosubstances by the high-iron diamine method. *Stain Technol.* 47:245–248.

Sorvari, T. E., and Arvilommi, H. S. (1973).   Some chemical, physical and histochemical properties of three diamine fractions obtained by gel chromatography from the high-iron diamine staining solution used for localizing sulphated mucosaccharides. *Histochem. J.* 5:119–130.

Sorvari, T. E., and Laurén, P. A. (1973).   The effect of various fixation procedures on the digestability of sialomucins with neuraminidase. *Histochem. J.* 5:405–412.

Spicer, S. S. (1961).   The use of various cationic reagents in histochemical differentiation of mucopolysaccharides. *Am. J. Clin. Pathol.* 36:393–407.

Spicer, S. S. (1965).   Diamine methods for differentiating mucosubstances histochemically. *J. Histochem. Cytochem.* 13:211–234.

Spicer, S. S., and Jarrel, M. H. (1961).   Histochemical reaction of an aromatic diamine with acid groups and periodate engendered aldehydes in mucopolysaccharides. *J. Histochem. Cytochem.* 9:368–379.

Spicer, S. S., Horn, R. G., and Leppi, T. J. (1967).   Histochemistry of Connective Tissue Mucopolysaccharides. In B. M. Wagner and D. E. Smith (Eds.), *The Connective Tissue.* Williams & Wilkins, Baltimore.

Spicer, S. S., Leppi, T. J., and Stoward, P. J. (1965).   Suggestions for a histochemical terminology of carbohydrate-rich tissue components. *J. Histochem. Cytochem.* 13:599–603.

Steedman, H. F. (1950).   Alcian Blue 8GS: A new stain for mucin. *Q. J. Microsc. Sci.* 91:477–479.

Stockwell, R. A., and Scott, J. E. (1965).   Observations on the acid glycosaminoglycan (mucopolysaccharide) content of the matrix of ageing cartilage. *Ann. Rheum. Dis.* 24:341–350.

Toch, E. P. C., and Pearse, A. G. E. (1965).   Preservation of tissue mucins by freeze-drying and vapour fixation. I. Light microscopy. *J. R. Microsc. Soc.* 84:519–537.

Troyer, H. (1974).   A microspectrophotometric study of metachromasia. *J. Histochem. Cytochem.* 22:1118–1121.

van Duijn, P., and Riddersma, S. H. (1973).   Purification of Pararosaniline and Atabrine by chromatography on lipophilic Sephadex LH-20. *Histochem. J.* 5:169–172.

Van Lenten, L., and Ashwell, G. (1971).   Studies on the chemical and enzymatic modification of glycoproteins: A general method for the titration of sialic acid–containing glycoproteins. *J. Biol. Chem.* 246:1889–1894.

Weber, P., Harrison, F. W., and Hof, L. (1975).   The histochemical application of dansylhydrazine as a fluorescent labeling reagent for sialic acid in cellular glycoconjugates. *Histochemistry* 45:271–277.

Weiss, C., and Mirow, S. (1972).   An ultrastructural study of osteoarthritic changes in the articular cartilage of human knees. *J. Bone Joint Surg.* [Am.] 54:954–972.

Wood, W. S., and Culling, C. F. A. (1975).   Perianal Paget disease: Histochemical differentiation utilizing the borohydride-KOH-PAS reaction. *Arch. Pathol.* 99:442–445.

Yarbo, C. L., Miller, B., and Anderson, C. E. (1954).   Purifying pararosanilin for use in colorless Schiff reagent. *Stain Technol.* 29:299–300.

7

LIPIDS

Tissue lipids are generally hydrophobic, a distinctive characteristic that gives rise to certain problems in fixation and staining of lipids. These problems are responsible for the fact that the histochemistry of lipids lags behind most other aspects of histochemistry.

CLASSIFICATION OF LIPIDS

A classification of biologically important lipids will include the following three major groups: (1) simple lipids, (2) compound lipids, and (3) steroids. These major groups, especially the latter two, can be further subdivided.

Simple Lipids

This group contains the common triglycerides, which consist of three moles of fatty acids esterified with 1 mole of glycerol. The fatty acids so combined with glycerol are mostly from 14 to 18 carbons in length. Some of the fatty acids also contain one or more unsaturated bonds. Table 7-1 lists some of the most common fatty acids found in nature. Most triglycerides contain two or three different types of fatty acids. Oils are those triglycerides that are liquid at room temperature, whereas the fats are those that are solids. Thus, there is no sharp distinction between the two. The melting point of the triglycerides is determined primarily by the degree of saturation of the fatty acids; an increased number of double bonds depresses the melting point. Highly unsaturated oils serve as drying oils in the traditional paints and varnishes. The lengths of the fatty acids play a somewhat less important role in determining the melting point. Generally, vegetable oils and fats contain greater percentages of unsaturated fatty acids than those derived from animals.

Waxes, which also belong to the group of simple lipids, are fatty acids that are esterified with alcohols other than glycerol, usually long-chained primary alcohols. Beeswax and wool wax are common types of animal waxes. Waxes are responsible for the natural gloss on polished apples and leaves. (Earwax, or cerumen, is not a true wax; it

Table 7-1. Some of the Most Common and Familiar Fatty Acids

Common Name	Systematic Name	Carbon Atoms	Double Bonds
Lauric acid	Dodecanoic acid	12	0
Myristic acid	Tetradecanoic acid	14	0
Palmitic acid	Hexadecanoic acid	16	0
Stearic acid	Octadecanoic acid	18	0
Arachidic acid	Eicosanoic acid	20	0
Oleic acid	9-octadecanoic acid	18	1
Linoleic	9,12-octadecadienoic	18	2
Linolenic	9,12,15-octadecatrienoic	18	3
Arachidonic	5,8,11,14-eicosatetraenoic	20	4

is, rather, a mixture of a brownish pigment, fat, sebaceous secretions, and cellular debris.) Unlike fats, waxes are rarely constituents of cells.

Compound Lipids

Three groups of compounds constitute the bulk of the compound lipids: phosphoglycerides, sphingomyelins, and glycosphingosides.

The phosphoglycerides are very important constituents of cell membranes. These compounds are similar to the triglycerides except that one of the fatty acids is replaced with a phosphate ester. Lecithin is the most abundant phosphoglyceride in nature, and it contains choline. Several cephalins exist; one contains ethanolamine and another serine. Another system of naming these compounds is to refer to them as phosphatidyl compounds. Thus, lecithin would be known as phosphatidyl choline and the kephalins as phosphatidyl ethanolamine and phosphatidyl serine.

Plasmalogens can be regarded chemically as derivatives of the phosphatidyl compounds that were briefly considered above. The fatty acid in the α-position is replaced by an unsaturated ether. By acid hydrolysis the ether bond can be easily converted to an aldehyde. This reaction is an important consideration for histochemists.

In the sphingomyelins, the compound sphingosine replaces glycerol, and the single fatty acid is attached by way of an amide bond. As in lecithin, the nitrogenous base on the phosphate radical is choline.

The glycosphingosides also contain sphingosine instead of glycerol; unlike the sphingomyelins, they lack a nitrogenous base and phosphate radicals. They contain one or more monosaccharide units that are attached by way of an ester bond. Cerebrosides contain one molecule of glucose or galactose, whereas gangliosides contain one molecule of neuraminic acid and three molecules of galactose.

Cardiolipin essentially consists of two molecules of phospholipid tied together through another glycerol molecule. It occurs in mitochondria, particularly in the heart.

The composition of some of the most common compound lipids is presented in Table 7-2.

Steroids

The basic structure of steroids is the cyclopento-perhydrophenanthrene ring system. It should be emphasized that this is an aliphatic ring system, although it is sometimes tempting to think of it as an aromatic ring system. Some of the steroid derivatives, however,

Table 7-2. Composition of Some of the Most Common Lipids

Type of Lipid	Composition		
Simple lipids			
Triglycerides	1 glycerol	3 fatty acids	—
Compound lipids			
Phospholipids			
Phosphatidyl choline (lecithin)	1 glycerol	2 fatty acids	1 choline
Phosphatidyl ethanolamine (cephalin)	1 glycerol	2 fatty acids	1 ethanolamine
Phosphatidyl serine (cephalin)	1 glycerol	2 fatty acids	1 serine
Sphingomyelin	1 sphingosine	1 fatty acid	1 choline
Glycosphingosides			
Cerebrosides	1 sphingosine	1 fatty acid	1 glucose or 1 galactose
Gangliosides	1 sphingosine	1 fatty acid	3 galactose and 1 neuraminic acid

may contain several double bonds. A series of highly important naturally occurring compounds of animals belong to this group, including vitamin D, the sex hormones, adrenocortical hormones, bile acids, and cholesterol. Cholesterol palmitate, or lanolin, belongs to this group; it is used a great deal in the manufacture of medicinal products. (Lanolin also qualifies as a wax and indeed is called wool wax.) It is widely distributed in the human body and is abundant in wool.

FIXATION AND PRESERVATION OF LIPIDS

Because of the physical properties of lipids, special problems are encountered in fixing and preserving lipids for histochemistry. Dehydration and paraffin embedding are not generally considered possible, because the organic solvents necessary for dehydration and clearing would dissolve nearly all the tissue lipids except those firmly bound to proteins. Nonetheless, several investigators have claimed that they were able to fix lipids sufficiently well to resist organic solvents. Elftman (1954) used dichromate oxidation (controlled chromation) as a way of fixing lipids. It would appear that lipids were demonstrated with a certain degree of success in paraffin sections following this method of fixation. Wigglesworth (1957) fixed tissues with osmium tetroxide in preparation for paraffin embedding, although he felt that xylene should still be avoided. Cedarwood oil was therefore substituted. Lipids appeared black but the intensity of the stain could be greatly enhanced by posttreating the sections with gallic acid derivatives. The Air Force Institute of Pathology Manual (Luna, 1968) also presents a method for preserving lipids with osmium tetroxide while embedding in paraffin. The osmium is removed with hydrogen peroxide following sectioning, and sections are then stained with oil red O. Fixing with either dichromate or osmium tetroxide greatly alters the chemistry of the lipids, and histochemical results are not always reliable (Holczinger, 1965). Fixation with osmium tetroxide is very common for electron microscopy, especially because osmium-fixed lipids are very electron opaque. It is very likely, however, that many lipids are lost even with osmium tetroxide fixation, and it seems that this type of fixation offers little significant value to histochemists.

It was once thought that glycol methacrylate embedding methods would offer a way of preserving lipids in embedded tissues (Feder, 1963). Because glycol methacrylate is water soluble, tissues can be embedded in it without the use of organic dehydrating solvents. However, such hopes were shattered when Cope and Williams (1968) established that while infiltrating glutaraldehyde-fixed tissues with glycol methacrylate, nearly half of the neutral lipid was extracted. Even osmium postfixation reduced lipid extraction only moderately. My own attempts at preserving lipids in glycol methacrylate–embedded tissues have been quite unsatisfactory.

Freezing the tissues assures the preservation of all lipids to the stage of sectioning (frozen sections). The reader should be reminded that the compound lipids have a somewhat hydrophobic group at one end of the molecule. It has been shown that some compound lipids, especially lecithin, are lost into water from frozen sections (Roozemond, 1967). Empirical observations have shown that fixation with Baker's formol-calcium preserves phospholipids (Adams, 1965). The observations have been confirmed with carefully designed analytical experiments (Roozemond, 1969a). However, it was also shown that the cephalins were destroyed or lost during aldehyde fixation (Roozemond, 1969b). The loss is not as rapid with whole rat brains, but nonetheless the phosphoglycerides are destroyed after prolonged formalin fixation (Deierkauf and Heslinga, 1962; Heslinga and Deierkauf, 1962). They are presumably hydrolyzed to their lyso derivatives. It is therefore necessary to use fixatives very judiciously.

Calcium has been shown empirically to have an important role in preserving certain lipids (Baker, 1958). Again, analytical experiments of Roozemond (1967, 1969a) have confirmed those empirical observations. Calcium probably forms complex coacervates with phospholipids, whereas formaldehyde may stabilize phospholipids by fixing closely associated proteins. Formol-calcium fixation for up to 24 hours is therefore recommended unless demonstration of cephalins is desired.

METHODS FOR DETECTING LIPIDS
Lipid-Soluble Pigments
A series of lipid-soluble pigments can be used for detecting lipids. These substances, especially the sudans and oil red O, have been

called stains or dyes even though they do not contain chromotropes and do not form salt linkages with the substances to which they impart color. They should therefore be regarded as pigments rather than stains or dyes.

These pigments are used as a saturated solution in a solvent in which they are barely soluble. When lipid-containing tissue sections are exposed to the solution, the pigment becomes dissolved in the lipid deposits because it is more soluble in the lipids than in the solvent. The solubility of these pigments in lipids depends on the long aliphatic chains of the fatty acids; therefore practically all lipids are demonstrated.

Currently the two most common of these pigments are oil red O and sudan black B. Sudan IV, oil red 4B, and Fettrot are used to a lesser extent.

The choice of solvents for the lipid soluble pigments is of critical importance. Chiffelle and Putt (1951) set forth three requirements for the solvent: (1) it should not extract any lipids from the tissue sections, (2) it should dissolve sufficient amounts of pigment to yield brilliant staining results, and (3) it should be easy, rapid, and economical to prepare and be stable in stock solution over long periods of time without deterioration, evaporation, or precipitation. Many different solvents have been used by various investigators, including 50 and 70 percent ethanol, 60 percent isopropanol, 60 percent triethylphosphate, and propylene and ethylene glycol. In their search for a suitable solvent, Chiffelle and Putt (1951) felt that either propylene glycol or ethylene glycol fulfilled the above-mentioned requirements most satisfactorily. It appeared that the most significant virtue of these two solvents was that they did not dissolve any lipids out of the tissue sections. Propylene glycol was preferred because of its ability to dissolve somewhat greater amounts of sudan black B and sudan IV. Results are therefore more brilliant. Propylene glycol and ethylene glycol thus are the solvents of choice for the sudan pigments.

Birefringence
Lipids that can be examined in their crystalline state will exhibit birefringence in polarized light, suggesting an isotropic nature of the solid lipids. The birefringence is usually seen in the form of a Maltese cross. (The Maltese cross is a distinctive eight-pointed white figure against a

black background displayed by the Hospitalers of the Order of St. John of God). Cholesterol and its esters are the most common lipids to give this distinctive birefringence pattern, although they are by no means the only ones.

Many other tissue components are of course also anisotropic and exhibit birefringence other than the Maltese-cross pattern. Some common examples are amyloid, collagen, myosin, myelin sheaths, silica in silicious granulomas, and crystals of phosphates in pseudo-gout.

Nile Blue Sulfate Method

The nile blue sulfate method offers the possibility of simultaneously staining neutral lipids and acidic lipids by different colors. The method was introduced by Smith and Mair (1909), but confidence in the claims to its specificity remained low for a long time. Cain (1947) studied the method thoroughly and established its specificity for neutral and acidic lipids. Histochemists now consider it a respectable method.

The staining solution contains three different molecular species. An oxazine salt forms the dominant species, which has the following structure:

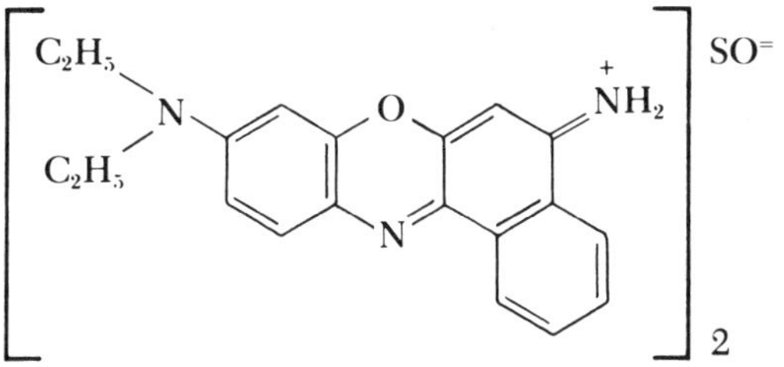

A small amount of ionization causes some free base to be present. Oxazone is the third component that is present as a result of hydrolysis:

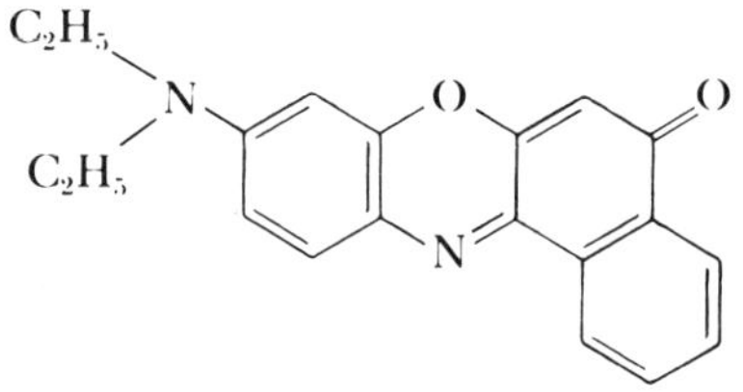

The salt is blue; the free base and oxazone are red. The oxazone can be readily extracted with certain organic solvents such as toluene.

The oxazone (and perhaps also the free base) is responsible for the red coloration of neutral lipids. Its mechanism of action is similar to that of the sudan pigments; it becomes preferentially dissolved by the lipid droplets. The free base, which is not formed to any appreciable extent in a 1 percent solution, combines with acidic lipids to form blue-colored compounds. A dilute solution contains more free base and less oxazone, and it stains all acidic lipids more readily. This is the reason for staining two slides in a 1 percent solution and restaining one of the slides in a more dilute solution.

Cain (1947) summarized his conclusions concerning the specificity of the nile blue sulfate method as follows: (1) lipids in their solid state are not stained; (2) triglycerides, if pure, are colored red by the oxazone and to a lesser extent by the free base; (3) fatty acids are colored blue by oxazine either in the dilute solution or in the more concentrated solution; (4) lecithin (and perhaps all phospholipids) stain deep blue when solid; and (5) cholesterol remains unstained.

Menschik (1953) introduced a modification of the nile blue sulfate method that is supposed to be specific for phospholipids. It is based on two steps of differentiation following staining. The first differentiation step is with hot acetone, which removes neutral lipids. The second step utilizes weak acids in order to destain proteins. Menschik's claim to specificity is no doubt fairly accurate, and the technique is easier to use than the acid-hematein method.

Acid-Hematein Method

It has long been known that tissues fixed or mordanted with chromic acid could be stained more readily with hematoxylin. With this knowledge, Baker (1946, 1947) set out to study the phenomenon and developed a staining method thought to be specific for phospholipids. The method involves a rigorous schedule of fixation, chromation, staining, and differentiation. A borax-ferricyanide mixture is used for differentiation.

Although it was originally thought that all phospholipids stain, it was since found that only the choline-containing lipids (lecithin and sphingomyelin) stain with this method (Bourgeois and Hack, 1962). The method has since been studied extensively by Adams (1965) and by Adams et al. (1965). From these studies, it would appear that cephalins and cerebrosides also give a weak reaction. Although choline is essential for a positive reaction, unsaturation in the fatty acid chains of the same molecule accentuates the staining intensity. This accentuation is not prevented by prior bromination, as might be expected, but it actually seems to enhance the intensity in an inexplicable manner.

Certain other tissue components, especially nucleoproteins and "mucins," also give a positive reaction. Baker (1947) therefore suggested using a parallel control section that was pretreated with hot pyridine. Only phospholipids are said to be extracted by hot pyridine, and areas rendered unstainable by extraction represent choline-containing lipids. However, the specificity of the extraction procedure has been severely criticized.

Iron Hematoxylin Method for Phospholipids

Elleder and Lojda (1973a) introduced a method for demonstrating phospholipids that they felt was more specific than other methods. The preparation of Weigert-Lillie iron hematoxylin (Lillie, 1965) was modified to eliminate ethanol; this forms the basis of the method. Other types of hematoxylin were tried but were unsatisfactory, because the staining solution was unstable, lipid artifacts appeared, or overstaining occurred. The procedure was coupled with preextraction of phospholipids by a mixture of chloroform and methanol. The sites where preextraction abolished hematoxylin are to be regarded as sites of phospholipids.

These authors also coupled the iron hematoxylin method with acetone and mild alkali preextractions to make it specific for sphingolipids (Elleder and Lojda, 1973b). Sphingolipids are the only alkali-resistant phospholipids. Acid hematoxylin could also be used with similar specificity, but the sensitivity is lower.

OTAN Method

Adams (1959) introduced the osmium tetroxide–α-naphthylamine (OTAN) method as a way to demonstrate normal and degenerating myelin simultaneously. The method depends on the principle that degenerating myelin reduces osmium tetroxide and is stained black. Normal myelin, on the other hand, binds unreduced osmium tetroxide, which is subsequently chelated with α-naphthylamine to form a red compound. The method is generally useful for distinguishing hydrophilic phospholipids (sphingomyelins, cephalins, lecithin, and cerebrosides) from hydrophobic lipids (cholesterol esters, triglycerides, and fatty acids). Degenerating myelin contains primarily phospholipids, which is the basis for using the method to detect sites of myelin degeneration.

A modification of this method was developed by Adams and Bayliss (1963) for demonstrating sphingomyelin specifically. The modification involves treatment with NaOH and is based on the fact that sphingomyelin is alkali-stable, whereas other phospholipids are alkali-labile.

The Plasmal Reaction

Ever since Feulgen and Voit (1924) described the plasmal reaction, it has engendered a great deal of controversy. They observed that in tissues treated with a dilute solution of mercuric chloride, aldehyde groups appear that can be demonstrated with Schiff's reagent. The reaction is supposed to be specific for plasmalogen phosphatides. It is uncertain whether vinyl groups in unsaturated fatty acids are hydrolyzed with mercuric chloride. The controversy about the plasmal reaction particularly concerned the appropriateness of various fixatives and the various types of controls.

The controversy has been adequately discussed by Cain (1949a). Space here will not permit a full discussion of the plasmal reaction, which is of limited importance. Anyone wishing to carry out the reaction should read the paper by Cain (1949a).

A false-positive reaction is known as the pseudoplasmal reaction. It results from the inadvertent formation of aldehyde groups at the double bonds in the fatty acid chains during some aspect of tissue preparation (Cain, 1949b). These aldehydes will of course react with Schiff's reagent and increase the color intensity.

Methods for Demonstrating Unsaturated Lipids

Unsaturated bonds in the fatty acid chains can be oxidized by a number of methods. Three of these methods provide the basis for important histochemical methods for demonstrating unsaturated fatty acids: reaction with osmium tetroxide, reaction with performic acid to form aldehydes, and bromination.

Osmium tetroxide has been used for many years as a fixative for cells and tissues. It has been thought to react with unsaturated lipids

(Criegee, 1936; Wigglesworth, 1957). It was shown more recently that osmium tetroxide reacts specifically with *cis*-double bonds of fatty acids but not with *trans*-double bonds. Osmium bridges, or bis-osmates, are formed between two fatty acid chains containing unsaturated bonds (Korn, 1966a, 1966b, 1967).

An extremely opaque black reaction product is formed, the nature of which is not understood. It is thought that the formation of osmium bridges occurs in two steps. The first would be the formation of an osmium–fatty acid monoester. Two of these monoesters would then react to form the bis-osmate bridge and a free molecule of osmium oxide. The osmium oxide was suspected of being the visible black product. However, Korn (1967) showed that the osmium–fatty acid ratio was 1:2, which renders the above explanation unlikely. It is more likely that the osmium incorporated into the bridge provides the opacity.

Lillie (1952) successfully oxidized unsaturated lipids with performic acid to generate aldehyde groups. The aldehyde groups could then be demonstrated with Schiff's reagent. Performic acid does not oxidize 1,2-glycols. Therefore, preexisting 1,2-glycols such as glycogen do not interfere with the reaction. The formation of aldehydes with performic acid probably does not involve 1,2-glycols as intermediates. It is possible, however, that 1,2-glycols are formed at some of the unsaturated bonds but are not further oxidized and make no contribution to the color. Lillie proposed that the reaction goes in two steps. First, peroxides and perhaps epoxides are formed. The epoxides would be hydrolyzed to form 1,2-glycols and the peroxides simply undergo rearrangement to form aldehyde groups:

$$-HC{=}CH- + HCO_3H \longrightarrow -HC\overset{O}{\underset{}{\diagup\diagdown}}CH- + HCOOH$$

$$-HC\overset{O}{\underset{}{\diagup\diagdown}}CH- + H_2O \longrightarrow -HC\overset{HO}{\underset{|}{|}}-\overset{OH}{\underset{|}{CH}}-$$

and

$$-HC{=}CH- + 2HCO_3H \longrightarrow -HC\overset{O}{\underset{O}{\diagup\diagdown\diagup\diagdown}}CH- \text{ or}$$

$$-HC\overset{O-O}{\underset{|\quad|}{\quad}}CH- \text{ rearrangement} \longrightarrow -HC\overset{O}{\underset{}{\|}} \quad \overset{O}{\underset{}{\|}}CH-$$

Prior bromination completely blocks the performic acid–Schiff reaction. The specificity for unsaturated lipids was confirmed by Adams (1965).

Not only can bromination be used for blocking the above reaction, but also it can serve as the basis for another histochemical method for demonstrating unsaturated lipids (Mukherji et al., 1960; Norton et al., 1962). When brominated tissue sections are treated with silver nitrate, silver bromide is formed. It can then be "developed" to form metallic silver granules in the same way that photographic paper is developed. Adams (1965) confirmed the specificity of this method for hydrophilic unsaturated lipids.

Schultz Method for Cholesterol

The Schultz (1924) method is based on an earlier biochemical method. The tissue sections are first oxidized with a buffered solution of ferric ammonium sulfate for as long as seven days. They are then treated with a mixture of equal parts of glacial acetic acid and concentrated sulfuric acid. This leads to the formation of a blue-green pigment, which is thought to be 7-hydroxycholesterol.

This method was studied and evaluated by Reiner (1953); it has several shortcomings, including low sensitivity and the formation of gas bubbles in the tissue sections. Furthermore, the color is short-lived, and the sections should be examined in the acid mixture. The specificity, however, is considered to be high.

Rossouw et al. (1976) found that the iron alum and the acid mixture could be premixed with certain advantages. The color was stable for several hours, and the gas bubble formation was minimized. Also, the premixed reagent was stable and could be stored.

Digitonin Method for Cholesterol

The digitonin method is also a modification of an earlier biochemical technique. The principle of the method is that digitonin forms a crystalline complex with cholesterol that is birefringent. Of course it must be viewed with polarized light. The complex is insoluble in lipid solvents, and the cholesterol esters can be extracted following the reaction. Digitonin must be dissolved in ethanol, which also dissolves cholesterol. This may cause diffusion of the reaction product (Adams, 1969) and is the main shortcoming of the method.

PAN Method for Cholesterol

The perchloric acid–naphthoquinone (PAN) method, which was introduced by Adams (1961), is free of the shortcomings of the former two methods. The tissue must be fixed for at least a week in formol-calcium to obtain maximum reaction. Perchloric acid reacts with cholesterol to form cholestra-3,5-diene:

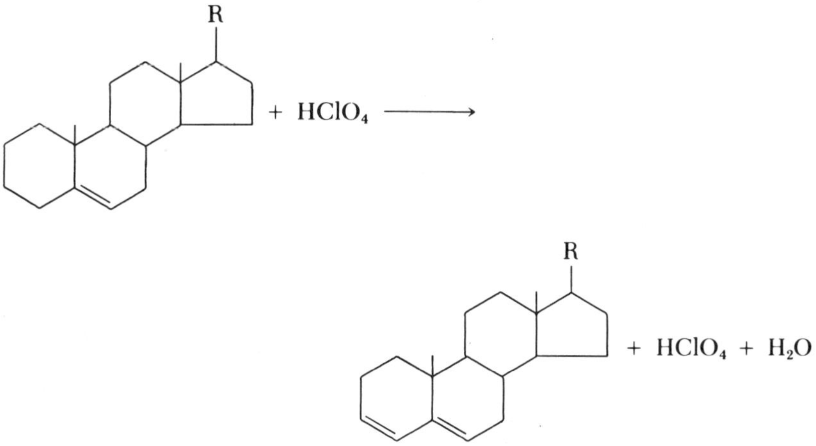

The diene then reacts with 1,2-naphthoquinone-4-sulfonic acid in an unknown manner. The result is a red pigment that turns blue when heated. Other tissue components also give colored reaction products, but only cholesterol and its esters form a dark blue color upon heating. The method has been further evaluated by Adams (1965), and it remains the method of choice for demonstrating cholesterol and its esters.

CHIFFELLE-PUTT METHOD FOR LIPIDS
(Chiffelle and Putt, 1951)

TISSUES:
Fix in formol-calcium and use frozen sections.

PREPARATION OF SUDAN BLACK B STAINING SOLUTION:
Suspend 1 gm of sudan black B in 100 ml of propylene glycol and heat

to 100°C for a few minutes. While it is hot, filter it through Whatman No. 2 paper. Cool and filter again through a coarse sintered-glass filter or through glass wool with vacuum.

PROCEDURE:
(1) Mount frozen sections on glass slides and air-dry.
(2) Wash in several changes of water to remove formalin.
(3) Dehydrate for 5 minutes in two changes of pure propylene glycol.
(4) Immerse in sudan black B staining solution for 5 to 10 minutes.
(5) Differentiate in warm 85 percent propylene glycol for 2 to 3 minutes.
(6) Rinse in 50 percent propylene glycol.
(7) Wash in water.
(8) Mount in glycerin jelly.

COMMENT:
The nuclei may be counterstained between steps 7 and 8 if desired. Stain with 50 percent Ehrlich's hematoxylin in distilled water for 5 to 10 minutes.

RESULTS:
Lipids are stained black. If the counterstaining step is included, the nuclei will stain blue.

OIL RED O METHOD FOR LIPIDS

TISSUES:
Fix tissues with formol-calcium or phosphate-buffered formalin. Use frozen sections.

PREPARATION OF OIL RED O STOCK SOLUTION:
Make a saturated solution of oil red O in 99 percent isopropanol.

PREPARATION OF OIL RED O WORKING SOLUTION:
Mix 6 parts of oil red O stock solution with 4 parts of distilled water. Prepare 1 hour before using and filter immediately before using.

PROCEDURE:
(1) Mount frozen sections on slides and allow to air-dry.
(2) Wash in water.

(3) Rinse in 70 percent ethanol.

(4) Stain in oil red O working solution for 15 minutes.

(5) Differentiate in 70 percent ethanol until no more dye diffuses out of the tissue.

(6) Wash well with water.

(7) Mount in glycerin jelly.

RESULTS:

Triglycerides, fatty acids, and cholesterol esters are colored strongly red. Phospholipids and cerebrosides are colored only faintly pink.

NILE BLUE SULFATE METHOD
(Cain, 1947)

TISSUES:

Fix tissues in formol-calcium. Use frozen sections, 8 or 10 microns thick. Mount sections on slides and air-dry them thoroughly.

PROCEDURE:

(1) Stain two sections (A and B) in 1 percent aqueous nile blue for 5 minutes at 60°C.

(2) Wash both sections in water at 60°C.

(3) Differentiate in 1 percent acetic acid for 3 seconds at 60°C.

(4) Restain section A in 0.02 percent nile blue for 10 to 15 minutes at 60°C.

(5) Wash both sections briefly in water at 60°C.

(6) Wash sections in water at room temperature.

(7) Mount in glycerin jelly.

NOTE:

Some workers stain a third section with the sudan black B method as a control.

RESULTS:

Neutral lipids stain red while acidic lipids stain blue. If sections A and B stain similarly, disregard B. If section A stains deeper than B, then the deeper-staining material is probably palmitic or stearic acid.

ACID HEMATEIN METHOD FOR PHOSPHOLIPIDS
(Baker, 1946, 1947; Adams, 1965)

TISSUES:
Fix tissues with formol-calcium. Use frozen sections.

PREPARATION OF ACID HEMATEIN SOLUTION:
Add 1 ml of 1 percent sodium periodate ($NaIO_4$) to 50 ml of 0.1 percent hematoxylin. Heat the solution to the boiling point and cool. Add 1 ml of glacial acetic acid. The solution should be made up fresh each day.

PROCEDURE:
(1) Mount frozen sections on slides and dry thoroughly in air.
(2) Treat sections with an aqueous solution of 5 percent potassium dichromate containing 1 percent calcium chloride for 18 hours at room temperature and then for another 18 to 24 hours at 60°C.
(3) Wash thoroughly with tap water.
(4) Stain with acid hematein solution for 5 hours at 37°C.
(5) Wash thoroughly in tap water.
(6) Differentiate with a solution containing 0.25 percent potassium ferricyanide and 25 percent sodium tetraborate for 18 hours at 37°C.
(7) Wash thoroughly in tap water.
(8) Mount with glycerin jelly.

RESULTS:
Lecithin and sphingomyelin (choline-containing lipids) stain blue-black.

IRON HEMATOXYLIN METHOD FOR PHOSPHOLIPIDS
(Elleder and Lojda, 1973a)

TISSUES:
Use fresh-frozen sections. Mount sections on clean glass slides and allow them to dry.

PREPARATION OF STOCK SOLUTION A:
$FeCl_3 \cdot 6H_2O$ 2.5 gm
$FeSO_4 \cdot 7H_2O$ 4.5 gm

Concentrated HCl   2 ml
Distilled water   298 ml

PREPARATION OF STOCK SOLUTION B:
Hematoxylin   1 gm
Distilled water   100 ml

Dissolve by heating gently.

PREPARATION OF STAINING SOLUTION:
Combine 3 parts of stock solution A with 1 part of stock solution B immediately before use.

STAINING PROCEDURE:
(1a) Treat one section with cold acetone (0° to 4°C) for 10 minutes and allow it to dry in air.
(1b) Treat another section with a chloroform-methanol mixture (2:1) for 10 minutes at room temperature and wash briefly with acetone and then in water.
(2) Fix both sections with formol-calcium for 10 minutes.
(3) Wash in distilled water.
(4) Stain for 6 to 8 minutes.
(5) Dip several times in 0.2 percent HCl.
(6) Wash well in tap water.
(7) Mount with Apathy's syrup *or* dehydrate with acetone, clear with xylene, and mount with Canada balsam.

RESULTS:
Phospholipids stain with a deep blue color in the acetone-extracted section. The chloroform-methanol–extracted section should have no stain.

IRON HEMATOXYLIN METHOD FOR SPHINGOMYELIN
(Elleder and Lojda, 1973b)

PROCEDURE:
This method is exactly like the previous method (iron hematoxylin method for phospholipids) with the exception that an alkaline prehy-

drolysis step is added. Following the acetone extraction of the first slide, hydrolyze it in 1N NaOH at room temperature for 1 to 2 hours. The chloroform-methanol extraction of the second slide is unmodified. Continue with all the steps of the previous method.

CONTROLS:
Elleder and Lojda (1973b) advised that the plasmal reaction be applied to a parallel section to check for the presence of plasmalogens. If present, they will also be stained with iron hematoxylin. If plasmalogens are troublesome, they can be removed by an acid prehydrolysis step prior to the alkaline prehydrolysis step.

RESULTS:
Sphingomyelins are stained blue in the acetone-extracted section. The chloroform-methanol–extracted section should have no color.

OTAN METHOD FOR PHOSPHOLIPIDS
(Adams, 1959)

TISSUES:
Fix tissues with formol-calcium. Use frozen sections, 5 to 15 microns thick. Do not mount on slides.

PREPARATION OF OSMIUM TETROXIDE SOLUTION:
Mix 1 part of a 1 percent solution of osmium tetroxide with 3 parts of a 1 percent solution of potassium chlorate ($KClO_3$). Carefully protect the eyes, nose, and other mucous membranes against the vapors of osmium tetroxide. Fill the container with the solution and close tightly to prevent volatilization of the osmium tetroxide.

PREPARATION OF α-NAPHTHYLAMINE SOLUTION:
Add enough α-naphthylamine to water heated to 40°C to make a saturated solution, and filter. Do not allow the naphthylamine to contact the skin; do not breathe the dust. The solution is used at 37°C.

PROCEDURE:
(1) Treat the free-floating sections with osmium tetroxide solution for 18 hours.

(2) Wash sections in distilled water for 10 minutes and mount on glass slides.
(3) Treat with α-naphthylamine solution for 15 to 20 minutes at 37°C.
(4) Wash in distilled water for 5 minutes.
(5) Counterstain in 1 percent alcian blue in 3 percent acetic acid for 15 to 30 minutes (this is optional).
(6) Wash in tap water.
(7) Mount in glycerin jelly.

RESULTS:
Unsaturated lipids (triglycerides and cholesterol esters) and fatty acids stain black. Phospholipids stain orange-red.

PLASMAL REACTION FOR PLASMALOGEN PHOSPHATIDES
(Hayes, 1949; Adams, 1969)

TISSUES:
Fix tissues briefly (3 to 6 hours) in formol-calcium. Use frozen sections.

PROCEDURE:
(1) Mount two frozen sections on slides and dry.
(2) Hydrolyze one section with 1 to 5 percent mercuric chloride for 10 minutes.
(3) Wash in three changes of distilled water.
(4) Treat both sections with Schiff's reagent for 20 minutes.
(5) Rinse in three changes of metabisulfite solution (10 percent $K_2S_2O_5$ in 0.05N HCl).
(6) Wash in running water for 20 minutes.
(7) Mount with glycerin jelly.

COMMENT:
Discard the Schiff's reagent following this procedure, because it is probably contaminated with mercury.

RESULTS:
Plasmalogen phosphatides are stained red-to-pink. The control section (in which the mercuric chloride is omitted) is necessary to evaluate the amount of pseudoplasmal reaction.

OSMIUM TETROXIDE METHOD FOR UNSATURATED LIPIDS
(Adams, 1959)

TISSUES:
Fix tissues in formol-calcium. Use frozen sections.

PROCEDURE:
(1) Mount frozen sections on slides and allow to dry in air.
(2) Treat with 1 percent osmium tetroxide for 2 to 18 hours.
(3) Wash in running tap water for 20 minutes.
(4) Mount in glycerin jelly.

COMMENT:
Osmium tetroxide is volatile, and its vapors are injurious to mucous membranes. Use it only in a fume hood. Exclude air from the container as much as possible by filling it to the top (step 2).

RESULTS:
Unsaturated lipids are stained black. Other lipids remain unstained.

PERFORMIC ACID–SCHIFF METHOD FOR UNSATURATED LIPIDS
(Lillie, 1952)

TISSUES:
Fix tissues in formol-calcium. Use frozen sections.

PREPARATION OF CHROME-GELATIN SLIDES:
Dip slides in a solution of 1 percent gelatin containing 0.05 percent chrome potassium alum. Allow the slides to drain and air-dry.

PREPARATION OF PERFORMIC ACID:
Add 4.5 ml of hydrogen peroxide (30 percent) and 0.5 ml of concentrated sulfuric acid to 45 ml of formic acid (98 percent). Allow the solution to "mature" for 1 hour and then stir well before using. Discard after 24 hours.

PROCEDURE:

(1) Mount frozen sections on chrome-gelatin slides and allow them to dry.
(2) Treat with performic acid for 10 to 20 minutes.
(3) Wash in tap water for 5 minutes.
(4) Treat with Schiff's reagent for 20 minutes.
(5) Wash in running tap water for 20 minutes.
(6) Mount in glycerin jelly.

RESULTS:

Unsaturated lipids stain red. The reagents are not fat-soluble; unsaturated lipids that may occur inside fat globules remain unstained.

BROMINE–SILVER NITRATE METHOD FOR UNSATURATED LIPIDS
(Norton et al., 1962)

TISSUES:

Fix tissues in formol-calcium. Use frozen sections.

PREPARATION OF BROMINE SOLUTION:

Add 1 ml of bromine to 390 ml of a 2 percent solution of potassium bromide.

PROCEDURE:

(1) Mount frozen sections on slides and air-dry.
(2) Treat with bromine solution for 1 minute.
(3) Wash in distilled water.
(4) Place in 1 percent solution of sodium bisulfide for 5 minutes.
(5) Wash in seven changes of distilled water.
(6) Treat with 1 percent silver nitrate in 1N nitric acid for 18 hours.
(7) Wash in seven changes of distilled water.
(8) Develop for 10 minutes in Kodak Dektol developer (or equivalent) diluted (1:1 v/v).
(9) Wash in water.
(10) Mount in glycerin jelly.

RESULTS:
Unsaturated lipids stain black or brownish black.

PAN METHOD FOR CHOLESTEROL
(Adams, 1961)

TISSUES:
Fix tissue blocks in formol-calcium. Use frozen sections.

PREPARATION OF THE SOLVENT:

Ethanol, 95 percent	50 ml
Perchloric acid, 60 percent	25 ml
Formalin (40 percent formaldehyde)	2.5 ml
Water	22.5 ml

PREPARATION OF THE NAPHTHOQUINONE SOLUTION:
Dissolve 10 mg of 1,2-naphthoquinone-4-sulfonic acid in 10 ml of the solvent. Use the same day.

PROCEDURE:
(1) Place free-floating sections in formol-calcium fixative for at least 1 week (preferably 3 or 4 weeks) in order to promote oxidation of cholesterol.
(2) Mount sections on slides and allow them to air-dry.
(3) Spread a thin layer of the naphthoquinone solution on the section and heat the section on a hot plate to 65° to 70°C for 5 to 10 minutes or until the red color that first forms turns completely blue.
(4) Place a drop of 60 percent perchloric acid on the section and apply a coverslip.

RESULTS:
Only cholesterol and related sterols are colored blue. The color is not stable in water or in any kind of mounting medium. The color is stable in perchloric acid for only a few hours.

MODIFIED SCHULTZ METHOD FOR CHOLESTEROL
(Weber et al., 1956)

TISSUES:
Fix tissues in formol-calcium for 2 to 3 days. Cut frozen sections at 20 to 30 microns. Use free-floating sections.

PREPARATION OF IRON ALUM SOLUTION:
Dissolve 2.5 gm of ferric ammonium sulfate in 100 ml 0.2M acetate buffer, pH 3. Upon addition of the sulfate, the pH of the solution will be lowered to about 2.0.

PREPARATION OF ACETIC ACID–SULFURIC ACID MIXTURE:
Add concentrated sulfuric acid slowly, with stirring, to precooled glacial acetic acid. Combine in a ratio of 1:1.

PROCEDURE:
(1) Wash the sections for 24 hours with several changes of distilled water.
(2) Treat the sections with the iron alum solution for 7 days at 37°C.
(3) Wash the sections in three changes of the acetate buffer (0.2M, pH 3.0), for 1 hour in each change.
(4) Rinse in distilled water.
(5) Fix for 10 minutes in 5 percent formalin.
(6) Mount the sections on slides and remove excess water from the edge with filter paper.
(7) Place a drop of the acetic acid–sulfuric acid mixture on a coverslip.
(8) Invert the slide and apply the section to the drop of acetic acid–sulfuric acid mixture.
(9) Turn the slide right side up and examine under the microscope.

RESULTS:
Areas with cholesterol or its ester will turn blue-green within a few seconds. The color is stable for 30 to 60 minutes.

MODIFIED SCHULTZ METHOD FOR CHOLESTEROL
(Rossouw et al., 1976)

TISSUES:

Fix tissues in formol-calcium for 7 days at 4°C and use 20-micron frozen sections, or use fresh-frozen sections.

PREPARATION OF IRON ALUM SOLUTION:

Dissolve 2.5 gm of ferric ammonium sulfate in 100 ml 0.2M acetate buffer, pH 3. Upon addition of the sulfate, the pH of the solution will be lowered to about 2.0.

PREPARATION OF FERRIC CHLORIDE SOLUTION:

Dissolve 2.5 gm of ferric chloride ($FeCl_3 \cdot 6H_2O$) in 85 percent phosphoric acid (H_3PO_4). Dilute to 100 ml with additional phosphoric acid.

PREPARATION OF PREMIXED COLOR REAGENT:

Combine 25 volumes of glacial acetic acid, 25 volumes of concentrated sulfuric acid, and 2 volumes of ferric chloride solution. Mix in a brown bottle at room temperature and allow the air bubbles to escape.

PROCEDURE:

(1) Mount the frozen sections on slides and allow to dry in air.
(2) Incubate sections in iron alum solution for 3 days at 37°C.
(3) Rinse in three changes of acetate buffer for 1 hour each.
(4) Rinse quickly in distilled water.
(5) Post-fix in 5 percent formalin for 10 minutes.
(6) Remove excess water and invert section on a drop of premixed color reagent placed on a coverslip.
(7) Turn right side up and examine.

RESULTS:

Cholesterol and its esters turn green. The color lasts for only a few hours.

REFERENCES

Adams, C. W. M. (1959). A histochemical method for the simultaneous demonstration of normal and degenerating myelin. *J. Pathol. Bacteriol.* 77:648–650.

Adams, C. W. M. (1961). A perchloric acid–naphthoquinone method for the histochemical localization of cholesterol. *Nature* 192:331–332.

Adams, C. W. M. (1965). Histochemistry of Lipids. In C. W. M. Adams (Ed.), *Neurohistochemistry*. Elsevier, Amsterdam. Pp. 6–66.

Adams, C. W. M. (1969). Lipid histochemistry. *Adv. Lipid Res.* 7:1–62.

Adams, C. W. M., and Bayliss, O. B. (1963). Histochemical observations on the localisation and origin of sphingomyelin, cerebroside and cholesterol in the normal and atherosclerotic human artery. *J. Pathol. Bacteriol.* 85:113–119.

Adams, C. W. M., Abdulla, Y. H., Bayliss, O. B., and Weller, R. O. (1965). The reaction of Baker's chromate reagent with lipids. *J. Histochem. Cytochem.* 13:410–411.

Baker, J. R. (1946). The histochemical recognition of lipine. *Q. J. Microsc. Sci.* 87:441–470.

Baker, J. R. (1947). Further remarks on the recognition of lipine. *Q. J. Microsc. Sci.* 88:463–465.

Baker, J. R. (1958). Fixation in cytochemistry and electron-microscopy. *J. Histochem. Cytochem.* 6:303–308.

Bourgeois, C., and Hack, M. H. (1962). Concerning the specificity of the dichromate-hematoxylin, dichromate–Sudan Black B techniques for the histochemical detection of phospholipids. *Acta Histochem.* 14:297–306.

Cain, A. J. (1947). The use of Nile Blue in the examination of lipoids. *Q. J. Microsc. Sci.* 88:383–392.

Cain, A. J. (1949a). A critique of the plasmal reaction, with remarks on recently proposed techniques. *Q. J. Microsc. Sci.* 90:411–426.

Cain, A. J. (1949b). On the significance of the plasmal reaction. *Q. J. Microsc. Sci.* 90:75–86.

Chiffelle, T. L., and Putt, F. A. (1951). Propylene and ethylene glycol as solvents for Sudan IV and Sudan Black B. *Stain Technol.* 26:51–56.

Cope, C. H., and Williams, M. A. (1968). Quantitative studies on neutral lipid preservation in electron microscopy. *J. R. Microsc. Soc.* 88:259–277.

Criegee, R. (1936). Osmiumsäureester als Zwischenprodukte bei Oxidation. *Liebig. Ann. Chem.* 522:75–96.

Deierkauf, F. A., and Heslinga, F. J. M. (1962). The action of formaldehyde on brain lipids. *J. Histochem. Cytochem.* 10:79–82.

Elftman, H. (1954). Controlled chromation. *J. Histochem. Cytochem.* 2:1–8.

Elleder, M., and Lojda, Z. (1973a). Studies on lipid histochemistry. XI. New, rapid, simple and selective method for the demonstration of phospholipids. *Histochemistry* 36:149–166.

Elleder, M., and Lojda, Z. (1973b). Studies on lipid histochemistry. XII. Histochemical detection of sphingomyelin. *Histochemistry* 37:371–373.

Feder, N. (1963). Histochemical demonstration of lipids and enzymes in tissue specimens embedded at low temperatures in glycol methacrylate. *J. Cell Biol.* 19:23A.

Feulgen, R., and Voit, K. (1924). Über einen weilverbreiteten festen Aldehyd. *Pfluegers Arch. Ges. Physiol.* 206:389–410.

Hayes, E. R. (1949). A rigorous re-definition of the plasmal reaction. *Stain Technol.* 24:19–23.

Heslinga, F. J. M., and Deierkauf, F. A. (1962). The action of formaldehyde solutions on human brain lipids. *J. Histochem. Cytochem.* 10:704–709.

Holczinger, L. (1965).   Über die unspezifische Sudanophilie. *Acta His-tochem.* 20:374–380.

Korn, E. D. (1966a).   Synthesis of bis(methyl 9,10-dihydroxystearate) osmate from methyl oleate and osmium tetroxide under conditions used for fixation of biological material. *Biochim. Biophys. Acta* 116:317–324.

Korn, E. D. (1966b).   Modification of oleic acid during fixation of amoebae by osmium tetroxide. *Biochim. Biophys. Acta* 116:325–335.

Korn, E. D. (1967).   A chromatographic and spectrophotometric study of the products of the reaction of osmium tetroxide with unsaturated lipids. *J. Cell Biol.* 34:627–638.

Lillie, R. D. (1952).   Ethylenic reaction of ceroid with performic acid and Schiff reagent. *Stain Technol.* 27:37–45.

Lillie, R. D. (1965).   *Histopathological Technic and Practical Histochemistry* (3rd ed.). McGraw-Hill, New York.

Luna, L. G. (1968).   *Manual of Histologic Staining Methods.* McGraw-Hill, New York.

Menschik, Z. (1953).   Histochemical comparison of brown and white adipose tissue in guinea pigs. *Anat. Rec.* 116:439–455.

Mukherji, M., Deb, C., and Sen, P. B. (1960).   Histochemical demonstration of unsaturated lipids by bromine silver method. *J. Histochem. Cytochem.* 8:189–194.

Norton, W. T., Korey, S. R., and Brotz, M. (1962).   Histochemical demonstration of unsaturated lipids by a bromine-silver method. *J. Histochem. Cytochem.* 10:83–88.

Reiner, C. B. (1953).   The Schultz histochemical reaction for cholesterol. *Lab. Invest.* 2:140–151.

Roozemond, R. C. (1967).   Thin layer chromatographic study of lipid extraction from cryostat sections of rat hypothalamus by some fixatives. *J. Histochem. Cytochem.* 15:526–529.

Roozemond, R. C. (1969a).   The effect of calcium chloride and formaldehyde on the release and composition of phospholipids from cryostat sections of rat hypothalamus. *J. Histochem. Cytochem.* 17:273–279.

Roozemond, R. C. (1969b).   The effect of fixation with formaldehyde glutaraldehyde on the composition of phospholipids extractable from rat hypothalamus. *J. Histochem. Cytochem.* 17:482–486.

Rossouw, D. J., Chase, C. C., Raath, I., and Engelbrecht, F. M. (1976).   The histochemical localization of cholesterol in formalin-fixed and fresh frozen sections. *Stain Technol.* 51:143–145.

Schultz, A. (1924).   Eine Methode des mikrochemischen Cholesterin-nachweiss am Gewebsschnitt. *Zentralbl. Allg. Pathol.* 35:414–417.

Smith, J. L., and Mair, W. (1909).   An investigation of the principles underlying Weigert's method of staining medullated nerve. *J. Pathol. Bacteriol.* 13:14–27.

Wigglesworth, V. B. (1957).   The use of osmium in the fixation and staining of tissue. *Proc. R. Soc. London* [Biol.] 147:185–199.

Weber, A. F., Phillips, M. G., and Bell, J. T. (1956).   An improved method for the Schultz cholesterol test. *J. Histochem. Cytochem.* 4:308–309.

8

NUCLEIC ACIDS

CHEMISTRY OF NUCLEIC ACIDS

Historical Aspects

Nucleic acids were discovered over a century ago by Mieschner, who isolated nuclei from pus cells for chemical analysis and discovered a material that contained much phosphorus. He called the material "nuclein," but it is now known as nucleoprotein (Davidson, 1969). Mieschner later discovered that salmon sperm also contain an acidic compound, later identified as nucleic acid, and a basic substance that he named "protamine." Chemical analysis of these substances, however, had to await the development of better methods for preparing nucleic acids.

Nucleic acid from the thymus was extensively studied, and it was discovered that on hydrolysis it yields a sugar, phosphoric acid, and the four nitrogenous bases—adenine, guanine, cytosine, and thymine. This was called thymonucleic acid. The sugar was soon identified as deoxyribose (desoxyribose) and thymonucleic acid gradually came to be more rationally termed *deoxyribonucleic acid* (DNA).

A different nucleic acid, isolated from yeast, contained ribose for a sugar and uracil in place of thymine. This substance was thus called yeast nucleic acid, but its name was changed to *ribonucleic acid* (RNA).

Because DNA was first isolated from animal tissue and RNA from plant tissues, it was believed that DNA was characteristic of animal tissues and that RNA was its equivalent in plant tissues. This belief began to be questioned during the 1920s, when it became increasingly evident that RNA was also a normal constituent of a number of animal tissues.

The next concept to be abandoned was that nucleic acids were confined to the nucleus. Histochemical and cell fractionation studies established that RNA was contained not only in the nucleus but also in the cytoplasm. It is particularly abundant in the nucleolus and in the rough endoplasmic reticulum. We now know that small amounts of DNA also occur in the cytoplasm, especially in the mitochondria.

Clarification of the early concepts about nucleic acids is necessary for our understanding of their structure and function.

Structure and Function of Nucleic Acids

The basic unit of DNA is the deoxyribonucleotide, or simply nucleotide. It is composed of a unit each of deoxyribose sugar, phosphoric acid, and a nitrogenous base. The base may be either a purine (adenine or guanine) or a pyrimidine (cytosine or thymine). The purines and the pyrimidines are present in an equimolar ratio.

The polymer of DNA is formed by the nucleotides becoming covalently linked end to end through the phosphates and sugars. The nitrogenous bases are not involved in forming the polymer; they are attached at the 1-positions of the sugar and are suspended laterally. The continuity of the polymer is therefore through alternating sugars and phosphates. The phosphate is bound to the 5-position of the same nucleotide and to the 3-position of the adjacent nucleotide.

DNA is double-stranded; two polymers run parallel to each other. Actually, one strand runs in the 3-5 direction while the other one runs in the opposite direction—that is, in the 5-3 direction. The two strands of DNA are held together by hydrogen bonds between the nitrogenous bases of the two strands. The stereochemical "fit" between two opposite bases is very specific; only adenine can be matched with thymine and only guanine can be matched with cytosine. It should be noted that in each case a purine matches a pyrimidine. The two strands are twisted upon each other so as to form a double helix. The structure derives some stability from the hydrogen bonding between the bases.

It is thought that a chromosome has a core of the double-stranded DNA helix, which becomes coiled in a very complex manner during mitosis. Histones are associated with DNA in the chromosomes and may also help determine the structure of the chromosome. There are two types of histones, those that are rich in arginine and those that are rich in lysine. It was very tempting to postulate that the histones are involved in determining the metabolic state of the nucleus. It was envisioned that histones would tie up certain segments of the DNA molecule to prevent genetic expression. It stood to reason that the nucleus of a metabolically active cell should have less histone than one that is metabolically quiescent. It has been shown, however, that there is no correlation between the amount of histone a nucleus contains and the metabolic state of the cell. All contain about the same amount of histone. There is evidence, however, that there is a correlation between the predominant type of histone (arginine-rich or lysine-rich) and the metabolic state.

The structure of RNA is different from that of DNA in several important respects. The pyrimidine uracil is substituted for thymine; RNA is generally of much lower molecular weight; and, with few exceptions, RNA is single-stranded. (In certain viruses, RNA is double-stranded, and it is in fact substituted for DNA.) RNA is found in the nucleus, primarily in the nucleolus, and in the cytoplasm, particularly in the rough endoplasmic reticulum.

HISTOCHEMICAL METHODS
Feulgen Method for DNA
The time-honored method for demonstrating DNA is the Feulgen method. Feulgen and Rossenbeck (1924) discovered that aldehyde groups could be generated in DNA by mild acid hydrolysis, and they could then be demonstrated with Schiff's reagent.

The specificity of the reaction has been intensely scrutinized, especially during the 1940s. The discussions and arguments that were generated occupy many more pages in the histochemical literature than is justifiable. A summary of the discussions and arguments can be found in Pearse (1968). The Feulgen reaction has survived the controversy and today enjoys a great deal of respectability. Perhaps the one single most important factor in establishing confidence in the specificity of the method was the use of deoxyribonuclease-treated sections for controls. Such controls are always completely negative.

An explanation of the reaction mechanism was offered by Stacey et al. (1946) and revised by Overend and Stacey (1949). It is clear that purine bases (adenine and guanine) are rapidly and completely removed by warm acid hydrolysis, while the main backbone of the DNA molecule—the alternating phosphates and sugars—remains largely intact. The removal of the purine base frees carbon 1 of the sugar while the furanose form is maintained. Hydrolysis occurs between carbons 1 and 2 and an aldehyde group is generated, probably on carbon 1.

When treated with Schiff's reagent, the fuchsin-sulfurous acid molecules condense on the aldehydes and are recolorized in the process. It is thought that the mechanism of condensation is through a bisulfite bond. One molecule undoubtedly attaches to two aldehyde groups of adjacent sugars.

It is not clear what happens to the RNA except that no aldehyde groups are produced. It is very likely hydrolyzed to such an extent that it is leached from the tissue.

It has been known for a long time that different fixatives have a profound effect on the hydrolysis time necessary for optimal Feulgen staining (Bauer, 1932). This is generally assumed to result from the effect of fixatives on nucleoproteins. Bouin's fixative is generally considered unsuitable because it tends to extract nucleic acids.

The usual hydrolysis conditions involve a 1N HCl solution at 60°C; the length of hydrolysis varies from 5 to 60 minutes depending on the fixative used. Overhydrolysis causes extraction of DNA (and incidentally of histones as well), and the amount of color that is produced by Schiff's reagent is correspondingly reduced. In the traditional method of hydrolysis, aldehyde formation and DNA extraction probably overlap somewhat.

In recent years, a number of workers have carried out the Feulgen hydrolysis at lower temperatures. Two authors (Murgatroyd, 1968; Vahs, 1973) have suggested that for quantitative work the hydrolysis should be done with 5N HCl at 20°C. The color intensity is constant over a long period of hydrolysis time, probably because extraction of DNA is prevented under these conditions. The amount of color was 20 to 30 percent higher with cold Feulgen hydrolysis (Vahs, 1973). Low temperature hydrolysis maximizes the margin of safety between underhydrolysis and extraction of DNA, thus increasing the reproducibility of the results. (For a fuller discussion and references on the Feulgen hydrolysis, the work of Kjellstrand and Andersson [1975] should be consulted.)

Bruchhaus and Geyer (1974) found DNA to be extremely resistant to extraction by alkaline hydrolysis, using ethanolic barium hydroxide. Aldehydes are produced in the same way as they are with the standard Feulgen hydrolysis. RNA, on the other hand, is readily oxidized and lost from the tissue section. Quantitative determinations indicated that the color produced after 5 hours of hydrolysis remained the same as that produced after 1 hour. This method therefore offers another method of hydrolysis suitable for quantitation.

Methyl Green–Pyronin Method

The methyl green–pyronin method offers the possibility of differentially demonstrating RNA and DNA in the same tissue section; RNA is stained purple and DNA green. The earliest versions of the method were introduced by several workers around the turn of the century. The method became very popular after Brachet (1940 and 1942) showed that, under controlled conditions, the method would stain the two nucleic acids differentially and specifically. The great interest in the method has also brought about a large number of variations of the method.

Different methods of fixation are responsible for causing a great deal of variation in the results of the method. Bouin's and Carnoy's fixatives are often used but there have been reports of significant losses of RNA during Carnoy fixation. Lillie (1965) stated that aqueous formalin and Zenker's fixative are unsuitable because the sections stain all red or all blue.

A problem exists in explaining the specificity of the two stains. Methyl green is a triarylmethane dye while pyronin is a quinoline type of dye, but both are basic stains. The selectivity must be based on something other than simply basophilia. The specificity of methyl green for DNA is undoubtedly based on the nature of the DNA polymers. It appears from the work of Kurnick (1949) and Vercauteren (1950) that DNA loses its affinity for methyl green during depolymerization. Vercauteren determined that the distance between two adjacent electronegative phosphate groups of a DNA polymer is the same as the distance between the two positive charges of the methyl green molecule. When the DNA becomes denatured, the distance between the two phosphate groups is altered. All this suggests that a stereochemical fit between DNA and the dye molecule is responsible for the specificity.

The specificity of pyronin for RNA is lower than that of methyl green for DNA. Other tissue components that stain with pyronin include cartilage, osteoid, keratin, and the granules of eosinophils and of mast cells. Even DNA stains with pyronin, although the affinity of pyronin for native DNA is low. In the presence of methyl green, evidently, pyronin cannot compete with methyl green for DNA. However, when DNA becomes denatured, the affinity of pyronin for DNA increases. In order to facilitate interpretation of results, RNase controls are recommended.

The nature and content of dye impurities varies from batch to batch, and this inconsistency may be responsible for some variability in the results. Methyl violet is consistently a contaminant of methyl green. Better results can be obtained if the methyl violet is removed before using methyl green, which can be done by washing an aqueous solution of methyl green with chloroform. Chloroform dissolves the contaminant but the methyl green remains in the aqueous phase.

Feulgen–Methylene Blue Sequence
Despite all the modifications of the methyl green–pyronin procedure, problems related to dye impurities and dye variability remain. To circumvent these problems, Spicer (1961) and Garvin et al. (1976) developed a different sequence that capitalized on the great reliability of the Feulgen method for demonstrating DNA. They modified the Feulgen procedure so that Bouin's fixative is substituted for the hydrochloric acid in the hydrolysis step. Satisfactory hydrolysis is achieved without RNA extraction. The RNA is subsequently stained with the basic dye methylene blue. However, this sequence does not increase the specificity for RNA and should probably also be used with RNase extraction controls.

Alkaline Fast Green Method for Histones
The alkaline fast green method was introduced by Alfret and Gerschwind (1953) as a technique for demonstrating basic nuclear proteins. Ordinarily, alkaline fast green stains only free basic groups of histones, which (coupled with nucleic acids) are masked and cannot be stained unless the nucleic acids are first removed. Alfret and Gerschwind (1953) hydrolyzed the nucleic acids with trichloroacetic acid at 90° to 100°C. Prentø and Lyon (1973) found that this procedure gave a very narrow margin of confidence between incomplete DNA removal (underhydrolysis) and extraction of histones (overhydrolysis). Slight variation in temperature is probably responsible for the erratic results sometimes observed with this procedure. Prentø and Lyon (1973) therefore suggested that hydrolysis should be carried out at 60°C, while compensating with a longer extraction period. The lower

hydrolysis temperature is easier to control accurately and the margin of confidence is greatly increased.

The method of Alfret and Gerschwind calls for dissolving the fast green stain in distilled water and adjusting it to pH 8.0 with sodium hydroxide. The absence of a buffer has been found objectionable because it is difficult to achieve and maintain the exact pH for any period of time. Chayen et al. (1973) found that phosphate-citrate buffer was more suitable as a solvent. Prentø and Lyon used an HCl-borate buffer for the same purpose but made it very weak (0.01M) in order to minimize ion competition.

The specificity of the method has been questioned. It is known that the method also stains other basic proteins, although no others occur in the nucleus. Other studies have also criticized the method on the basis that the dye-protein interaction either is not quantitative (Noeske, 1973) or is quantitative only in the absence of other proteins (Cohn, 1973).

Ammoniacal-Silver Method for Histones

A method was developed by Black and Ansley (1964, 1965, 1966) for selectively demonstrating histones with ammoniacal silver. The method was developed originally to be used with smears, but Myśliwski (1970) adapted it to use with paraffin sections, evidently with equal success. This method has the advantage over the fast green method in that it is unnecessary to remove DNA prior to staining.

The tissue sections (or smears) are exposed to buffered formalin just before exposure to ammoniacal silver. This is an absolute requirement, regardless of the original method of fixation. The necessity for this step suggests that the formaldehyde molecules so introduced into the tissue are somehow responsible for the reaction with silver. However, the reaction mechanism is poorly understood.

STANDARD FEULGEN METHOD FOR DNA

TISSUES:

Fix tissues with either formalin or Carnoy's fixative. Use paraffin sections.

PROCEDURE:

(1) Deparaffinize sections and bring to water.
(2) Hydrolyze sections in 1N HCl, preheated to 60°C, for 10 minutes.
(3) Rinse in 1N HCl at room temperature.
(4) Transfer directly into Schiff's reagent for 30 to 60 minutes.
(5) Wash in three changes of 0.5 percent sodium metabisulfite (or potassium metabisulfite) for 2 minutes each.
(6) Wash in water.
(7) Dehydrate through graded alcohols, clear in xylene, and mount with synthetic medium.

RESULTS:
DNA stains red to reddish purple.

MODIFIED FEULGEN REACTION
(Vahs, 1973)

TISSUES:
Fix tissues with alcohol–formalin–acetic acid solution and use paraffin sections.

PROCEDURE:

(1) Bring sections to water.
(2) Treat sections with 5N HCl at 20°C for 40 minutes.
(3) Transfer directly to Schiff's reagent (see p. 106) for 30 to 90 minutes.
(4) Wash in three changes of 0.5 percent sodium metabisulfite for 2 minutes each.
(5) Wash in distilled water.
(6) Dehydrate, clear, and mount.

RESULTS:
DNA will stain red or reddish purple.

METHYL GREEN–PYRONIN METHOD FOR DNA AND RNA

TISSUES:
Fix tissues with alcoholic formalin, acetic alcohol formalin, or Carnoy's fixative. Use paraffin sections. Alternatively, fresh-frozen sections can be used.

PREPARATION OF METHYL GREEN STOCK SOLUTION:
Prepare a 2 percent solution of methyl green; pour it into a separatory funnel and add an equal amount of chloroform. Shake well and then allow the chloroform to form a separate layer, which is removed and discarded. Repeat this operation six or eight times or until no more violet is removed.

PREPARATION OF PYRONIN Y STOCK SOLUTION:
Prepare a 2 percent solution of pyronin Y.

PREPARATION OF STAINING SOLUTION:
Combine 7.5 ml of the methyl green stock solution and 12.5 ml of a 2 percent solution of Pyronin Y with 30 ml of $0.2M$ acetate buffer, pH 4.8.

PROCEDURE:
(1) Bring sections to water. (Fresh-frozen sections may be placed directly into the staining solution.)
(2) Stain in staining solution for 6 minutes.
(3) Blot dry with filter paper.
(4) Place in two changes of n-butyl alcohol for 5 minutes each.
(5) Place in xylene for 5 minutes.
(6) Place in cedar oil for 5 minutes.
(7) Mount with Permount.

RESULTS:
Nuclear chromatin is stained green; nucleoli and cytoplasmic RNA are stained red.

FEULGEN–METHYLENE BLUE METHOD FOR NUCLEIC ACIDS
(Garvin et al., 1976)

TISSUES:
Fix the tissues with formol-calcium, phosphate-buffered formalin, Zenker's fluid, or Carnoy's fluid for 24 hours. Use paraffin sections.

PREPARATION OF SCHIFF'S REAGENT:
Sodium sulfite, 5 gm; basic fuchsin, 4 gm; distilled water, 192 ml

Stir for 3 to 4 hours or until the solution is pale brown. Add 0.5 gm activated charcoal and filter through double filter paper. (Note: This Schiff's reagent has a considerably higher concentration of basic fuchsin than other preparations, making it preferable for this procedure.)

PREPARATION OF METHYLENE BLUE STOCK SOLUTION:
Add 1 gm methylene blue to 500 ml distilled water.

PREPARATION OF METHYLENE BLUE STAINING SOLUTION:
Methylene blue stock solution, 48 ml; citric acid, $0.1M$, 1.4 ml; dibasic sodium phosphate, $0.2M$, 0.6 ml

Adjust the pH to 3.5 with additional citric acid or dibasic sodium phosphate, added drop by drop.

PROCEDURE:
(1) Deparaffinize sections and bring to water.
(2) Hydrolyze in Bouin's fluid for 1 hour at 60°C.
(3) Wash in running water for 5 minutes.
(4) Treat with Schiff's reagent for 20 minutes.
(5) Wash in three changes of 0.5 percent sodium metabisulfite for 2 minutes each.
(6) Wash in running water for 5 minutes.
(7) Stain with methylene blue staining solution for 3 minutes.
(8) Dehydrate for 1 minute each in 95 percent ethanol, in two changes of 100 percent ethanol, in xylene-ethanol (1:1) and in three changes of xylene.
(9) Mount with synthetic resin.

RESULTS:
DNA (i.e., heterochromatin) is stained red-to-purple and RNA is stained blue.

ALKALINE FAST GREEN FCF METHOD FOR HISTONES
(Prentø and Lyon, 1973)

TISSUES:
Fix tissues for 12 hours in phosphate-buffered formalin. Use paraffin sections. (Avoid temperatures exceeding 50°C while drying the sections.)

PREPARATION OF ALKALINE FAST GREEN STAINING SOLUTION:
Dissolve 0.5 gm fast green FCF in 450 ml distilled water. Add 50 ml 0.1*M* borate-HCl buffer, pH 8. The final pH should be 8.1 to 8.2. Add a crystal of thymol and store in a stoppered 500-ml bottle.

PROCEDURE:
Use two sections, A and B. Section A is subjected to the entire procedure, while section B is subjected to all but steps 2, 3, and 4.

(1) Bring sections to water.
(2) Immerse in distilled water at 60°C (this step is optional).
(3) Extract with 5 percent trichloroacetic acid at 60°C for 80 to 90 minutes.
(4) Wash in tap water for 5 minutes, or in five changes of distilled water for 2 minutes each.
(5) Stain for 30 minutes in alkaline fast green solution in a vessel with a tight-fitting lid.
(6) Rinse in tap water or in three changes of distilled water.
(7) Dehydrate through two changes of 95 percent ethanol and then with absolute ethanol. (After staining at pH 8, the chromatin-bound dye is not appreciably differentiated by either water or ethanol.)
(8) Clear in toluene and mount in synthetic resin.

RESULTS:

In section B, all unblocked proteins with a pI above 8 stain green. Section A stains as section B but there is a pronounced staining of the nuclei and endoplasmic reticulum because of the increase in the chromatin pI.

AMMONIACAL-SILVER METHOD FOR HISTONES
(Black and Ansley, 1965; Myśliwski, 1970)

TISSUES:

Fix tissues in alcoholic formalin and use paraffin sections, or make fresh smears and allow them to air-dry.

PREPARATION OF AMMONIACAL-SILVER SOLUTION:

Add a 10 percent solution of silver nitrate drop by drop to concentrated ammonium hydroxide until the first permanent turbidity appears.

PROCEDURE:

(1) Bring sections to water or use smears.
(2) Treat sections with sodium acetate–neutralized 10 percent formalin for 1 hour. (If controls are desired, parallel slides may be treated with a nitrite solution for 5 to 15 minutes between steps 1 and 2, followed by a wash in water. The nitrite solution is prepared by mixing equal parts of 10 percent trichloroacetic acid and 10 percent sodium nitrite.)
(3) Wash vigorously in five changes of distilled water.
(4) Treat with ammoniacal-silver solution for 10 seconds with vigorous agitation.
(5) Wash vigorously in five changes of distilled water.
(6) Treat with 3 percent formalin for 2 minutes (dilute 15 ml of the fixative to 50 ml with water; neutralize with sodium acetate).
(7) Wash thoroughly with water.
(8) Dehydrate, clear, and mount with a synthetic mountant.

RESULTS:

Histones will stain yellowish brown. Histones will fail to stain in the controls.

ENZYMATIC EXTRACTION OF RNA

PROCEDURE:

(1) Bring two sections to water.

(2) Treat the one section with a 0.05 to 0.1 percent solution of RNase in 0.05M tris buffer, pH 7, for 1 hour at room temperature. Incubate the other section in buffer solution containing no enzyme.

(3) Wash both sections in distilled water.

(4) Stain with the methyl green–pyronin method.

RESULTS:

Any areas staining red in the control and not in the enzyme-treated section can be presumed to be RNA.

REFERENCES

Alfret, M., and Gerschwind, I. I. (1953).   A selective staining method for the basic proteins of cell nuclei. *Proc. Natl. Acad. Sci. U.S.A.* 39:991–999.

Bauer, H. (1932).   Die Feulgensche Nuklealfärbung in ihrer Anwendung auf cytologische Untersuchungen. Z. *Zellforsch.* 15:225–247.

Black, M. M., and Ansley, H. R. (1964).   Histone staining with ammoniacal silver. *Science* 143:693–695.

Black, M. M., and Ansley, H. R. (1965).   Antigen-induced changes in lymphoid cell histones. I. Thymus. *J. Cell Biol.* 26:201–208.

Black, M. M., and Ansley, H. R. (1966).   Histone specificity revealed by ammoniacal silver staining. *J. Histochem. Cytochem.* 14:177–181.

Bruchhaus, H., and Geyer, G. (1974).   Studies on the alkaline hydrolysis of nucleic acid in tissue sections. V. Cytophotometric evaluation of the effect of an alkaline hydrolysis on the Feulgen reaction. *Histochem. J.* 6:579–581.

Chayen, J., Bitensky, L., and Butcher, R. G. (1973).   *Practical Histochemistry*. Wiley, London.

Cohn, N. S. (1973).   A model system analysis of the parameters in histone staining I. Alkaline Fast Green. *Histochem. J.* 5:529–545.

Davidson, J. N. (1969).   *The Biochemistry of Nucleic Acids* (6th ed.). Methuen, London.

Feulgen, R., and Rossenbeck, H. (1924).   Mikroskopisch-chemischer Nachweis einer Nukleinsaüre vom Typus der Thymonucleinsaüre und die darauf beruhende elektive Färbung von Zellkernen in mikroskopischen Präparaten. *Hoppe Seylers Z. Physiol. Chem.* 135:203–248.

Garvin, A. J., Hall, B. J., Brissie, R. M., and Spicer, S. S. (1976).   Cytochemical differentiation of nucleic acids with a Schiff–methylene blue sequence. *J. Histochem. Cytochem.* 24:587–590.

Kjellstrand, P. T. T., and Andersson, G. K. A. (1975).   Histochemical properties of spermatozoa and somatic cells. I. Relationship between the Feulgen hydrolysis pattern and the composition of the nucleoproteins. *Histochem. J.* 7:563–573.

Kurnick, N. B. (1949).   Methyl Green–Pyronin. I. Basis of selective staining of nucleic acids. *J. Gen. Physiol.* 33:243–264.

Lillie, R. D. (1965).   *Histopathologic Technic and Practical Histochemistry* (3rd ed.). Blakiston, New York.

Murgatroyd, L. B. (1968).   A quantitative investigation into the effect of fixative, temperature and acid strength upon the Feulgen reaction. *J. R. Microsc. Soc.* 88:133–139.

Myśliwski, A. (1970).   Reaction on histones with ammoniacal silver in the developing rat retina. *Acta Histochem.* 38:88–91.

Noeske, K. (1973).   Discrepancies between cytophotometric alkaline Fast Green measurements and nuclear histone protein content. *Histochem. J.* 5:303–311.

Overend, W. G., and Stacey, M. (1949).   Mechanism of the Feulgen nucleal reaction. *Nature* 163:538–540.

Prentø. P., and Lyon, H. (1973).   Nucleoprotein staining: An analysis of a standardized trichloroacetic acid–Fast Green FCF procedure. *Histochem. J.* 5:493–501.

Pearse, A. G. E. (1968).   *Histochemistry: Theoretical and Applied* (3rd ed.), Vol. 1. Little, Brown, Boston.

Spicer, S. S. (1961).   Differentiation of nucleic acids by staining at controlled pH and by a Schiff–Methylene Blue sequence. *Stain Technol.* 36:337–340.

Stacey, M., Deriaz, R. E., Teece, E. G., and Wiggins, L. F. (1946).   Chemistry of the Feulgen and Dische nucleal reactions. *Nature* 157:740–741.

Vahs, W. (1973).   Die Bedeutung der Hydrolyse-Art in der Feulgen-Cytophotometrie von Kernen mit unterschiedlichen Ploidiegraden. *Histochemistry* 33:341–348.

Vercauteren, R. (1950).   The structure of desoxyribose nucleic acid in relation to the cytochemical significance of the methyl green-pyronin staining. *Enzymologia* 14:134–140.

9

PIGMENTS AND MINERALS

Some minerals that are of the greatest physiological importance—for example, sodium, potassium, and chlorine—cannot be reliably demonstrated by currently available histochemical methods. Reliable methods do exist for demonstrating iron, copper, calcium, zinc, and a few other minerals that are nevertheless of considerable physiological importance. Minerals that readily form insoluble compounds are of course more readily demonstrated than minerals that do not form such compounds.

The pigments to be considered in this chapter represent only a few of the most important pigments for which histochemical methods exist.

PIGMENTS
Lipofuscin

Many cells of older subjects contain inclusions of lipofuscin, which occurs most commonly in heart muscle, liver, adrenal glands, and ganglion cells. In unstained tissue sections, it has a light tan or brown color. Because it is often found in association with lysosomal enzymes, lipofuscin is thought to be a residue that remains following lysosomal degradation of other cell organelles or phagocytosed material. It is seen more frequently in old age, a fact suggesting that it accumulates with age. It is therefore often called the "wear and tear" pigment.

Lipofuscin inclusions are thought to contain some lipid material because they stain somewhat with lipid stains. They are insoluble in organic solvents as well as in acids or bases. In electron micrographs, they appear very heterogeneous. The chemistry of lipofuscin inclusions is poorly understood and probably is very complex.

The Schmorl reaction is frequently used to demonstrate lipofuscins. In this method, a fresh solution of ferric ferricyanide (potassium ferricyanide dissolved in a ferric chloride solution) is reduced by lipofuscin to form a blue pigment. There was general agreement that the pigment was ferric ferrocyanide (Prussian blue), but Adams (1956) felt that it must be ferrosoferric ferricyanide, which is a greenish pigment. That may be the pigment formed at the less strongly reducing sites;

however, ferric ferrocyanide is almost certainly the pigment produced at the more strongly reducing sites.

The Schmorl method is not specific for lipofuscin, because melanin, argentaffin granules, and sulfhydryl groups also stain. It is therefore essential to use this method in conjunction with other methods if positive identification is required.

Lipofuscins are highly colored with sudan black B and they usually yield positive results in the periodic acid–Schiff (PAS) method. Since lipofuscin is acid-fast, the Ziehl-Neelsen method generally gives positive results also, especially if prolonged staining is used. The bleaching method of Hueck (1912) has often been used to distinguish between melanin and lipofuscin. With this method, the tissue section is first stained with the nile blue sulfate method, which stains both components equally (as well as other components). The section is then subjected to hydrogen peroxide bleaching, after which only the lipofuscin coloration remains.

Melanin

Melanin is a pigment that occurs in various body tissues such as the substantia nigra and the pigmented epithelium of the retina and the iris. However, it is most evident in the skin of those human races that traditionally live in geographical regions of high solar radiation. Special cells—called melanocytes—elaborate melanin, which is formed as small, homogeneous granules that are yellow, brown, or black.

The intrinsic coloration of melanin is sufficiently intense to be readily seen under the microscope. A histochemical method for demonstrating melanin serves no important purpose in demonstrating the pigment in normally occurring sites. Histochemical methods for melanin are useful primarily for making positive identification in abnormal conditions. It is at least as important to know what conditions abolish the pigmentation as it is to be able to employ specific histochemical reactions that alter or intensify the coloration.

Melanin is completely insoluble in organic solvents; this is undoubtedly related to the fact that it is strongly bound to protein, which makes it possible to distinguish melanin readily from lipid pigments. Melanin is soluble in strong alkali whereas lipofuscins are not, although this does not provide a practical method of differentiating between the two.

Melanin can be readily bleached with certain strong oxidizing agents; however, the process is quite slow with weak oxidants. Pearse (1972) prefers to use a 10 percent solution of hydrogen peroxide, which is convenient but requires 24 to 48 hours. Potassium permanganate, potassium chlorate, peracetic acid, and performic acid have also been used successfully. (For further details, see Lillie [1956, 1965].) Many other endogenous pigments are also bleached by the methods used to decolorize melanin. Thus, bleaching has limited value in differentiating melanin from other pigments. It is sometimes useful in revealing cellular detail in areas where it is normally obscured by heavy pigmentation.

Melanin is also argentophilic; that is, it has the ability to reduce silver nitrate to form metallic silver. Because other tissue pigments—particularly hematoidin and lipofuscin—are also argentophilic, the silver staining methods are not specific for melanin. However, Lillie (1956) and Lillie et al. (1957) found that the specificity of the silver methods for melanin could be greatly increased by maintaining a pH of 4.0 and limiting the incubation time to 10 minutes. The Fontana (1925) modification of the Masson (1914) silver technique is often used to demonstrate melanin. Many workers prefer to use Gomori's (1946) methenamine silver solution in this procedure.

Hemosiderin

Hemosiderin is a breakdown product of hemoglobin. Red blood cells live only 120 days on the average; when they die, they are phagocytosed by the reticuloendothelial phagocytes of the liver, spleen, and bone marrow. Within the phagocytes, hemosiderin is formed and deposited in the form of golden brown granules or inclusion bodies. Ferritin is a constituent of hemosiderin granules. In diseases involving increased red cell breakdown, or in conditions of hemochromotosis, the number of hemosiderin granules is greatly increased and their distribution becomes more widespread.

Hemosiderin is generally demonstrated with a modified version of Perls' (1867) method that is based on the iron content of hemosiderin. The iron is not readily available to react with ferrocyanide but it can readily be freed with dilute hydrochloric acid. The method of Gomori (1952) is preferred for demonstrating hemosiderin.

Hematoidin and Bile Pigments

The discoloration at the site of an old hemorrhage or infarct is the result of a yellowish brown pigment called hematoidin. Chemically, the pigment is identical with bilirubin, but the morphological distribution is different.

The traditional method of demonstrating bile pigments is by the Gmelin method. Nitric acid is applied to the tissue sections while the section is being observed under the microscope. If the pigment is bilirubin or hematoidin, it will go through a whole spectrum of colors in a period of only a few seconds to a few minutes. No permanent sections can be made, and the sensitivity may be too low to demonstrate the smallest of the granules. Many other oxidants have been tried with the aim of increasing the sensitivity and obtaining a stable color, but only one was successful enough to warrant consideration here. Glenner (1957) used potassium dichromate in his procedure, which seems to be quite reliable. The pH of the oxidizing solution must be carefully controlled, because a pH that is too low might result in solubilization of the pigment, whereas too high a pH might result in overoxidation.

MINERALS

Iron

The importance of iron in the body is well known. Iron is a major constituent of the proteins (namely, hemoglobin and myoglobin) involved with oxygen transport and storage. It has a similar position of importance in certain key proteins (such as the cytochromes and certain flavoproteins) of the electron transport system. Iron is stored in the liver as ferritin.

The time-honored method for demonstrating iron in tissue is Perls' (1867) Prussian blue reaction, which depends on the formation of ferric ferrocyanide when ferrocyanide reacts with ferric ions in the tissue.

$$4FeCl_3 + 3K_4Fe(CN)_6 \rightarrow Fe_4[Fe(CN)_6]_3 + 12KCl$$

Many other methods have been developed for demonstrating iron but none of them seem to be as specific, sensitive, and dependable as the

Perls method. This method will demonstrate free inorganic ferric iron as well as hemosiderin, and it will also demonstrate hemoglobin if the "masked" iron of hemoglobin is released by pretreating the tissue sections with hydrogen peroxide.

Zinc

Zinc is an important element in plant and animal tissues. The human body contains from 1.5 to 4.5 gm of zinc, which places it next to iron in abundance. The first major function to be recognized for zinc was its involvement in the production of insulin; its presence in the islets of Langerhans is frequently studied. Zinc was first recognized as an important constituent of enzymes when it was found in carbonic anhydrase. Since then zinc has been found to be an integral part of arginase, alkaline phosphatase, aminopeptidase, carboxypeptidase, and lactate dehydrogenase.

The method of choice for demonstrating zinc is that developed by Mager et al. (1953); it employs dithizon (diphenylthiocarbazone), which forms a red salt complex with zinc. This compound had been used earlier to demonstrate zinc but not with any great specificity, because dithizon binds with metals other than zinc as well. Mager et al. (1953) made the method very specific by employing a complex-forming buffer. The buffer prevents all metals but zinc from binding with dithizon.

Calcium

Calcium represents about 1.5 percent of the human body and thus is the most abundant mineral. However, 99 percent of it is incorporated into teeth and bone in the form of a calcium phosphate salt similar to hydroxyapatite. Calcium is also a functionally important component of serum, interstitial fluid, and soft tissues. About 60 percent of the serum calcium is ionized, while the rest is protein-bound. Soft tissues may contain no free ionized calcium at all. Only the physiological and abnormal deposits of calcium salts can be demonstrated histochemically.

Fixatives that contain metal or acid should be avoided when tissues are intended for calcium demonstration. Best results have been obtained by using 95 percent ethanol, either alone or saturated with sodium carbonate (Dahl, 1952). McGee-Russell (1958) preferred a fixative made of equal portions of concentrated formaldehyde and absolute ethanol.

von Kossa method. The von Kossa (1901) silver method is the oldest, although no longer the most popular, method for demonstrating calcium in tissue. It is based on the principle of metal substitution. If a tissue containing calcium salts is treated with a suitable cation solution, a new salt is formed. Thus the reaction depends more on the anion than calcium. If the transformation is to be successful, the new salt must be less soluble than the original calcium salt. In the von Kossa method, silver is used to replace calcium. The silver is then reduced with hydroquinone or some other photographic developer, so that it becomes visible as metallic silver.

This method is not specific for calcium; other metals are also detected, because it is actually the anions associated with calcium that are demonstrated. Cytological details of the sections suffer a great deal. Despite these drawbacks, the technique is useful and has been exploited extensively.

Other metals, such as cobalt and lead, have been used successfully in place of silver. However, because they have to be visualized as their sulfides, these methods are inferior to the von Kossa method.

Whether calcium oxalate, such as found in renal calculi, is visualized with the von Kossa method has been a matter of controversy. The method, in its original form, is certainly not recommended for that purpose. The results of Chaplin and Grace (1975) indicate that in order to obtain positive results, the concentration of silver nitrate should be increased or the incubation time prolonged, or both.

An interesting variation of the von Kossa method was developed by Yasue (1969a, 1969b) specifically for demonstrating calcium oxalate. The variation involves a silver nitrate–rubeanic acid sequence, which seems to be quite satisfactory for that purpose. Calcium phosphates and calcium carbonates are removed with acids and "leucinelike crystals" are removed with an alkaline wash before the silver nitrate treatment.

Alizarin red S method. Another method for demonstrating calcium offers a totally different approach. Alizarin red S (sodium alizarin sulfonate) can form a lake with calcium; this is the basis of many histochemical methods that have been developed for calcium. The dye has the following structure:

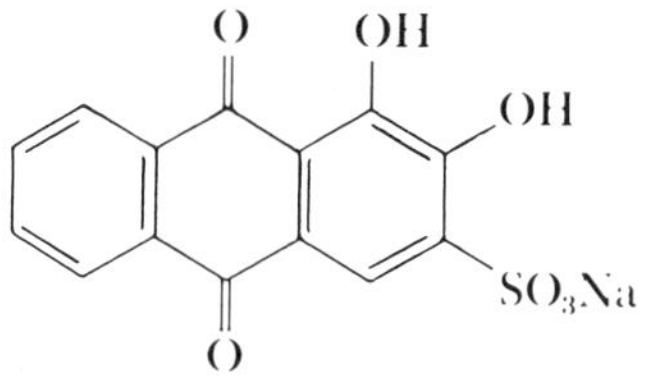

The structure shows an anthraquinone ring system with substituents O, OH, OH, SO₃Na.

The alizarin red S method of Dahl (1952) offered a great improvement over the older methods. Dahl emphasized the importance of controlling the pH, since lakes are formed with calcium only between pH 4 and pH 8; Dahl recommended a pH between 6.3 and 6.5. McGee-Russell (1958) improved Dahl's procedure in the following ways: (1) Albuminized slides are used (albumin had to be avoided in Dahl's procedure). (2) The differentiation step is eliminated. (3) By lowering the pH to 4.1 to 4.3, the dye solution is brown and contrasts satisfactorily with the forming calcium lake. The McGee-Russell method (1958) is therefore the method of choice.

Alizarin red S also forms lakes with other metals. However, the crimson red lake formed with calcium can be readily differentiated visually from the darker shades of red formed with aluminum, barium, magnesium, and mercury.

Nuclear fast red method. The nuclear fast red method of McGee-Russell (1955) is also based on the principle that the dye forms a lake with calcium. Nuclear fast red (Kernechtrot) has been confused with Kernechtrotsalz B, a totally unrelated compound. For this reason, McGee-Russell (1955) preferred the term *calcium red* to *nuclear fast red*.

The virtue of this method lies with its simplicity. Pearse (1972) found the method satisfactory only for invertebrate tissues, but many others have found it to be a good and useful method.

In addition to staining calcium phosphates and calcium carbonates, the dye also forms lakes with Pb^{2+}, Fe^{3+}, Cu^{2+}, K^+, Sn^{4+}, and Sr^{2+}, but not with Ba^{2+} or Mg^{2+}.

Copper

The physiological importance of copper has not always been appreciated. The amount of copper ordinarily present in mammalian tissues is in fact very low, and its presence is difficult to detect histochemically despite the fact that many oxidase enzymes have copper ligands. The better-known of these enzymes are tyrosinase, cytochrome oxidase, and uricase, which are widely distributed. Some non-mammalian tissues have sufficiently high concentrations of copper that it can be demonstrated histochemically. Human embryonic tissues contain several times the amount of copper that is present in adult tissues; that amount is also sufficient for histochemical detection. Wilson's disease (hepatolenticular degeneration) involves aberrant copper metabolism, so that abnormally high amounts of copper are found in the liver and in the lentiform nuclei of the brain (ceruloplasm, a copper-containing plasma protein, is also deficient in this disease).

Many methods have been developed for demonstrating copper in tissue, but the only method considered here is that of Okamoto and Utamura (1938). This method is highly sensitive and easy to apply. It utilizes rubeanic acid (dithiooxamide):

$$H_2N-C-C-NH_2$$
$$\underset{S}{\parallel}\ \underset{S}{\parallel}$$

Tissue copper reacts with rubeanic acid to form copper rubeanate, which is a greenish black pigment. Since the color is stable for at least 3 years (Howell, 1959), permanent slides are quite satisfactory.

Copper sometimes forms complexes with tissue proteins, preventing it from reacting with rubeanic acid. Complexed copper can be released by treating the tissue sections with hydrogen peroxide (Gomori, 1952) or with HCl fumes (Uzman, 1956). Rubeanic acid also reacts with cobalt and nickel, but the colors formed are yellowish brown and blue-violet, respectively; and can be easily distinguished from the color of copper rubeanate.

SCHMORL REACTION FOR LIPOFUSCINS

TISSUES:
Use paraffin sections of formalin-fixed tissues.

PREPARATION OF FERROCYANIDE SOLUTION:
Combine 3 parts of a 1 percent solution of ferric chloride or ferric sulfate with 1 part of a 1 percent solution of potassium ferrocyanide (freshly prepared). Prepare just before use.

PROCEDURE:

(1) Deparaffinize sections and bring to water.

(2) Treat with ferrocyanide solution for 5 minutes.

(3) Wash in running tap water.

(4) Counterstain with 1 percent neutral red for 3 minutes (this step is optional).

(5) Dehydrate rapidly through graded alcohols, clear in xylene, and mount with synthetic medium.

RESULTS:

Melanin granules, argentaffin granules, and lipofuscin pigment are stained deep blue. Nuclei (if counterstained) are red.

ZIEHL-NEELSEN METHOD FOR LIPOFUSCIN

TISSUES:

Use paraffin sections of formalin-fixed tissues.

PREPARATION OF CARBOL-FUCHSIN STAIN:

Basic fuchsin	10 gm
Phenol	50 gm
Alcohol	100 ml
Distilled water	1 liter

PROCEDURE:

(1) Deparaffinize sections and bring to water.

(2) Stain with carbol-fuchsin stain for 3 hours at 60°C.

(3) Wash in running water.

(4) Differentiate in acid-alcohol (1 percent HCl in 70 percent ethanol) until red blood cells are light pink.

(5) Counterstain lightly with 0.5 percent toluidin blue.

(6) Wash in running water.

(7) Dehydrate through graded alcohols, clear in xylene, and mount with synthetic resin.

RESULTS:

Lipofuscins and ceroid are stained bright red and the nuclei are light blue.

METHENAMINE SILVER METHOD FOR MELANIN
(Masson, 1914; Fontana, 1925; Gomori, 1946)

TISSUES:
Formalin fixatives are preferred, although any fixative except those containing dichromate may be used. Use paraffin sections.

PREPARATION OF GRAM'S IODINE:

Iodine crystals	1 gm
Potassium iodide	2 gm
Distilled water	300 ml

PREPARATION OF METHENAMINE SILVER STOCK SOLUTION:
Add 5 ml of a 5 percent silver nitrate solution to 100 ml of a 3 percent methenamine (hexamine) solution. Shake the mixture until the initial precipitate disappears. This solution can be stored in a refrigerator for several months.

PREPARATION OF METHENAMINE SILVER STAINING SOLUTION:
Mix 50 ml of the methenamine silver stock solution with 25 ml boric acid–borate buffer, pH 8, and 25 ml distilled water.

PROCEDURE:
(1) Deparaffinize sections and bring to water.
(2) Treat with Gram's iodine for 10 minutes.
(3) Treat with 5 percent sodium thiosulfate for 2 minutes.
(4) Wash with several changes of distilled water.
(5) Incubate with methenamine silver staining solution in a closed vessel for 18 to 24 hours.
(6) Wash with several changes of distilled water.
(7) Fix with 5 percent sodium thiosulfate for 2 minutes.
(8) Wash with running tap water for 2 minutes.
(9) Counterstain with 1 percent neutral red for 3 minutes.
(10) Rinse in distilled water.
(11) Dehydrate through graded alcohols, clear with xylene, and mount with a synthetic resin.

RESULTS:
Melanin and argentaffin granules are stained black and nuclei are stained red.

MODIFIED PERLS' METHOD FOR HEMOSIDERIN
(After Gomori, 1952)

TISSUES:
Fix tissues with a neutral formalin fixative. Avoid fixatives that contain acid or chromate. Use paraffin sections.

PROCEDURE:
(1) Deparaffinize sections and bring to water.
(2) Place slides in a Coplin jar containing 30 ml of a 5 percent solution of potassium ferrocyanide. (Prepare fresh and filter immediately before use.) After 5 minutes, add 15 ml of 10 percent HCl and mix. Incubate for an additional 30 minutes.
(3) Wash in distilled water.
(4) Counterstain with 1 percent aqueous neutral red for 3 minutes (this step is optional).
(5) Wash in tap water.
(6) Dehydrate through graded alcohols, clear with xylene, and mount with synthetic medium.

RESULTS:
Hemosiderin and ferric iron are stained deep blue, and nuclei (if counterstained) are red.

METHOD FOR BILIRUBIN, HEMOSIDERIN, AND LIPOFUSCIN
(Glenner, 1957)

TISSUES:
Use fresh-frozen cryostat sections. Mount on slides and allow them to dry at room temperature.

PREPARATION OF ACETIC ACID–FERROCYANIDE SOLUTION:
Combine equal volumes of 2 percent potassium ferrocyanide and 5 percent acetic acid. Prepare just prior to use.

PREPARATION OF BUFFERED DICHROMATE SOLUTION:
Combine equal volumes of 3 percent aqueous potassium dichromate and phosphate buffer. The phosphate buffer is prepared by adding 8 ml $0.1N$ HCl and 17 ml $0.1M$ potassium dihydrogen phosphate.

PREPARATION OF OIL RED O:
Add 24 ml distilled water to 16 ml of 1 percent oil red O in isopropyl alcohol. Allow it to stand for 10 minutes and filter.

PROCEDURE:
(1) Treat the sections for 5 minutes with a 2 percent aqueous solution of potassium ferrocyanide.
(2) Treat for 20 minutes with acetic acid–ferrocyanide solution.
(3) Rinse briefly in running water.
(4) Treat for 15 minutes in buffered dichromate solution.
(5) Wash in running water.
(6) Stain for 20 minutes with oil red O.
(7) Wash in running water.
(8) Mount with Apathy's syrup.

RESULTS:
Lipofuscin is colored dark red-orange, hemosiderin is stained dark blue, and oxidized bilirubin is stained dark emerald green.

PERLS' METHOD FOR IRON

TISSUES:
Fix tissues with a neutral formalin fixative. Avoid fixatives that contain acid or chromate. Use paraffin sections.

PREPARATION OF FERROCYANIDE SOLUTION:
Combine equal parts of 2 percent potassium or sodium ferrocyanide and 0.25N HCl (or 2 percent). Prepare immediately before use.

PROCEDURE:
(1) Deparaffinize sections and bring to water.
(2) Treat for 30 to 60 minutes with ferrocyanide solution.
(3) Wash in distilled water.
(4) Counterstain with 1 percent aqueous neutral red for 3 minutes (this step is optional).
(5) Wash in tap water.
(6) Dehydrate through graded alcohols, clear in xylene, and mount with synthetic medium.

RESULTS:

Deposits of ferric iron are stained deep blue and nuclei (if counterstained) are red.

DITHIZON METHOD FOR ZINC
(Mager et al., 1953)

TISSUES:

Fix tissues with two changes of cold absolute ethanol for 1 hour each and use paraffin sections. Alternatively, fresh-frozen sections or paraffin sections of freeze-dried tissues can be used.

PREPARATION OF COMPLEX-FORMING BUFFER:

Dissolve the following reagents in approximately 1 liter of zinc-free water:

Sodium thiosulfate·5H$_2$O	550 gm
Sodium acetate·3H$_2$O	90 gm
Potassium cyanide	10 gm

Adjust the pH to 5.5 with glacial acetic acid and add zinc-free water to make 2 liters. Extract the solution with dithizon in carbon tetrachloride until all zinc is removed.

PREPARATION OF DITHIZON STAINING SOLUTION:

Mix 24 ml 0.01 percent dithizon in acetone with 18 ml distilled water. Adjust the pH to 3.7 with 1N acetic acid. Add 5.8 ml of the complexing buffer and 0.2 ml 20 percent potassium tartrate. Add these reagents slowly while stirring.

PROCEDURE:

(1) Deparaffinize sections and bring to water.
(2) Place in dithizon staining solution for 10 minutes.
(3) Wash with chloroform.
(4) Rinse quickly with water.
(5) Mount with Karo syrup.

RESULTS:

A red-to-purple granular stain is considered positive for zinc. Tissues have a diffuse yellow-to-pink background stain. Fat (in frozen sections) stains from green to yellow.

SILVER NITRATE–RUBEANIC ACID METHOD FOR CALCIUM OXALATE
(Yasue, 1969a, 1969b)

TISSUES:

Fix with any formalin fixative and use paraffin sections.

RUBEANIC ACID SOLUTION:

Prepare a saturated solution of rubeanic acid in 100 ml of 70 percent ethanol containing 2 drops of concentrated NH_4OH.

PROCEDURE:
(1) Deparaffinize sections and bring to water.
(2) Treat with 5 percent acetic acid for 30 minutes.
(3) Wash in distilled water.
(4) Treat with 5 percent KOH for 30 minutes.
(5) Wash in distilled water.
(6) Treat with 5 percent aqueous silver nitrate for 10 to 20 minutes.
(7) Wash thoroughly with running distilled water.
(8) Treat with rubeanic acid solution for 1 minute.
(9) Rinse in 50 percent ethanol.
(10) Rinse in water.
(11) Counterstain with 2 percent methyl green for 2 minutes.
(12) Dehydrate through graded alcohols, clear with xylene, and mount with synthetic mountant.

RESULTS:

Calcium oxalate deposits are stained brown or black while nuclei are stained green.

ALIZARIN RED S METHOD FOR CALCIUM
(McGee-Russell, 1958)

TISSUES:

Fix the tissues with a 1:1 ratio of full-strength formalin and absolute ethanol or any nonacid, nonmetallic formalin fixative. Use paraffin sections.

STAINING SOLUTION:

Prepare a 2 percent solution of alizarin red S and adjust the pH to 4.1 to 4.3 using dilute ammonium hydroxide. The solution is stable, and it should be a deep iodine color.

PROCEDURE:
(1) Deparaffinize sections and bring to 50 percent ethanol.
(2) Rinse rapidly in distilled water.
(3) Cover section with the staining solution and observe it through a microscope. Incubate for 30 seconds to 5 minutes.
(4) Shake off stain and blot carefully with filter paper.
(5) Plunge directly into acetone for 10 to 20 seconds.
(6) Place in a solution with a 1:1 ratio of acetone and xylene for 10 to 20 seconds.
(7) Clear in xylene and mount with synthetic mountant.

RESULTS:

Sites of calcium are indicated by heavy orange-red deposits of dye. The remainder of the tissue is covered by a diffuse pink color.

NUCLEAR FAST RED METHOD FOR CALCIUM
(McGee-Russell, 1955)

TISSUES:

Use paraffin sections of formalin-fixed tissues.

PREPARATION OF STAINING SOLUTION:

Wash 2 gm of commercial nuclear fast red with 100 ml distilled water and repeat the process. Make a saturated aqueous solution of the residue (about 0.25 gm per 100 ml water) to be used as the staining solution.

PROCEDURE:
(1) Deparaffinize sections and bring to water.
(2) Stain for 5 to 10 minutes.
(3) Wash in distilled water.
(4) Dehydrate through graded alcohols, clear with xylene, and mount with a synthetic medium.

RESULTS:
Calcium deposits are indicated by a red dye lake.

RUBEANIC ACID METHOD FOR COPPER
(Howell, 1959)

TISSUES:
Fix tissues with formalin or Bouin's fixative and use paraffin sections.

STOCK SOLUTION OF RUBEANIC ACID:
Dissolve 0.1 gm rubeanic acid in 100 ml absolute ethanol.

WORKING SOLUTION OF RUBEANIC ACID:
Add 2 ml of the stock solution to 40 ml of a 10 percent solution of sodium acetate.

PROCEDURE:
(1) Deparaffinize sections and bring to water.
(2) Place sections in a Coplin jar of working solution of rubeanic acid and seal it. Incubate the jar at 37°C for 12 to 24 hours.
(3) Wash in distilled water.
(4) Mount with Farrant's medium.

RESULTS:
Deposits of copper or copper-containing protein will be stained black.

REFERENCES

Adams, C. W. M. (1956). A stricter interpretation of the ferric ferrocyanide reaction with particular reference to the demonstration of protein-bound sulfhydryl and disulphide groups. *J. Histochem. Cytochem.* 4:23–35.

Chaplin, A. J., and Grace, S. R. (1975). Calcium oxalate and the von Kossa method with reference to the influence of citric acid. *Histochem. J.* 7:451–458.

Dahl, L. K. (1952). A simple and sensitive histochemical method for calcium. *Proc. Soc. Expl. Biol. Med.* 80:474–479.

Fontana, A. (1925). Über die Silberdarstellung des *Treponema pallidum* und anderer Mikroorganismen in Ausstrichen. *Dermatol. Z.* 46:291–293.

Glenner, G. G. (1957). Simultaneous demonstration of bilirubin, hemosiderin, and lipofuscin pigments in tissue sections. *Am. J. Clin. Pathol.* 27:1–5.

Gomori, G. (1946). A new histochemical test for glycogen and mucin. *Am. J. Clin. Pathol. (Tech. Bull.)* 7:177–179.

Gomori, G. (1952). *Microscopic Histochemistry.* University of Chicago Press, Chicago.

Howell, J. S. (1959). Histochemical demonstration of copper in copper fed rats and in hepatolenticular degeneration. *J. Pathol. Bacteriol.* 77:473–484.

Hueck, W. (1912). Pigmentstudien. *Beitr. Pathol. Anat. Allg. Pathol.* 54:68–232.

Lillie, R. D. (1956). A Nile Blue staining technic for the differentiation of melanin and lipofuscins. *Stain Technol.* 31:151–153.

Lillie, R. D. (1957). Metal reduction reactions of the melanins: Histochemical studies. *J. Histochem. Cytochem.* 5:325–333.

Lillie, R. D. (1965). *Histopathological Technic and Practical Histochemistry* (3rd ed.). McGraw-Hill, New York.

Lillie, R. D., Greco-Henson, J. P., and Burtner, H. C. J. (1957). Metal reduction reactions of the melanins. Silver and ferric ferrocyanide reduction by various reagents in vitro. *J. Histochem. Cytochem.* 5:311–324.

Mager, M., McNary, W. F., and Lionetti, F. (1953). The histochemical detection of zinc. *J. Histochem. Cytochem.* 1:493–504.

Masson, M. P. (1914). La glande endocrine de l'intestin chez l'homme. *C. R. Acad. Sci.* 158:59–61.

McGee-Russell, S. M. (1955). A new reagent for the histochemical and chemical detection of calcium. *Nature* 175:301–302.

McGee-Russell, S. M. (1958). Histochemical methods for calcium. *J. Histochem. Cytochem.* 6:22–42.

Okamoto, K., and Utamura, M. (1938). Biologische Untersuchungen des Kupfers; über die histochemische Kupfernachweismethode. *Acta Sch. Med. Univ. Kioto* 20:573–580.

Pearse, A. G. E. (1972). *Histochemistry: Theoretical and Applied* (3rd ed.), Vol. 2. Williams and Wilkins, Baltimore.

Perls, M. (1867). Nachweis von Eisenoxyd in gewissen Pigmenten. *Virchows Archiv.* 39:42.

Uzman, L. L. (1956). Histochemical localization of copper with rubeanic acid. *Lab. Invest.* 5:299–305.

von Kossa, J. (1901).   Ueber die im Organismus kuenstlich erzeugbaren Ver-
kalkungen. *Beitr. Pathol. Anat.* 29:163–202.
Yasue, T. (1969a).   Histochemical identification of calcium oxalate. *Acta His-
tochem. Cytochem.* 2:83–95.
Yasue, T. (1969b).   Renal crystalline deposition and its pathogenesis. *Acta
Histochem. Cytochem.* 2:96–111.

10

IMMUNOHISTOCHEMISTRY

The development of immunohistochemical techniques has made possible the histochemical localization of a wide variety of indigenous proteins and peptides, as well as foreign pathogenic agents such as bacteria, rickettsia, and viruses. Any antigenic substance or agent that can be maintained in situ during tissue preparation has the potential for being localized immunohistochemically.

As early as 1933, Hopkins and Warmall conjugated isocyanate with proteins; this accomplishment was essential to the development of immunohistochemistry. An independent study showed that antibodies conjugated with dye compounds would retain some of their "agglutinating power" (Marrack, 1934). Coons et al. (1942) demonstrated that fluorescein isocyanate could serve as a label for identifying adsorbed antibody. Many studies have reported using this compound or its successor fluorescein isothiocyanate for labeling antibodies. The historical beginnings of fluorescence immunohistochemistry are delightfully recounted in an essay by Coons (1961). Enzyme labels have now largely replaced the fluorescent compounds; the most widely used enzyme for this purpose is horseradish peroxidase.

TISSUE PREPARATION

Few generalizations can be made concerning the best methods of preparing tissues for immunohistochemistry. A method that works well in one application may be totally unsuitable in another. Coons and Kaplan (1950) first used cryostat sections of fresh-frozen tissues, and Coons and his associates have used frozen sections extensively since then. With this method, the tissue components substantially retain their immunological properties. Morphology, however, is difficult to preserve with unfixed frozen sections. Usually sections of greater than optimal thickness are obtained, precluding good histological detail. Ice crystals may cause cellular damage and distortion. During incubation, diffusion of the immunologically active substances may occur, particularly in the lower-molecular-weight compounds such as the peptide hormones. For some applications, frozen sections are totally unsuitable (Arnold et al., 1974). Despite these drawbacks, frozen sections are probably suitable for more immunohistochemical procedures than is any other method of preparing tissues (at any rate, it is the most frequently used method).

Freeze-dried tissues have been used extensively also. If the tissues are frozen instantly, excellent histological detail can be preserved. Following the freeze-drying procedure, the tissues are embedded in preparation for sectioning. Paraffin, polyester wax, and glycol methacrylate have been used as embedments. The heat from paraffin embedding may have a harmful effect on the immunological properties of proteins, although the effect is always difficult to evaluate. Paraffin-embedded freeze-dried tissues have been found unsuitable in some cases (Arnold et al., 1974; Gervais, 1972).

Some investigators (Balfour, 1961; Sutherland, 1970; Arnold et al., 1974) have reported favorable results with freeze-substitution. At the low temperature used for the substitution process, alcohol fixation does not take place; it occurs only at higher temperatures. The process is often looked upon unfavorably because of the length of time required for substitution.

It is often possible to fix tissues while preserving immunological activity adequately. Sainte-Marie (1962) fixed tissues in 95 percent ethanol, and then dehydrated further in absolute ethanol, cleared in xylene, and embedded in paraffin. Up to the paraffin step, the procedure is performed at 4°C. This method gave adequate preservation of bovine immunoglobulin G (IgG), influenza virus A, and antibody in mouse spleen, but it did not adequately preserve certain other components. Numerous investigators have found this method satisfactory in many applications, but it is by no means always satisfactory.

A great need exists for fixatives that will preserve good cellular and tissue morphology without destroying the immunological properties of tissue components. Glutaraldehyde is an excellent fixative but antigenicity is often completely destroyed. Paraformaldehyde is less destructive to antigens but it is still often unacceptable. Adequate fixation together with preservation of antigenicity is especially difficult in immunoelectron microscopy.

There is evidently no correlation between the preservation of enzyme activity and antigenicity of the same enzyme molecule. Aldehydes will preserve esterases rather well for histochemical purposes but will totally destroy their antigenicity (Vladutiu et al., 1973).

The future of immunohistochemical fixatives seems to lie with bifunctional fixatives, which act on select functional groups of proteins. The diimidoesters, already studied as fixatives for enzyme histochemistry (Hassell and Hand, 1974; Hand and Hassell, 1976), show the greatest potential as immunohistochemical fixatives. They react only with ε-amino groups to form crosslinks. Enzymatic activity is preserved remarkably well. According to several reports (Dutton et al., 1966; McLean and Singer, 1970; Wofsy and Singer, 1963) there is virtually no loss of immunological activity of proteins treated with diimidoesters.

Two short reports from Pearse's laboratory (Kendall et al., 1971; Polak et al., 1972) have shown promising results with two carbodiimidates as fixatives. The degree of antigenicity preserved in six polypeptide hormones was excellent. Pearse and his associates found that short periods of fixation (1 hour) gave better ultrastructural preservation than longer periods (24 hours), suggesting that the process of fixation is a biphasic action. Following the initial phase of fixation, the lysosomal hydrolases, which were no doubt well preserved, were probably activated, accounting for the deterioration of ultrastructure.

The carbodiimidates evidently first react with carboxyl groups to form O-acylisourea groups, which then react with nearby amino groups to form amide crosslink bonds. The carbodiimidate molecules themselves do not become incorporated into the linkage.

A totally new principle of fixation was developed for electron microscopic immunohistochemistry by McLean and Nakane (1974). The principle is similar to one developed in the same laboratory for peroxidase-antibody conjugation (Nakane and Kawaoi, 1974). The fixative contains periodate, lysine, and paraformaldehyde. The periodate oxidizes the carbohydrate moieties to form aldehyde groups, while lysine, which is a diamine, reacts with the periodate-generated aldehydes to form crosslinks between tissue components. Paraformaldehyde (2 percent) was added to achieve some stability of proteins and lipids. This fixative still falls short of being ideal, because some antigenicity is lost. However, it merits consideration because it preserves ultrastructure as well as glutaraldehyde does and preserves antigenicity as well as paraformaldehyde does. McLean and Nakane (1974) suggested that, in order to achieve better preservation of antigenicity, as little as 0.5 percent paraformaldehyde could be used. The ultrastructure would of course not be as well preserved, but it would probably be adequate for light microscopy.

BASIC PRINCIPLES

The specificity of immunohistochemical methods is based on the fact that an antibody will bind only with the antigen that stimulated the production of the antibody. Tagging an antibody permits localizing of the corresponding antigen. Similarly, tagging an antigen allows the corresponding antibody to be localized in tissue sections, although this latter procedure has far fewer applications than the former.

It is important to realize that antibodies can be prepared not only against a wide variety of antigens, but also against other antibodies. Antibodies against other antibodies are used in the indirect methods. Antibodies are prepared against the whole γ-globulin fraction of a different species, rather than against the specific antibody of interest. For instance, if the γ-globulin fraction of rabbit serum is repeatedly injected into a sheep, the sheep will produce an antibody (in its γ-globulin fraction) against rabbit γ-globulin. This would be known as sheep antirabbit γ-globulin.

An antibody can be rendered visible by chemically conjugating it with a fluorescent dye, or by conjugating it with an enzyme that can then be demonstrated with a conventional histochemical method.

IMMUNOFLUORESCENCE METHODS
Specific Techniques

Direct method. In the direct method (Fig. 10-1), an appropriate antibody is first prepared against the antigen to be localized. The antibody must be conjugated with a fluorescent dye such as fluorescein isothiocyanate. A slide with the tissue section is then flooded with a solution of the antibody for a period of from 10 minutes to several hours. The antibody is usually dissolved in phosphate-buffered saline (PBS) at a slightly alkaline pH. Following the incubation, the slide is carefully washed to remove all the excess antibody. The slide is studied with a fluorescence microscope and appropriate photomicrographs are obtained; this should be done within a matter of a few hours because the fluorescence fades.

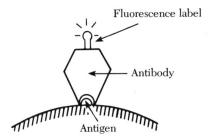

Figure 10-1. The direct immunofluorescence method. The fluorescence-labeled antibody is layered directly onto the antigen.

Indirect method. The indirect method (Fig. 10-2), developed by Weller and Coons (1954), uses two antibodies in sequence. The first one is specific against the antigen of interest. It is equivalent to the antibody of the direct method except that it has no fluorescent label. After washing away the unbound antibody, the tissue section is incubated with the second antibody, prepared against the first antibody. This antibody is previously conjugated with a fluorescent dye, which finally renders the complex visible.

The indirect method is not appreciably more difficult to perform than the direct method, because the second antibody is generally available commercially, already labeled. This method is often favored because it is somewhat more sensitive. Also, if demonstration of more

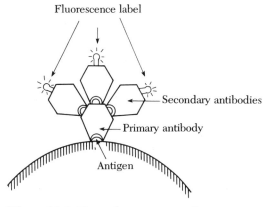

Figure 10-2. The indirect immunofluorescence method. First, the unlabeled primary antibody is applied. Then the labeled secondary antibody, which is specific for the primary antibody, is layered onto the primary antibody.

than one antigen is desired, only one labeled antibody is required, provided of course that the first antibody in each case is of the same species of animal.

Fluorescent Dyes
Coons and co-workers (1942) found fluorescein isocyanate to be the most suitable dye available at that time. It is easily conjugated with proteins and has a brilliant fluorescence in the yellow-green range, which contrasts well with most naturally occurring fluorescence of tissues. The compound is unstable, however, so that it is generally necessary to synthesize it as needed. Moreover, its synthesis requires the use of phosgene gas, which is highly toxic.

A more stable fluorescent compound, fluorescein isothiocyanate, was introduced by Riggs et al. (1958). This compound is stable as a powder and can be stored as such. It can be merely dissolved and used when needed. It is widely available commercially.

Conjugation
The original method of conjugation was to add a solution of fluorescein isocyanate, drop by drop, to a cold solution of antibody. The solvent for the dye was a mixture of dioxane and acetone. The antibody was dissolved in saline and buffered with a carbonate-bicarbonate buffer (Coons et al., 1942; Coons and Kaplan, 1950). The unconjugated fluorescein compound was dialyzed away. The conjugated protein was then concentrated by precipitating with ammonium sulfate and by redialysis. The method of conjugating the isothiocyanate compounds was not essentially different from the method of conjugating the isocyanate compounds (Riggs et al., 1958). Marshall et al. (1958) found that fluorescein isothiocyanate could be added in its powdery form directly to the buffered solution of protein, thus avoiding organic solvents. Denaturation of antibodies was reduced, permitting the use of antisera of lower antibody titer. This has become the standard method for conjugating fluorescein isothiocyanate with antibodies.

Disadvantages of the Immunofluorescence Methods
Despite the usefulness of the immunofluorescence methods, there remain certain disadvantages: (1) It is necessary to have a special

fluorescence microscope, which may be quite expensive. (2) Localization of the fluorescence-stained elements may be difficult because the tissue morphology is hard to distinguish. (3) Permanent preparations cannot be made because the fluorescence fades. These disadvantages provided the incentive to search for better methods.

ENZYME-LABELED ANTIBODY METHODS

Replacement of the fluorescence-labeled antibody methods was brought about because of the greater advantages and additional utility offered by the enzyme-labeled antibody methods. The principle of these methods is to adsorb the enzyme to the antigenic sites with the antibody. The enzyme is then localized according to an appropriate conventional histochemical technique.

The first enzyme to be used for this purpose was acid phosphatase (Ram et al., 1966). This enzyme is fairly stable and is easy to detect histochemically. However, Ram and his associates found it difficult to obtain consistent results from the conjugation step. It was for that reason that Nakane and Pierce (1966, 1967) investigated the use of horseradish peroxidase as an antibody label.

The horseradish peroxidase-labeled antibody methods have the following advantages over the immunofluorescence methods: (1) The sections can be examined with an ordinary bright-field microscope. (2) The sections can be counterstained so that the component being localized can be studied in relation to the morphology of the tissue. (3) Because the final reaction product is stable, permanent slide preparations can be made. (4) The final reaction product can be made electron-dense and the same tissue blocks can be used to prepare sections for electron microscopy.

Horseradish peroxidase has been used very successfully in immunohistochemical methods. However, more recently another type of peroxidase of lower molecular weight and higher specific activity has been used. These are short peptide fragments of cytochrome c (called microperoxidase). The first such fragment to be used was an 11-amino acid fragment (Feder, 1971). The fragment can be reduced to an octapeptide (8-MP), which has 300 times the peroxidase activity of cytochrome c (Ryan et al., 1976).

Specific Techniques

Direct method. The direct method is similar to the immunofluorescence direct method except, of course, that the antibody must be conjugated with peroxidase. The sites of antibody absorption are visualized by carrying out the peroxidase reaction of Graham and Karnovsky (1966). The tissue sections can be counterstained and mounted as a permanent preparation.

Indirect method. The indirect method is likewise similar to the indirect immunofluorescence method. The second antibody is labeled with peroxidase, which is similarly demonstrated with the Graham and Karnovsky method. The second antibody is usually commercially available already conjugated with peroxidase.

Unlabeled antibody enzyme method. The unlabeled antibody enzyme method (Fig. 10-3), developed by Mason et al. (1969), does not require chemical conjugation between peroxidase and the antibody. A very similar method was developed by Sternberger et al. (1970). The peroxidase is adsorbed onto the antigenic sites by an antiperoxidase antibody. Four components are used sequentially in this method: (1) An unconjugated antibody is adsorbed onto the antigen. This is the primary antibody and is the same as the first antibody that would be used in the indirect method. (2) An antibody against the primary antibody is applied next. This is actually an antiserum prepared against

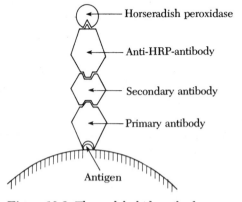

— Horseradish peroxidase

— Anti-HRP-antibody

— Secondary antibody

— Primary antibody

Antigen

Figure 10-3. The unlabeled antibody enzyme method. The four components are layered sequentially. Alternatively, the peroxidase and antiperoxidase components may be allowed to react before being layered on as a complex. (HRP = horseradish peroxidase.)

the gamma globulin fraction of the species from which the primary antibody is made. It is the same as the second antibody used in the direct method except that it is not labeled. (3) Antiperoxidase, an antibody against horseradish peroxidase, is the third component. It is prepared in the same species as the primary antibody; thus, the second antibody is specific against it also. (4) Horseradish peroxidase is finally adsorbed onto the antiperoxidase. Peroxidase, adsorbed onto its antibody, has substantially unaltered enzymatic activity. The peroxidase and antiperoxidase may be allowed to form a peroxidase-antiperoxidase (PAP) complex prior to being layered onto the tissue section.

It is crucial to understand that the primary antibody and the antiperoxidase (the first and third components) are prepared from the same species; that is, they are homologous. The second component is an antibody prepared against the gamma globulin fraction of the first species. We can think of the second component as being sandwiched between the primary antibody and the antiperoxidase. This procedure therefore is often called the "sandwich technique."

This method obviates the need for chemical conjugation and therefore circumvents the injurious effect of conjugation on the immunoreactivity of the antibodies. It also eliminates the problem of residual unconjugated antibody competing with the enzyme-labeled antibody for the antigenic sites.

The requirement for using homologous antisera for the first and third components of this system was investigated by Erlandsen et al. (1975). This requirement is evidently not absolute; despite the fact that the primary antibody and the antiperoxidase were prepared from different species, Erlandsen and associates were able to demonstrate glucagon, insulin, and growth hormone. As long as the coupling antibody (the second component) shows even a low cross reactivity with the primary antibody and the antiperoxidase, the reaction can be successfully carried out. This finding should facilitate the use of specific antisera derived from less commonly used antisera-producing species and the antisera of higher primates, including man.

Inhibition of Endogenous Peroxidase

The final step in the peroxidase immunohistochemical procedures is always the peroxidase reaction of Graham and Karnovsky (1966); however, this procedure also localizes endogenous peroxidase. It is therefore essential to inhibit or destroy endogenous peroxidase, or somehow to distinguish it from the adsorbed peroxidase. Various substances have been suggested for inhibiting endogenous peroxidase. Robinson and Dawson (1975) found that sodium azide, potassium cyanide, sodium nitroferricyanide, and 3-amino-1,2,4-triazole inhibited all endogenous peroxidase except that in macrophages, neutrophils, eosinophils, and erythrocytes. Other inhibitors (methanol, acetic acid in methanol, and phenylhydrazine) inhibited all endogenous peroxidase, but also destroyed the antigenicity of the antigen to be demonstrated (gastrin in this case), rendering the inhibitors useless. Wier et al. (1974) claimed total inhibition of all endogenous peroxidase with ethanolic HCl with no loss of antigenicity.

Robinson and Dawson (1975) found that the most satisfactory approach to the problem was to demonstrate endogenous peroxidase differentially. Prior to the immunohistochemical procedure, endogenous peroxidase was demonstrated with the method of Nakane (1968) using 4-chloro-1-naphthol. This produced a gray-blue reaction product that contrasted adequately with the brown color of the diaminobenzidine reaction product of horseradish peroxidase.

Conjugation of Antibodies with Peroxidase

Conjugation is accomplished with some type of bifunctional reagent that reacts with proteins. In the original work of Nakane and Pierce (1966, 1967), p,p'-difluoro-m,m'-dinitrodiphenyl sulfone (FNPS) was used, but carbodiimides can also be used for this purpose (McGuigan, 1968); both compounds have been used extensively in protein chemistry.

A more widely used method of conjugation was presented by Avrameas (1969) using glutaraldehyde. Glutaraldehyde was added slowly to a solution containing the antibody and peroxidase. An improved version of the method was published by Avrameas and Ternynck (1971). Glutaraldehyde was first allowed to react with peroxidase for 18 hours. The product remained in solution; it is not known what prevented excessive crosslinking and precipitation. The activated peroxidase is separated on a Sephadex G-25 column. The antibody is then allowed to react with the activated peroxidase for 24 hours in the cold. The conjugated antibody is then precipitated with ammonium sulfate, redissolved, and exhaustively dialyzed against PBS. Still other methods have been used for conjugating microperoxidases. Kraehenbuhl et al. (1974) used p-formyl-benzoyl-N-OH-succinamide. Ryan et al. (1976) used the bifunctional ester, bis-succinyl-succinate, in a two-step reaction sequence.

A totally novel method has been used to accomplish conjugation (Kawaoi and Nakane, 1973; Nakane and Kawaoi, 1974; Kawaoi, 1975). With periodate oxidation, aldehyde groups were introduced on the carbohydrate moieties of the horseradish peroxidase, and the antibodies were then allowed to react with the aldehyde-laden peroxidase. Kawaoi and Nakane (1973) claimed 95 percent efficiency of coupling with no loss of biological activity. Their method seems therefore to have a considerable advantage over the glutaraldehyde methods, which have a rather low efficiency.

REFERENCES

Arnold, W., Mitrenga, D., and Mayersbach, H. (1974). The preservation of substance and immunological activity of intravenously injected human IgG in the mouse liver. *Acta Histochem.* 49:161–175.

Avrameas, S. (1969). Coupling of enzymes to proteins with glutaraldehyde. Use of the conjugates for the detection of antigens and antibodies. *Immunochemistry* 6:43–52.

Avrameas, S., and Ternynck, T. (1971). Peroxidase labeled antibody and Fab conjugates with enhanced intracellular penetration. *Immunochemistry* 8:1175–1179.

Balfour, B. M. (1961). Immunological studies on a freeze-substitution method of preparing tissue for fluorescent staining. *Immunology* 4:206–218.

Coons, A. H. (1961). The beginnings of immunofluorescence. *J. Immunol.* 87:499–503.

Coons, A. H., and Kaplan, M.H. (1950). Localization of antigen in tissue cells. II. Improvements in a method for the detection of antigen by means of fluorescent antibody. *J. Exp. Med.* 91:1–13.

Coons, A. H., Creech, H. J., Jones, R. N., and Berliner, E. (1942). The demonstration of pneumococcal antigen in tissues by the use of fluorescent antibody. *J. Immunol.* 45:159–170.

Dutton, A., Adams, M., and Singer, S. J. (1966). Bifunctional imidoesters as cross-linking reagents. *Biochem. Biophys. Res. Commun.* 23:730–739.

Erlandsen, S. L., Parsons, J. A., Burke, J. P., Redick, J. A., Van Orden, D. E., and Van Orden, L. S. (1975). A modification of the unlabeled antibody enzyme method using heterologous antisera for the light microscopic and ultrastructural localization of insulin, glucagon and growth hormone. *J. Histochem. Cytochem.* 23:666–677.

Feder, N. (1971). Microperoxidase. An ultrastructural tracer of low molecular weight. *J. Cell Biol.* 51:339–343.

Gervais, A. G. (1972). Localization of transplantation antigens in tissue sections: Effects of various fixatives and use of tissue preparations other than frozen sections. *Experientia* 28:342–343.

Graham, R. C., and Karnovsky, M. J. (1966). The early stages of absorption of injected horseradish peroxidase in the proximal tubules of mouse kidney: Ultrastructural cytochemistry by a new technique. *J. Histochem. Cytochem.* 14:291–302.

Hand, A. R., and Hassell, J. R. (1976). Tissue fixation with diimidoesters as an alternative to aldehydes. II. Cytochemical and biochemical studies of rat liver fixed with dimethylsuberimidate. *J. Histochem. Cytochem.* 24:1000–1011.

Hassell, J. R., and Hand, A. R. (1974). Tissue fixation with diimidoesters as an alternative to aldehydes. I. Comparison of cross-linking and ultrastructure obtained with dimethylsuberimidate and glutaraldehyde. *J. Histochem. Cytochem.* 22:223–239.

Hopkins, S. J., and Warmall, A. (1933). Phenyl isocyanate protein compounds and their immunological properties. *Biochem. J.* 27:740–753.

Kawaoi, A. (1975). New method for peroxidase-protein conjugation. *Acta Histochem. Cytochem.* 8:41.

Kawaoi, A., and Nakane, P. K. (1973). An improved method of conjugation of peroxidase with proteins. *Fed. Proc.* 32:840.

Kendall, P. A., Polak, J. M., and Pearse, A. G. E. (1971). Carbodiimide fixation for immunohistochemistry: Observations on the fixation of polypeptide hormones. *Experientia* 27:1104–1106.

Kraehenbuhl, J. P., Galardy, R. E., and Jamieson, J. D. (1974). Preparation and characterization of an immunoelectron microscope tracer consisting of a heme-octapeptide coupled to Fab. *J. Exp. Med.* 139:208–223.

Marrack, J. (1934). Nature of antibodies. *Nature* 133:292–293.

Marshall, J. D., Eveland, W. C., and Smith, C. W. (1958). Superiority of fluorescein isothiocyanate (Riggs) for fluorescent-antibody technic with a modification of its application. *Proc. Soc. Exp. Biol. Med.* 98:898–900.

Mason, T. E., Phifer, R. F., Spicer, S. S., Swallow, R. A., and Dreskin, R. B. (1969). An immunoglobulin-enzyme bridge method for localizing tissue antigens. *J. Histochem. Cytochem.* 17:563–569.

McGuigan, J. E. (1968). Immunochemical studies with synthetic human gastrin. *Gastroenterology* 54:1005–1011.

McLean, I. W., and Nakane, P. K. (1974). Periodate-lysine-paraformaldehyde fixative: A new fixative for immunoelectron microscopy. *J. Histochem. Cytochem.* 22:1077–1083.

McLean, J. D., and Singer, S. J. (1970). A general method for the specific staining of intracellular antigens with ferritin-antibody conjugates. *Proc. Natl. Acad. Sci. U.S.A.* 65:122–128.

Nakane, P. K. (1968). Simultaneous localization of multiple tissue antigens using the peroxidase-labelled antibody method: A study on pituitary glands of the rat. *J. Histochem. Cytochem.* 16:557–560.

Nakane, P. K., and Kawaoi, A. (1974). Peroxidase-labeled antibody: A new method of conjugation. *J. Histochem. Cytochem.* 22:1084–1091.

Nakane, P. K., and Pierce, G. B. (1966).   Enzyme-labeled antibodies: Preparation and application for the localization of antigens. *J. Histochem. Cytochem.* 14:929–931.

Nakane, P. K., and Pierce, G. B. (1967).   Enzyme-labeled antibodies for the light and electron microscopic localization of tissue antigens. *J. Cell Biol.* 33:307–318.

Polak, J. M., Kendall, P. A., Heath, C. M., and Pearse, A. G. E. (1972). Carbodiimide fixation for electron microscopy and immunoelectron cytochemistry. *Experientia* 28:368–370.

Ram, J. S., Nakane, P. K., Rawlinson, D. G., and Pierce, G. B. (1966). Enzyme-labeled antibodies for ultrastructural studies. *Fed. Proc.* 25:732.

Riggs, J. L., Seiwald, R. J., Burckhalter, J. H., Downs, C. M., and Metcalf, T. G. (1958).   Isothiocyanate compounds as fluorescent labeling agents for immune serum. *Am. J. Pathol.* 34:1081–1097.

Robinson, G., and Dawson, I. (1975).   Immunochemical studies of the endocrine cells of the gastrointestinal tract. I. The use and value of peroxidase-conjugated antibody techniques for the localization of gastrin-containing cells in the human pyloric antrum. *Histochem. J.* 7:321–333.

Ryan, J. W., Day, A. R., Schultz, D. R., Ryan, N. S., Chung, A., Marlborough, D. I., and Dorer, F. E. (1976).   Localization of angio-tensin converting enzyme (kinase II). I. Preparation of antibody-heme-octapeptide conjugates. *Tissue Cell* 8:111–124.

Sainte-Marie, G. (1962).   A paraffin embedding technique for studies employing immunofluorescence. *J. Histochem. Cytochem.* 10:250–256.

Sternberger, L. A., Hardy, P. H., Cuculis, J. J., and Meyer, H. G. (1970).   The unlabeled antibody enzyme method of immunohistochemistry. Preparation and properties of soluble antigen-antibody complex (horseradish peroxidase-antihorseradish peroxidase) and its use in identifying spirochetes. *J. Histochem. Cytochem.* 18:315–333.

Sutherland, L. E. (1970).   A fluorescent antibody study of juxtaglomerular cells using the freeze-substitution technique. *Nephron* 7:512–523.

Vladutiu, G. D., Bigazzi, P. E., and Rose, N. R. (1973).   Localization of a primate-specific esterase using immunofluorescence and immunoperoxidase techniques. *J. Histochem. Cytochem.* 21:559–567.

Weir, E. E., Pretlow, T. G., Pitts, A., and Williams, E. E. (1974).   Destruction of endogenous peroxidase activity in order to locate cellular antigens by peroxidase-labeled antibodies. *J. Histochem. Cytochem.* 22:51–54.

Weller, T. H., and Coons, A. H. (1954).   Fluorescent antibody studies with agents of Varicella and Herpes Zoster propagated *in vitro*. *Proc. Soc. Exp. Biol. Med.* 86:789–794.

Wofsy, L., and Singer, S. J. (1963).   Effects of the amidination reaction on antibody activity and on the physical properties of some proteins. *Biochemistry* 2:104–116.

11

PRINCIPLES OF ENZYME HISTOCHEMISTRY

GENERAL ENZYMOLOGY

Nature of Enzyme Reactions

Within a given cell, a myriad of chemical reactions take place, the totality of which in concert are the essence of life. The sum of all the chemical reactions that occur within a cell or tissue is the metabolism. Some of these metabolic reactions are involved in building up cellular components—that is, they are anabolic; others are involved in degrading cellular components and are known as catabolic reactions. Some reactions are "uphill" or energy-requiring reactions, which is somewhat analogous to an engine "driving" a car up a grade. Other reactions are "downhill" or energy-yielding reactions. Most of this energy yielded from the downhill reactions is harnessed in the form of high energy compounds, primarily adenosine triphosphate (ATP). This chemical energy can be used to "drive" uphill reactions.

Theoretically, all these reactions can take place without the aid of a catalyst such as an enzyme, although in reality most of these reactions are negligibly slow. For a compound to undergo a chemical reaction, it must first acquire a certain amount of kinetic energy. The molecules of a given population do not possess an equal amount of kinetic energy. Only a very few may possess sufficient energy to undergo the reaction; this quantity of energy is called the energy of activation. The fewer the molecules that possess that amount of energy, the slower the reaction is. It is as if an energy hurdle exists; molecules have to get over the hurdle if the reaction is to take place. It is possible to increase the rate of a reaction by two methods. (1) Energy can be added in the form of heat, which is a very common practice in the organic chemistry laboratory. To add heat to a biological system would be devastating, however. (2) The energy of activation can be lowered, which is precisely what enzymes do; enzymes help substrates get over the energy hurdle.

It must be clearly understood, however, that enzymes do not drive reactions in the sense that they supply the energy; they merely facilitate or increase the velocity of the reactions by lowering the amount of kinetic energy required. The function of enzymes might be compared to the function of the wheels of a car; they do not supply energy but, rather, make it possible for the energy that is developed by the engine to drive the car upgrade. The energy for reactions is provided by high energy compounds such as ATP. Similarly, just as wheels are necessary for a car to coast downhill, enzymes are necessary to facilitate downhill reactions.

Different enzymes are involved in a wide variety of metabolic activities. Cells are frequently specialized to carry out specific functions and are therefore endowed with a certain set of enzymes, presumably those necessary to carry out the cells' functions. However, we do not always understand exactly how all of the enzymes are involved. Cells with a very active aerobic metabolism might show high cytochrome oxidase activity, while those with high anaerobic metabolism would be expected to show high lactate dehydrogenase activity. Cells actively synthesizing glycosaminoglycans would probably show high uridine diphosphogalactose-4-epimerase activity. Cells involved in either calcification or membrane transport usually have high alkaline phosphatase activity. Many other examples could be listed, but it should be obvious that it is frequently important to assess the enzymatic activity of certain cells or groups of cells. Enzyme histochemistry offers a method of obtaining this kind of information.

Enzyme Kinetics

For an enzyme to catalyze or promote a chemical event, the substrate must come in intimate contact with the enzyme. If the substrate concentration is kept low, it can be shown that the likelihood that a substrate molecule will be in close association with an enzyme molecule at a given instant is directly proportional to the concentration. Therefore it follows that when the enzyme concentration is held constant, the reaction rate is proportional to the substrate concentration. This is called the initial velocity (V).

This relationship holds only for low concentrations of substrate, however. It is possible to increase the substrate to such a concentration that the active site of the enzyme is constantly in contact with a substrate molecule—that is, all the active sites are always saturated. At that point, the enzyme is working as hard as it can, and any further increase in substrate will not increase the reaction rate. This point is called the maximum velocity (V_{max}) (Fig. 11-1).

At low concentrations of substrate, where the reaction rate is directly

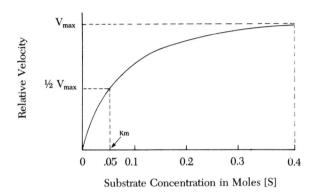

Figure 11-1. The enzyme velocity is plotted against the substrate concentration to show the relationship of ½ V_{max} to K_m.

proportional to the substrate concentration, the reaction is termed first order kinetics. At high substrate concentrations, where the reaction rate is independent of substrate concentration, the reaction involves zero order kinetics. Reactions are also known that are of second order kinetics, or directly proportional to the square of the substrate concentration. It should be emphasized that at first order kinetics, the enzyme reaction rate is directly proportional to two parameters: the concentration of substrate and the concentration of enzyme. On the other hand, at zero order kinetics the enzyme reaction rate varies only with enzyme concentration and is independent of substrate concentration.

Histochemists aim to work with zero order kinetics while demonstrating enzymes in situ. However, it is sometimes difficult to achieve sufficiently high concentrations of substrate for zero order kinetics with certain synthetic substrates.

It is assumed that the substrate and the enzyme temporarily form a complex. The enzyme-substrate complex can then either form reaction products and the freed enzyme or simply dissociate into the original substrate and the enzyme. The reaction can be written thus:

$$E + S \underset{K_2}{\overset{K_1}{\rightleftarrows}} ES \overset{K_3}{\rightarrow} E + P \tag{1}$$

E = enzyme, S = substrate, ES = enzyme-substrate complex, P = products.

Theoretically, the enzyme and product can also reassociate so that the reaction could be reversed. Indeed this is frequently the case in nature; in vitro, however, the product is not allowed to build up to any significant concentration, and the reverse reaction can be ignored.

The rate of reaction of an enzyme at a high concentration of substrate (where the enzyme is saturated) is the V_{max}; however, we have seen that the velocity at that point is independent of substrate concentration. It would therefore be useful if we had an expression of enzyme velocity in terms of substrate concentration.

It is pertinent to consider the dynamics of equation 1 in terms of ES. The rate of *formation* of ES is derived from a consideration of the law of mass action:

$$V_f = K_1([E] - [ES])[S] \tag{2}$$

The factor $[E - ES]$ is an expression of the uncombined enzyme.

The rate of *disappearance* of ES is:

$$V_d = K_2[ES] + K_3[ES] \tag{3}$$

$K_2[ES]$ expresses the formation of the original substrate and enzyme, and $K_3[ES]$ expresses the formation of the product and freed enzyme.

Equations 2 and 3 are equal ($V_f = V_d$) when the formation of ES is equal to the disappearance, or when equation 1 is in steady state. Then we can write:

$$K_1([E] - [ES])[S] = K_2[ES] + K_3[ES] \tag{4}$$

With appropriate algebraic manipulation, the equation can be rewritten as follows:

$$\frac{K_2 + K_3}{K_1} = \frac{([E] - [ES])[S]}{[ES]} \tag{5}$$

The factors on the left, including all the constants, can be written as a single constant, K_m.

The following relationships are known:

$$V = K_3[ES] \tag{6}$$

$$V_{max} = K_3[E] \tag{7}$$

Then by appropriate substitution and algebraic manipulation of equation 5, the following equation can be derived:

$$V = \frac{V_{max}}{1 + \frac{K_m}{[S]}} \tag{8}$$

This is known as the Michaelis-Menten equation. With this equation it can be seen that when K_m is equal to substrate concentration $[S]$, then $V = \frac{1}{2} V_{max}$; K_m is therefore defined as the concentration of substrate at which $V = \frac{1}{2} V_{max}$. This relationship is an extremely useful concept in understanding kinetics of enzymes; it is an expression of the kinetics of an enzyme that is independent of substrate concentration or enzyme concentration. However, one should remember that the K_m of an enzyme can be influenced by temperature, pH, activators, and inhibitors. Furthermore, one can determine that when $[S]$ becomes very large, such that the factor $K_m/[S]$ becomes small, then $V = V_{max}$ and zero order kinetics is achieved.

Histochemists aim to work in zero order kinetics while demonstrating enzymes in situ. It is desirable to have V as high as possible in order to keep incubation time as short as possible. What is more important, it is only when the enzymes are saturated with substrate that the amount of colored reaction product will reflect the relative concentration of enzyme. However, it may be difficult to achieve sufficiently high substrate concentrations with certain synthetic substrates to operate at zero order kinetics. In that case, a compromise must be accepted.

When studying the kinetics of an enzyme, it is useful to measure the velocity of the reaction at a wide range of substrate concentrations and to plot the reciprocal of the velocity against the reciprocal of the substrate concentration. The best-fitting straight line is then drawn through the points (see Fig. 11-2). The point at which the line intersects with the y axis $(1/v)$ is taken as $1/V_{max}$. If the line is further extrapolated to intersect with the negative area of the x axis, then the point of intersection is taken as $-1/K_m$. This is known as the Limeweaver-Burk plot, after the authors who established the method.

Effect of Temperature
It is well known that when the temperature is raised, the velocity of an enzyme increases. As a rule of thumb, the velocity is increased twofold

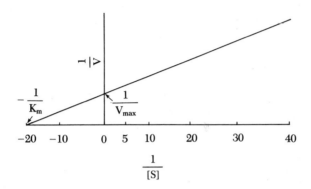

Figure 11-2. The reciprocal of the enzyme velocity is plotted against the reciprocal of the substrate concentration. This represents a convenient way of determining both V_{max} and K_m.

to threefold for every 10°C that the temperature is raised. The exact value can be readily established for each enzyme, and the factor by which the velocity is raised for every 10°C increase is known as the Q_{10} of the enzyme. If the temperature continues to be increased, a point will be reached at which the enzyme activity will drop off precipitously.

Effect of pH
The pH has an extremely important effect on the rate of enzyme reactions. Each enzyme has a maximum rate of action at a particular pH, called the pH optimum; the reaction rate gradually falls off on each side of the pH optimum. The pH optimum of most enzymes is within the range of the pH of the milieu in which the enzymes are naturally found. Trypsin and carboxypeptidase have pH optima of 7.8 and 7.5, respectively, which is similar to the pH of pancreatic juice. Pepsin has a very low pH optimum, which is in the vicinity of the pH of the gastric juice. Lysosomal hydrolases act to digest injured and dead cells and tissues, and they have pH optima of 3 to 5.5, which is also the pH of dead or dying cells and tissues.

Classification of Enzymes
Enzymes are classified according to the type of reaction that they catalyze; most of these reactions can be grouped into one of seven main classes of reactions. Thus enzymes that catalyze hydrolysis reac-

tions are called hydrolases, while hydrolases can be further divided into subcategories such as esterases, carbohydrases, and peptidases according to the type of bond that is involved in the hydrolysis reaction. Each of these types of enzymes can be subdivided at two more levels; each time the substance being hydrolyzed is more narrowly specified. The International Union of Biochemists has assigned a numbering system to enzymes for the purpose of classification. Thus, the number of each oxidoreductase begins with 1 and is succeeded by three more numbers, which further specify the substance. Hydrolases all begin with 3: esterases are 3.1, carbohydrases are 3.2, and peptidases are 3.4. Phosphoesterases (or phosphatases) are designated 3.1.3. Phosphatases with an alkaline pH optimum are designated 3.1.3.1, while those with an acid pH optimum are designated 3.1.3.2. Each succeeding number, therefore, attempts to classify the enzyme more specifically.

Naming of Enzymes

In the early days, when only a few enzymes were known, they were given trivial names that often give no suggestion as to what the function of the enzyme might be. Some of these names that have persisted—such as trypsin, renin, and papain—refer to how and where the enzymes were obtained. Nowadays, enzymes are named by adding the suffix *ase* to the natural substrate of the enzyme. Thus the enzyme that hydrolyzes arginine is arginase; however, the name says nothing about its action on arginine. The names of many enzymes are derived from their substrate as well as from their action. For instance, consider the terms *aspartate aminotransferase, succinic dehydrogenase,* and *glutamate decarboxylase,* which all suggest a substrate and an action. It should be obvious that systematic naming of enzymes prevents the chaotic situation that would result from using only trivial names. For a more extensive discussion about the naming of enzymes, the work of Dixon and Webb (1964) should be consulted.

PRACTICAL ASPECTS OF ENZYME HISTOCHEMISTRY
Processing Tissues

Preparation of tissues for enzyme histochemistry presents problems not encountered in preparation of tissues for other purposes. Many aspects of tissue handling militate against preservation of enzymes. The aims of fixation for enzyme histochemistry are threefold: (1) to preserve morphology as faithfully as possible; (2) to preserve the maximum amount of enzyme activity; and (3) to prevent dislocation of enzymes. The first two are difficult to achieve simultaneously; indeed, to achieve maximum enzyme activity while fixing is a paradox, since it will be recalled that one of the aims of fixation is to inactivate enzymes. It is therefore necessary to compromise on both aspects.

Different enzymes show different degrees of sensitivities toward fixatives. Some are almost instantly inactivated while others retain a considerable amount of activity even after prolonged exposure to fixatives. For instance, cytochrome oxidase and succinic dehydrogenase are both totally inactivated by 2 to 4 hours exposure to formalin at 4°C, whereas only 20 percent of acid phosphatase and nonspecific esterase activities is inactivated under similar conditions (Pearse, 1960). In general, the oxidoreductases can tolerate little or no fixation, whereas the hydrolases can generally tolerate extensive fixation.

Some fixatives are more destructive of enzymes than others are. Janigan (1964) studied the effects of various aldehydes on enzymes under identical conditions. After 7 hours of fixation with formaldehyde at 0° to 2°C, 60 percent of the β-glucuronidase activity was retained, whereas with glutaraldehyde 24 percent was retained and with acrolein no activity was retained. Osmium tetroxide is also extremely detrimental to enzyme activity (Holt and Hicks, 1961).

It appears that the rate of enzyme inactivation is much greater during the first hour or two than thereafter (Seligman et al., 1951), but it is not entirely clear why this should be so. Part of the explanation may lie in the loss of the soluble fraction of the enzyme into the fixative solution. The existence of soluble and insoluble fractions, known as the lyoenzyme and the desmoenzyme, respectively, has been long recognized. Nachlas et al. (1956) measured the amount of enzyme that diffused from unfixed tissue blocks into distilled water in an hour. One-third of the alkaline phosphatase and leucine aminopeptidase activities was found in the diffusate, while one-half of the acid phosphatase activity and two-thirds of the esterase activity had also diffused out.

The earlier studies of the effects of fixatives on enzyme activity were hardly ever done under comparable conditions, and the results were often highly inconsistent. Christie and Stoward (1974) therefore set out to study the factors that affect the results. They used as a model the effect of formaldehyde fixation on liver and kidney acid phosphatase. Previous studies had used liver, which was assumed to be homogeneous. Christie and Stoward found that assumption unjustified and preferred to cut the organ into uniform 1-cubic-millimeter blocks. The blocks are then randomly divided into lots of 9 to 12 blocks each.

The following are some other important points made by Christie and Stoward (1974). (1) Washing after fixation is extremely important. Even though acid phosphatase activity was reduced to less than 50 percent immediately after fixation, 24 hours of washing restored most of the original activity. (2) Enzyme activity should be expressed in terms of units of dry mass. (3) Buffers used as vehicles for the fixative affect the enzyme. Christie and Stoward found that a 5-norbornene-2,3-dicarboxylic acid–sodium hydroxide buffer system preserved more enzyme activity than cacodylate or phosphate buffers. (4) Homogenization, using an efficient homogenizer, is essential. Four 10-second bursts of homogenization are preferred, after which the temperature is returned to 2° to 5°C.

If the tissues are to be embedded in paraffin, it is necessary to cope with both the deleterious effects of fixatives and the effects of organic solvents and heat during embedding. The effects of heat on enzymes are well known. Stafford and Atkinson (1948) found that after 50 hours of cold ethanol fixation, 88 percent of alkaline phosphatase and 20 percent of acid phosphatase remained. After embedding in paraffin (2 hours at 56°C) the percentages were 20 and 7, respectively, for alkaline and acid phosphatase. It is important to use small pieces of tissue, so that infiltration time can be minimized. Low-temperature paraffins have been developed but they have not gained wide acceptance, probably because the meager benefit did not outweigh the increased difficulty of cutting softer paraffin.

The problems of fixing and embedding are totally avoided by using frozen sections; this is the only practical method available for demonstrating enzymes that are highly sensitive to fixatives. Frozen sections are practically always used, even for the hydrolases, although the tissue blocks are sometimes briefly fixed before freezing. Freezing the tissue guards against diffusion of cellular components, autolysis, and microorganismal decomposition. In addition, frozen blocks provide support from which to cut sections. All cellular activity is "frozen in its steps" and remains so until it is thawed again.

With frozen sections the problem of diffusion of the lyoenzyme can still be a serious problem during incubation. The problem can be partially remedied either by very short formalin fixation or by the addition of neutral polymers such as polyvinyl alcohol or polyvinyl pyrrolidine to the medium.

Nature of Enzyme Histochemical Reactions

An enzyme is not stained or visualized directly in any way; rather it is detected through one of its reaction products. The principle of demonstrating enzymes is to provide the conditions whereby the enzyme to be demonstrated will catalyze a reaction that results in a visible product. The enzyme product can then be studied and related to the histology of the section.

Most enzyme histochemical reactions fall into one of three general categories: simultaneous coupling, postincubation coupling, or substrate film techniques.

Simultaneous coupling. Simultaneous coupling is by far the most commonly used principle. The incubation medium contains the appropriately buffered substrate and the coupler. As the substrate is hydrolyzed by the enzyme, it is rapidly coupled and converted into a precipitate that either is visible or can be converted into a visible precipitate.

The diffusion of the reaction product is of great concern. The rate of coupling, which should be very rapid, is the most important factor in preventing diffusion; the rate of reaction product formation also obviously affects the degree of diffusion. In addition, if the substrate can be designed in such a way that the reaction product will be insoluble or have substantivity, diffusion will be limited. In general, coupling is slower and less efficient when the pH is low.

Postincubation coupling. With the postincubation coupling method, the reaction product must be sufficiently insoluble to remain in situ during the course of incubation. Following incubation the tissue section is subjected to a coupling agent (the incubation medium does not contain the coupling reagent).

To require that the substrate be sufficiently soluble to achieve zero order kinetics and yet give rise to a highly insoluble product is a rather stringent requirement. Only one class of substrates has been successfully used with this method, namely, β-naphthols substituted at the 6-position with benzoyl or bromo groups. If the reaction product can be made more substantive, then the insolubility requirement is not so stringent.

The postincubation coupling method has been used to demonstrate acid phosphatase, sulfatase, and β-glucuronidase. The substrates are made by esterifying β-naphthols with the appropriate compound.

The postincubation coupling method is suitable to use when the pH is relatively low, because at a low pH it would be difficult to obtain efficient coupling with the simultaneous coupling method. This method also has an advantage when long incubation periods are required, because the diazo coupling salts are frequently rather unstable. Another distinct advantage of postincubation coupling is that the deleterious effects of the diazo coupling salts on the enzymes are avoided.

Substrate film techniques. In the substrate film techniques, the substrate is supplied in the form of a very thin film, which becomes depleted at the enzyme sites. Following an appropriate incubation period, the remaining substrate is stained, and the presence of the enzymes is detected by the absence of the substrate in the film. (These techniques are further considered on page 255.)

Nachlas et al. (1957) recognized two additional general methods for demonstrating enzymes—insolubilization of a colored substrate and intermolecular rearrangement. However, these methods are used so rarely that they merit no consideration here.

Enzyme Selectivity

In order to demonstrate a particular enzyme, it is usually only necessary to supply the appropriate substrate under the proper conditions. However, there are frequently several enzymes that will use the same substrate. How then does one know which enzyme is being demonstrated? Three examples will be dealt with that are of histochemical interest.

Isoenzymes. In the case of hydrolases, there are frequently several isoenzymes. One type of acid phosphatase is associated with smooth endoplasmic reticulum and another is associated with lysosomes; the two have somewhat different but widely overlapping characteristics.

There are many esterase isoenzymes that are electrophoretically separable, but these isoenzymes often have very similar enzymatic characteristics and cannot be distinguished separately in tissue sections. It is generally impossible to know which form of the enzyme is being demonstrated; rather, one demonstrates total esterase or phosphatase activity.

Malic acid oxidation. The oxidation of malic acid is somewhat unusual because it can be accomplished by two distinctly different dehydrogenases, catalyzing two distinctly different reactions. One of the enzymes, called malic enzyme, catalyzes the interconversion of malic acid to pyruvic acid. The other enzyme, malate dehydrogenase, catalyzes the formation of α-ketoglutarate from malic acid. Fortunately, these two enzymes have mutually exclusive coenzyme requirements—the former is dependent on nicotinamide-adenine dinucleotide (NAD) and the latter is dependent on nicotinamide-adenine dinucleotide phosphate (NADP). This difference in requirements makes it possible, therefore, to demonstrate these two enzymes selectively.

Uridine diphosphoglucose conversion. Uridine diphosphoglucose (UDP-glucose) can also serve as the substrate of two different enzymes: UDP-glucose dehydrogenase can oxidize UDP-glucose to UDP-glucuronic acid, while UDP-galactose-4-epimerase can convert (or epimerize) UDP-glucose to UDP-galactose. Both reactions are dependent on NAD. The histochemical methods given for these two enzymes call for almost identical incubation media. The epimerase reaction is reversible, and the equilibrium is toward the formation of UDP-glucose. The fact that the equilibrium favors the formation of UDP-glucose allows the epimerase to be demonstrated even more readily with UDP-galactose and solves the problem of which enzyme is being demonstrated with respect to the epimerase. However, when one attempts to demonstrate UDP-glucose dehydrogenase, it is not possible to exclude some epimerase activity as well.

Factors Affecting Localization of Enzymes

The following discussion is adapted and updated from the important works of Nachlas et al. (1957) and Pearse (1968).

Requirements of the substrate. A fundamental requirement of the substrate is that it must be readily hydrolyzed by the enzyme. It is well known that an enzyme will show different kinetics with different synthetic substrates.

High solubility of the substrates is also necessary in order to achieve zero order kinetics. The naphthol AS compounds are sparingly soluble in water; however, they can be first solubilized in a small volume of a high polar water-soluble organic solvent such as dimethyl sulfoxide or dimethylformamide. These solvents may also serve to produce decreased solubility of the reaction product, which can cause diffusion.

The substrate should be as pure as possible. Unesterified naphthol compounds will be coupled and precipitated the instant the coupling salt is added, resulting in an indeterminable lowering of the concentration of both the substrate and the coupling salt. In some cases impurities may inhibit the enzyme to be demonstrated.

There are other attributes for the substrate that are desirable but not strict requirements. For instance, for the sake of convenience, a substrate should be stable so that it can be stored in the form in which it is used. It should also be readily available and economical—substrates that are very expensive and those that are not commercially available are often a source of discouragement for investigators unprepared to synthesize their own.

Requirements of the reaction product. It is rather self-evident that the reaction product must be insoluble in aqueous solution. It is perhaps less evident that it should be insoluble in lipid also; if the reaction product were soluble in lipid, it might be encouraged to translocate or diffuse to lipid droplets. It should have substantivity (or an affinity for proteins), although not selectively for just certain proteins. If a reaction product exhibits substantivity, the requirement for insolubility is less stringent. Finally, the reaction product must be highly colored (that is, it must have a high color index), or it must be capable of being converted to a highly colored product.

Requirements of coupling reagent. It is imperative that the coupling reagent have a high coupling rate with the reaction production. If the reaction product is not instantly coupled, it may diffuse from the site of enzyme activity. It should couple efficiently at the pH optimum of the enzyme. Actually, both lead (in the methods of phosphatases) and the diazonium salts couple very inefficiently at a pH of less than 5. For the sake of achieving efficient coupling it is therefore sometimes necessary to compromise somewhat on the pH. The coupling agent should be stable in aqueous solution for the duration of the incubation period. Some coupling agents, such as fast blue B (tetrazotized *o*-dianisidine), are stable for no more than 5 minutes. The coupling agent should not yield false-positive results. The potential danger can be appreciated when one considers that diazonium salts have been used to couple with and demonstrate phenol groups in tissue. However, the color of a false-positive reaction product will be different from the enzyme reaction product and will appear as diffuse background staining. Finally, the coupling agent should not significantly inhibit the enzyme to be demonstrated. Diazonium salts do often inhibit enzymes to a certain degree. Some of this effect is primarily due to the effect of stabilizing metal ions associated with the diazonium compounds, although some of the inhibition is almost certainly due to the diazonium compound itself.

REFERENCES

Christie, K. N., and Stoward, P. J. (1974).   A quantitative study of the fixation of acid phosphatase by formaldehyde and its relevance to histochemistry. *Proc. R. Soc. Lond.* [Biol.] 186:137–164.

Dixon, M., and Webb, E. C. (1964).   *Enzymes* (2nd ed.). Academic Press, New York.

Holt, S. J., and Hicks, R. M. (1961).   Studies on formalin fixation for electron microscopy and cytochemical staining purposes. *J. Biophys. Biochem. Cytol.* 11:31–45.

Janigan, D. T. (1964).   Tissue enzyme fixation studies. I. The effects of aldehyde fixation on β-glucuronidase, β-galactosidase, N-acetyl-β-glucosaminidase, and β-glucosidase in tissue blocks. *Lab. Invest.* 13:1038–1050.

Nachlas, M. M., Prinn, W., and Seligman, A. M. (1956).   Quantitative estimation of lyo- and desmoenzymes in tissue sections, with and without fixation. *J. Biophys. Biochem. Cytol.* 2:487–502.

Nachlas, M. M., Young, A. C., and Seligman, A. M. (1957).   Problems of enzymatic localization by chemical reactions applied to tissue sections. *J. Histochem. Cytochem.* 5:565–583.

Pearse, A. G. E. (1960).   *Histochemistry: Theoretical and Applied* (2nd ed.). Little, Brown, Boston.

Pearse, A. G. E. (1968).   *Histochemistry: Theoretical and Applied* (3rd ed.), Vol. 1. Little, Brown, Boston.

Seligman, A. M., Chauncey, H. H., and Nachlas, M. M. (1951).   Effect of formalin fixation on the activity of five enzymes of rat liver. *Stain Technol.* 26:19–23.

Stafford, R. O., and Atkinson, W. B. (1948).   Effect of acetone and alcohol fixation and paraffin embedding on activity of acid and alkaline phosphatase in rat tissues. *Science* 107:279–281.

12

PHOSPHATASES

The phosphatases are a group of enzymes that liberate orthophosphate from organic phosphates. They are widely distributed in mammalian tissues and have been studied extensively.

Phosphatases exhibit varying degrees of specificity. The nonspecific phosphatases hydrolyze virtually any organic phosphate and show only limited preferences. 5'-Nucleotidase shows limited specificity, hydrolyzing nearly all nucleotides; other phosphatases, such as adenosine triphosphatase (ATPase), show rather high specificity for their substrates.

Each of the nonspecific phosphatases is classified as an alkaline or an acid phosphatase, depending on whether its pH optimum is in the alkaline or acid range. It is strange that there are no nonspecific phosphatases with pH optima near neutrality. However, most of the specific phosphatases have pH optima in the vicinity of neutrality, usually somewhat on the alkaline side.

ALKALINE PHOSPHATASE
Biochemistry
Alkaline phosphatases have been divided into three categories based on kinetics, response to inhibitors, stability, and electrophoretic mobility (Moss, 1969a). The first type, placental alkaline phosphatase, occurs in the human placenta but generally not in adult tissues. However, tumor tissues sometimes elaborate a kind of alkaline phosphatase that resembles placental alkaline phosphatase. The second type is intestinal alkaline phosphatase, which also somewhat resembles placental alkaline phosphatase. The third category comprises the alkaline phosphatases, which occur in bone, kidney, and liver. The placental enzyme is clearly under separate genetic control, which differentiates it from the others. Alkaline phosphatases from other sources are fairly heterogeneous, and it is uncertain whether they represent separate genetic groups or whether the differences result from carbohydrate prosthetic groups. Alkaline phosphatase from an osteosarcoma has been shown to be completely different from liver,

intestinal, kidney, and placental alkaline phosphatase on immuno-logical grounds (Singh and Tsang, 1975).

Alkaline phosphatase contains zinc as an integral part of the mole-cule and is therefore considered a metalloenzyme. Its active center contains a serine residue, and the mechanism of action involves the formation of phosphoryl serine residues. The enzyme contains a number of carbohydrate prosthetic groups including sialic acid, but the amount varies from one type to another.

High concentrations of alkaline phosphatase are often associated with absorptive cells, especially in the brush borders of the intestinal mucosa and the proximal tubules of the kidney. Alkaline phosphatase is always present wherever calcification occurs, particularly in os-teoblasts and in chondrocytes of cartilage that is about to give way to endochondral bone formation.

In addition to being able to hydrolyze a wide variety of monophos-phates, alkaline phosphatase is also quite capable of hydrolyzing pyrophosphates (Cox and Griffin, 1965). It has also been reported to hydrolyze ATP (Moss and Walli, 1969).

It has long been known that most types of alkaline phosphatases are activated by magnesium ions, but Moss (1969b) showed that only the orthophosphatase activity of alkaline phosphatase is activated; mag-nesium ions actually inhibit pyrophosphatase activity. Zinc ions in-hibit alkaline phosphatase of bone and cartilage (Takada et al., 1968). Highly specific inhibitors of alkaline phosphatase, L-tetramisole, and related compounds were introduced by Borgers (1973). These in-hibitors make it possible to distinguish clearly between alkaline phos-phatase and other phosphatases whose specificities overlap with al-kaline phosphatase (Borgers and Thoné, 1975 and 1976).

Histochemical Demonstration

Metal precipitation methods. Alkaline phosphatase was first demon-strated histochemically by Gomori (1939) and Takamatsu (1939). These two investigators developed essentially the same method simultaneously and independently. The method was based on the precipitation of calcium phosphate at the site of enzyme activity. Sec-tions were incubated with the organic phosphate ester as a substrate along with a high concentration of Ca^{2+} at a pH of 9 or above. As the orthophosphate was enzymatically released, it was immediately precipitated with the Ca^{2+} as calcium phosphate. However, calcium phosphate was an invisible white precipitate and had to be converted to a different form to be seen microscopically. Both Gomori and Takamatsu used the von Kossa silver method for this purpose, but most workers since then have preferred to convert the calcium salt into a cobalt salt and blacken it with ammonium sulfide. Theoretically, any histochemical method for calcium could be used for the last step. The reactions involved in demonstrating alkaline phosphatase are summarized:

$$\text{Glycerophosphate} + Ca^{2+} \xrightarrow[\text{pH 9.2}]{\text{phosphatase}} \text{glycerol} + \text{calcium phosphate}$$

$$\text{Calcium phosphate} + \text{cobalt nitrate} \longrightarrow \text{cobalt phosphate}$$

$$\text{Cobalt phosphate} + \text{ammonium sulfide} \longrightarrow \text{cobalt sulfide (black)}$$

The original method was modified and improved several times. Gomori's revised (1952) method seems to have found most favor among histochemists. The methods based on calcium phosphate precipitation are still useful although they have largely given way to methods using substituted naphthol phosphate compounds as substrates.

Naphthyl methods. Methods based on precipitation and visualization of the organic portion of the hydrolyzed substrate started with the work of Menten et al. (1944), who used β-naphthyl phosphate as the substrate. The liberated naphthol was coupled with diazotized α-naphthylamine, which formed a colored precipitate at the site of enzyme activity. Naphthyl compounds were soon found to be much more satisfactory because the precipitates were much more insoluble than the lead or calcium precipitates.

Other naphthyl compounds were tried in modifications of the method of Menten et al.; the main drawback was the requirement always to prepare fresh diazotized coupling compounds. Later, through the work of Manheimer and Seligman (1948), histochemists learned to use stabilized diazoates, which could be stored.

The methods utilizing the substituted naphthols always showed essentially the same distribution patterns as were obtained with the calcium phosphate precipitation methods. The naphthol methods are convenient to use and have three additional advantages. (1) Because the reaction product is much more insoluble, localization is sharper and the diffusion problems are essentially solved. (2) The naphthol compounds are more rapidly hydrolyzed, which allows shorter incubation periods. (3) The reaction product is immediately visible and the reaction can therefore be monitored and terminated when optimal color intensity is developed.

The method of Burstone (1962) is recommended for demonstrating alkaline phosphatase using naphthol AS-MX-phosphate. The reaction proceeds as follows:

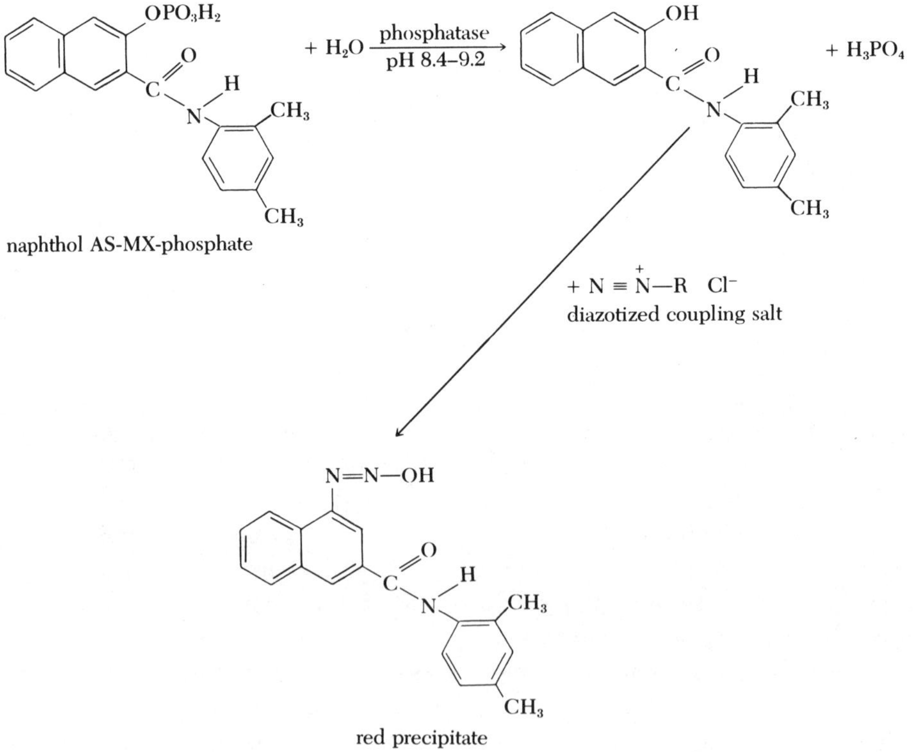

naphthol AS-MX-phosphate

+ N ≡ N—R   Cl⁻
diazotized coupling salt

red precipitate

A number of other naphthol AS compounds are available commercially and can be substituted for naphthol AS-MX-phosphate. Many stabilized diazoates are also commercially available. However, care should be exercised in substituting the coupling salt. Pearse (1968) studied the suitability of various diazo compounds to use with alkaline phosphatase on the basis of color, decomposition, coupling speed, and enzyme inhibition. Out of the 22 compounds tested, only 5 were satisfactory: fast blue RR, fast red TR, fast violet B, fast blue VRT, and fast black B.

ACID PHOSPHATASE

The enzyme acid phosphatase is one of the multitude of acid hydrolases that have been shown to occur in lysosomes. Knowledge of acid phosphatase considerably predates the knowledge of lysosomes. (For a review of lysosomes, see DeDuve and Wattiaux [1966].) Acid phosphatase from the prostate gland was described by Kutscher and Wolbergs (1935) and that from erythrocytes by King et al. (1945).

Biochemistry

Biochemical studies have clearly demonstrated the presence of more than one type of acid phosphatase. Even within the same organ (rat liver), there may exist several types of acid phosphatase (Goodlad and Mills, 1957). One might wonder then if all types of acid phosphatases are lysosomal or if perhaps only one type is lysosomal. Through the cell fractionation studies of Lin and Fishman (1972) we now know that, at least within the kidney, there is a lysosomal acid phosphatase and an acid phosphatase that is associated with the smooth endoplasmic reticulum. Rosenbaum and Rolon (1962) presented evidence that the lead precipitation method and the azo dye method demonstrated different enzymes. Indeed, Lin and Fishman (1972) found that their lysosomal acid phosphatase preferred β-glycerophosphate for a substrate, whereas their microsomal enzyme preferred naphthyl phosphates.

Few generalizations can be made concerning activators and inhibitors of acid phosphatase—undoubtedly because of the existence of multiple unrelated forms of acid phosphatase. The reader should refer to the studies of Goodlad and Mills (1957) and Lin and Fishman (1972).

A zinc-activated acid phosphatase, which has been described in the coroid plexus (Felicetti and Rath, 1975) and in rat brain tumors (Rath and Felicetti, 1975), is chromatographically and electrophoretically different from other phosphatases. There is no indication which subcellular organelle this acid phosphatase is associated with.

Histochemical Demonstration

Metal precipitation methods. Gomori (1941) first localized acid phosphatase in tissues, using a modification of his method for alkaline phosphatase. Instead of using calcium nitrate as a capture agent, he used lead nitrate. The incubation medium was buffered at pH 4.7 to 5.0. At this pH lead phosphate is still rather insoluble, whereas calcium phosphate would have been quite soluble.

Early studies showed this method to be quite unsatisfactory, however, because results were variable and because certain tissue components, especially nuclei, showed an affinity for lead ions. Gomori (1950) then offered an improved version of his original method by keeping the pH at 5.0 while adjusting the buffer strength and increasing the substrate concentration tenfold. The results were considerably better but still did not satisfy all workers.

Naphthyl methods. The naphthyl phosphate compounds at first were also not completely satisfactory; they were less satisfactory for acid phosphatase than for alkaline phosphatase, because the rate of coupling was slower at a low pH than at a higher pH. However, at least from a retrospective viewpoint, the use of naphthyl phosphate compounds offered a great deal more potential than the lead precipitation method. These compounds permitted the coupling step to be carried out after incubation (the postcoupling method). This offered certain advantages, and for a time the postcoupling method was thought to be the method of choice.

Highly satisfactory methods for demonstrating acid phosphatase were developed by Burstone (1958a, 1958b, 1958c, 1961, and 1962) by using naphthol AS phosphate compounds. The advantages outlined earlier for alkaline phosphatase also hold for acid phosphatase. The coupling compounds found most suitable by Pearse (1968) were fast garnet GBC, fast red LTR, fast blue VRT, and hexazotized pararosanilin. Many workers, including Burstone (1961), have found red violet LB to be quite satisfactory. Hexazotized pararosanilin has been particularly recommended as a coupler at acid pH levels because it couples extremely rapidly, even at a low pH (Davis and Ornstein, 1959).

5'-NUCLEOTIDASE
Biochemistry

The enzyme 5'-nucleotidase catalyzes the dephosphorylation of mononucleotides. All mononucleotides are hydrolyzed, although some are hydrolyzed more rapidly than others. Deoxyribonucleotides are also hydrolyzed, according to Hardonk (1968), but not according to Goff and Miliare (1974). The enzyme is widely distributed in mammalian tissues but is generally absent from rapidly growing tissues. The presence of 5'-nucleotidase is thought to be correlated with functional competence, leading Hardonk and Koudstaal (1968) to postulate that the enzyme has a catabolic function in the cell; they relegated it to a role of degrading residual nucleotides. This has become the generally accepted role for the enzyme (Goldberg, 1973).

The Hardonk and Koudstaal concept has been challenged recently on the grounds that a few rapidly growing tissues have been found to contain high concentrations of the enzyme (Klaushofer and Böck, 1974). It is possible that the Hardonk and Koudstaal concept is inadequate; however, if it is successfully discredited, it will be difficult to replace that concept with another that is equally attractive.

The pH optimum of 5'-nucleotidase varies considerably depending on the species of origin, although generally it is somewhat on the basic side of neutrality; human 5'-nucleotidase has a pH optimum of 7.8. The pH optimum is influenced by cationic activators as well as by the nature of the substrate. (For a short discussion see Hardonk [1968].) It is curious that Otte (1958) found a strong reaction for 5'-nucleotidase in the superficial chondrocytes of calves' articular cartilage at a pH of 9.2. When β-glycerophosphate was substituted for the substrate, no activity was observed.

Magnesium is known to activate the enzyme and is always added in the incubation medium. Other divalent cations are generally less effective as activators, although Hardonk (1968) found Mn^{2+} to be at least as effective as Mg^{2+}. No cation is known to inhibit the enzyme consistently; however, Ni^{2+} is very effective as an inhibitor of the enzyme from certain species.

Histochemical Demonstration

The calcium phosphate precipitation method has been used to demonstrate 5'-nucleotidase (Gomori, 1952). Two problems that were associated with this method led workers to use lead for precipitating the phosphate. First, if the pH was dropped to near neutrality, where there was no longer any overlap with alkaline phosphatase activity, then calcium phosphate was no longer sufficiently insoluble and a diffusion problem would be encountered. Second, at the required concentrations of calcium, significant enzyme inhibition occurred. The lead precipitation method of Wachstein and Meisel (1957) is most commonly used for demonstrating 5'-nucleotidase.

If 5'-nucleotidase occurs in areas where alkaline phosphatase may be expected, a control section should be used, with the substitution of α-glycerophosphate for the nucleotide substrate. The amount of 5'-nucleotidase can be estimated by comparing the two sections. Alternatively, a specific inhibitor of alkaline phosphatase, L-tetramisole, may be used to suppress all alkaline phosphatase activity.

ADENOSINE TRIPHOSPHATASE
Biochemistry

There are three distinctly different types of ATPase, each of which hydrolyzes the terminal group of adenosine triphosphate:

$$\text{Adenosine} - O - HPO_2 - O - HPO_2 - O - HPO_2 - OH + H_2O \longrightarrow$$
$$\text{(ATP)}$$

$$\text{Adenosine} - O - HPO_2 - O - HPO_2 - OH + H_3PO_4$$
$$\text{(ADP)}$$

The three types occur at different subcellular sites and are distinguished on the basis of activators and inhibitors. (1) Mitochondrial ATPase is located in the mitochondral wall; it is activated by Mg^{2+} and inhibited by Ca^{2+}. A good correlation exists between the amount of mitochondrial ATPase and the intensity of metabolic activity. (2)

Myosin ATPase is located in striated muscle; it is activated by Ca^{2+} and inhibited by Mg^{2+} (exactly the reverse of the situation seen in mitochondrial ATPase). (3) The third type is Na^+/K^+-activated ATPase, in which the requirement for Na^+ and K^+ is absolute; no activity is expressed in the absence of these cations. This ATPase is often associated with membranes engaged in active transport, and it is thought to be linked with the Na^+/K^+ pump.

Histochemical Demonstration

Mitochondrial ATPase. Histochemical studies are usually concerned with the mitochondrial ATPase; for this purpose, the method of Wachstein and Meisel (1956) or their own modification (Wachstein et al., 1960) is generally used. The liberated phosphate is trapped and precipitated with Pb^{2+} in a manner similar to Gomori's method (1952) for acid phosphatase.

During the late 1960s an intense controversy raged over the validity of the method of Wachstein and Meisel (1956) for demonstrating ATPase. The dispute was slowly resolved in favor of the validity of the method. There is no doubt, however, that the interpretation of the results is generally less casual as a result of the controversy. First, Rosenthal et al. (1966) reported that ATP could be hydrolyzed by Pb^{2+} under normal histochemical incubation conditions, undermining the validity of the method. Novikoff (1967), however, felt that the amount of hydrolysis by Pb^{2+} was negligible because (1) the staining was dependent on divalent cations, (2) it is abolished by various inhibitors, and (3) most important, it is site-specific for various substrates. Then came the report that the lead phosphate resulting from the enzymatic hydrolysis of ATP is preferentially bound to certain tissue sites irrespective of the sites of enzyme activity (Ganote et al., 1969). Evidence was also presented to indicate that the reaction product is not simply lead phosphate but, rather, a complex compound containing significant amounts of the nucleotide substrate. However, Koenig and Vial (1973) could find no evidence of nonenzymatic precipitation of ATP, even at high concentrations of ATP and Pb^{2+}. Many of the parameters were again studied extensively by Goff and Milaire (1974); on the basis of inhibition experiments, heat inactivation experiments, and a comparison of the enzymatic hydrolysis of ATP with that of AMP, they concluded that "it is unlikely that the hydrolysis of ATP was a nonenzymatic catalysis." This attitude toward the method is shared by most workers, and only a few die-hard critics might still disagree. The method of Wachstein et al. (1960) can be confidently recommended for mitochondrial ATPase.

Myosin ATPase. Methods have been devised for demonstrating the myosin ATPase as well as for the Na^+/K^+-activated ATPase. Padykula and Herman (1955a and 1955b) devised a method for myosin ATPase using Ca^{2+} to precipitate the phosphate. Calcium also serves to activate the enzyme. This method has also been severely criticized (Gillis and Page, 1967) because controls in which the substrate was omitted also showed some precipitation. The procedure has been modified specifically for muscle ATPase by Guth and Samaha (1969 and 1970), and it has been found satisfactory by some workers.

The method of Guth and Samaha utilizes either an acid or an alkaline preincubation step. Myosin ATPase is alkali-stable and acid-labile, whereas nonmyosin ATPase (interfibrillar ATPase) is alkali-labile and acid-stable. By using the alkaline preincubation step, the method can be rendered specific for myosin ATPase, and only the white (alpha) muscle fibers will show the reaction product. However, by using the acid preincubation step, the red (beta) fibers are preferentially stained.

Sodium/potassium-activated ATPase. Modifications of the Wachstein and Meisel method (1957) have been used for demonstrating Na^+/K^+-activated ATPase in the ciliary body (McClurkin, 1964) and in red cell ghosts (Marchesi and Palade, 1967). This method has also been criticized. Tormey (1966) claimed that Na^+/K^+-activated ATPase was not demonstrated at all because neither the presence of ouabain (a specific inhibitor of Na^+/K^+-activated ATPase) nor the absence of Na^+ and K^+ affected the intensity of the reaction. Guth and Albers (1974) devised a new method for demonstrating this enzyme using p-nitrophenyl phosphate as the substrate. Biochemical studies using purified Na^+/K^+-activated ATPase have shown that p-nitrophenyl phosphate is hydrolyzed by this enzyme. Moreover, if ATP is used as the substrate, a heavy divalent cation would have to be used to trap the liberated phosphate ions, and such divalent cations strongly inhibit the enzyme. Using this method, the enzyme is optimally active at pH 7.9. However, when dimethyl sulfoxide is added, the pH optimum of the enzyme shifts to 9.0, increasing the specific activity of the enzyme severalfold. The liberated inorganic phosphate is thought to bind to tissue components at pH 9—which can then be visualized with a cobalt chloride–ammonium sulfide sequence.

GLUCOSE 6-PHOSPHATASE

The distribution of glucose 6-phosphatase is more limited than that of most other phosphatases. Chiquoine (1953, 1955) surveyed a wide variety of tissues but found this enzyme only in the liver, kidney, and small intestine—tissues that are capable of freezing glucose into the circulatory system. Since then, however, glucose 6-phosphatase has been localized in some other tissues. (For a review, see Rosen [1974].) It is an enzyme primarily of the smooth endoplasmic reticulum and is found in microsomal fractions of cells; indeed, it is used as a biochemical marker for microsomes in cell fractionation studies. However, approximately 15 to 20 percent of the hepatic glucose 6-phosphatase is present in the nuclear membrane. Glucose 6-phosphatase is tightly bound to the membrane, which renders purification difficult.

Biochemistry

Glucose 6-phosphatase has a fairly high specificity for glucose 6-phosphate, but evidently it also hydrolyzes ribose 5-phosphate to a limited extent (Hardonk, 1968). Besides its hydrolytic activity, the enzyme can also form glucose 6-phosphate by transferring a phosphate group from another compound to the 6-position of free glucose. Thus, the enzyme can also serve as a transferase.

The pH optimum of glucose 6-phosphatase is about 6.7. The enzyme does not seem to be activated by magnesium ions, as are many other phosphatases, but it is in fact inhibited by $0.1M$ magnesium ions. 1,5-Sorbitan 6-phosphate is a specific inhibitor of glucose 6-phosphatase. It is also inhibited by fluoride, zinc, and cyanide ions (Chiquoine, 1955). The activity is also somewhat impaired by trismaleate and cacodylate buffers. The biological significance of glucose 6-phosphatase has been reviewed by Nordlie (1979).

Histochemical Demonstration

The method of Chiquoine (1953) is generally used for demonstrating glucose 6-phosphatase. This method is based on the same principle as that for Gomori's method for acid phosphatase. Liberated phosphate ions are precipitated with lead ions and the resulting lead phosphate is converted to a black precipitate with ammonium sulfide. Fresh cryostat sections are required, because the enzyme is readily inactivated by either formalin fixation or paraffin embedding. The sensitive nature of the enzyme is useful, however, in differentiating it from the nonspecific acid and alkaline phosphatases.

CALCIUM METHOD FOR ALKALINE PHOSPHATASE
(After Gomori, 1952)

TISSUES:
Fix small pieces of tissue in cold (−20°C) acetone for 24 hours. Place in two further changes of acetone (room temperature) for an additional hour each, clear in two changes of benzene for 30 minutes each, and embed in paraffin as quickly as possible. Cut thin paraffin sections.

PREPARATION OF INCUBATION MEDIUM:

Sodium β-glycerophosphate, 2 percent	2.5 ml
Sodium veronal (barbitone)	2.5 ml
$CaCl_2$, 2 percent (or $CaCl_2 \cdot 2H_2O$, 2.7 percent)	4.5 ml
$MgSO_4$, 2 percent	0.2 ml
Distilled water	0.3 ml

The pH should be approximately 9.2; adjust if necessary. This medium is fairly stable and can be stored at 0° to 4°C for several months.

PROCEDURE:
(1) Deparaffinize sections and bring to water.
(2) Incubate for 1 to 3 hours at 37°C.
(3) Rinse quickly in distilled water.
(4) Treat in 2 percent cobalt nitrate (or acetate) for 2 minutes.
(5) Wash for 1 minute to remove all excess cobalt chloride.
(6) Treat with 1 percent ammonium sulfide for 1 minute.
(7) Wash for 2 or 3 minutes.
(8) Dehydrate through graded alcohols, clear with xylene, and mount.

CONTROL:

Prepare an incubation medium in which the sodium β-glycero-phosphate is replaced with an equal amount of water.

RESULTS:

Black deposits of cobalt sulfide represent the sites of enzyme activity. Controls should be completely negative.

SUBSTITUTED NAPHTHOL METHOD FOR ALKALINE PHOSPHATASE

(After Burstone, 1962)

TISSUES:

Use fresh-frozen cryostat sections and fix the sections for 5 minutes in cold acetone before incubation. Alternatively, fix the tissues in formol-calcium for 3 to 4 hours at 0° to 4°C and use cryostat sections.

PREPARATION OF INCUBATION MEDIUM:

Dissolve 10 mg naphthol AS-MX-phosphate in 0.25 ml dimethylfor-mamide. (Alternatively, naphthol AS-BI phosphate or naphthol AS-TR phosphate may be used.) Dilute with 25 ml distilled water. Add 25 ml tris buffer, $0.2M$, pH 8.4 to 8.6. Just before use, add 30 mg fast red violet LB or fast red TR and filter.

PROCEDURE:

(1) Incubate sections for 15 minutes to 2 hours at room temperature.
(2) Wash in water for 5 minutes.
(3) Counterstain with Mayer's hematoxylin if desired.
(4) Wash in water for 5 minutes.
(5) Mount with glycerin jelly.

RESULTS:

The red azo dye indicates alkaline phosphatase activity, and the nuclei are black is counterstain is used.

LEAD PRECIPITATION METHOD FOR ACID PHOSPHATASE
(Gomori, 1952)

TISSUES:
Fix tissues in cold acetone and embed rapidly in paraffin at a temperature not exceeding 56°C. Use paraffin sections.

INCUBATION MEDIUM:
Prepare the following solution:

Acetate buffer, 0.05M, pH 5.0	50 ml
Lead nitrate	60 mg
Na β-glycerophosphate, 3 percent, 0.1M	5 ml

The solution will become turbid. Keep it in an incubator at 37°C for 24 hours and filter it. Add 3 ml distilled water to prevent precipitation upon evaporation. The incubation medium is ready to use and can be stored in a refrigerator for several months. Discard it if it becomes cloudy.

PROCEDURE:
(1) Deparaffinize sections and bring to water.
(2) Incubate for 1 to 24 hours.
(3) Rinse in distilled water.
(4) Rinse in 1 or 2 percent acetic acid.
(5) Rinse in distilled water.
(6) Immerse in 1 percent ammonium sulfide solution.
(7) Mount with glycerin jelly.

RESULTS:
Dark brown deposits of lead sulfide indicate sites of acid phosphatase.

SUBSTITUTED NAPHTHOL METHOD FOR ACID PHOSPHATASE
(Burstone, 1961)

TISSUES:
Fresh-frozen cryostat sections are satisfactory but Burstone recommended paraffin sections of freeze-dried tissues.

PREPARATION OF INCUBATION MEDIUM:
Dissolve 10 mg naphthol AS-BI phosphate with 0.25 ml dimethylformamide. Dilute with 25 ml distilled water. Add 25 ml of 0.2M acetate buffer, pH 5.2. Add 2 drops of 10 percent $MnCl_2$. Just before use, add 30 mg fast violet LB salt and filter.

PROCEDURE:
(1) Incubate sections for 30 minutes to 2 hours.
(2) Wash in distilled water.
(3) Counterstain with Mayer's hematoxylin if desired.
(4) Wash in water.
(5) Mount in glycerin jelly.

RESULTS:
The red azo dye indicates acid phosphatase activity, and the nuclei are black if counterstain is used.

LEAD METHOD FOR 5'-NUCLEOTIDASE
(Wachstein and Meisel, 1957)

TISSUES:
Use free-floating cryostat sections. Cut directly into incubation medium.

PREPARATION OF INCUBATION MEDIUM:
Combine the following:

Adenosine 5-phosphate, 0.125 percent	4.0 ml
Tris buffer, 0.2M, pH 7.2	4.4 ml
Magnesium sulfate, 0.1M	1.0 ml
Lead nitrate, 2 percent	0.6 ml

All solutions should be made with CO_2-free water (boiled). A small amount of precipitate may form when the lead nitrate is added, and it should be filtered just before use.

PROCEDURE:
(1) Incubate free-floating cryostat sections at 37°C for 30 to 60 minutes.

(2) Transfer to formol-saline fixative for 30 minutes.
(3) Transfer through two changes of CO_2-free distilled water.
(4) Treat with 2 percent ammonium sulfide for 2 minutes.
(5) Rinse in distilled water.
(6) Mount on slides and allow them to become nearly dry.
(7) Mount in glycerin jelly.

RESULTS:
5′-Nucleotidase activity appears brown to black.

LEAD METHOD FOR ADENOSINE TRIPHOSPHATASE
(Wachstein et al., 1960)

TISSUES:
Use cryostat sections of either unfixed or formalin-fixed tissues. Sections may be used either as free-floating or mounted.

PREPARATION OF INCUBATION MEDIUM:
Combine the following:

Disodium adenosine triphosphate, 0.125 percent	4.0 ml
Tris buffer, 0.2M, pH 7.2	4.4 ml
Magnesium sulfate, 0.1M	1.0 ml
Lead nitrate, 2 percent	0.6 ml

All solutions should be made with CO_2-free water (boiled). A precipitate may be formed when the lead nitrate is added, and the medium should be filtered just before use.

PROCEDURE:
(1) Incubate sections for 10 to 60 minutes at 37°C.
(2) Transfer through two changes of CO_2-free water.
(3) Treat with 2 percent ammonium sulfide for 2 minutes.
(4) Rinse in distilled water.
(5) Mount with glycerin jelly.

RESULTS:
Adenosine triphosphatase activity appears brown to black.

METHOD FOR CALCIUM-ACTIVATED (ACTOMYOSIN) ATPase
(Guth and Samaha, 1970)

TISSUES:
Use fresh-frozen cryostat sections. Allow the sections to dry at room temperature for 30 minutes to 3 hours.

PREPARATION OF FIXATIVE SOLUTION:

Commercial formaldehyde, 40 percent	50 ml
Sodium cacodylate	31 gm
$CaCl_2$	10 gm
Sucrose	115 gm
Water, to make 1 liter	

PREPARATION OF TRIS BUFFER:

Tris	12.1 gm
$CaCl_2$, 0.18M	100 ml
Distilled water	850 ml

Adjust the pH to 7.8 with HCl (1N to 6N) and bring the volume to 1 liter with distilled water.

PREPARATION OF ALKALINE PREINCUBATION SOLUTION:

2-Amino-2-methyl-1-propanol, 1.5M	3.35 ml
$CaCl_2$, 0.18M	10.0 ml
Distilled water	35 ml

Adjust the pH to 10.4 with KOH (1N to 10N) and bring the volume to 50 ml with distilled water. Prepare immediately before use.

PREPARATION OF INCUBATION MEDIUM:

2-Amino-2-methyl-1-propanol, 1.5M	3.35 ml
$CaCl_2$, 0.18M	5.0 ml
KCl	185 mg
Disodium adenosine triphosphate	76 mg

Adjust the pH to 9.4 with 6N HCl and bring the volume to 50 ml with distilled water. Prepare immediately before use.

PREPARATION OF ALKALINE WASHING SOLUTION:

2-Amino-2-methyl-1-propanol, 1.5M	13.4 ml
Distilled water	160.0 ml

Adjust the pH to 9.4 with HCl (1N to 6N) and bring the volume to 200 ml with distilled water.

PREPARATION OF ACID PREINCUBATION SOLUTION:

$CaCl_2$, 0.18M	100 ml
Glacial acetic acid	3.0 ml
Distilled water	875 ml

Adjust the pH to 4.35 with KOH (1N to 5N) and bring the volume to 1 liter with distilled water.

PROCEDURE:

(1) Expose the sections to the fixative solution for 5 minutes at room temperature.
(2) Rinse in tris buffer for 1 minute with agitation and blot excess on filter paper.
(3) Place sections in alkaline preincubation solution for 15 minutes.
(4) Rinse in two changes of tris buffer for 1 minute each and drain the excess buffer.
(5) Incubate for 15 to 60 minutes at 37°C.
(6) Wash in three changes of a 1 percent solution of $CaCl_2$ for 30 seconds each.
(7) Place in a 2 percent solution of cobalt chloride solution for 3 minutes. (Make up fresh each time.)
(8) Wash in four changes of alkaline washing solution for 30 seconds each.
(9) Place in a 1 percent solution of ammonium sulfide solution for 3 minutes.
(10) Wash in running tap water for 3 to 5 minutes.
(11) Dehydrate through graded alcohols, clear in xylene, and mount with synthetic mounting medium.

RESULTS:

Areas containing actomyosin are yellowish brown to black. A heavy reaction should be seen in white (alpha) muscle fibers, while little reaction should be seen in red (beta) fibers. In order to demonstrate the red fibers preferentially, the alternative incubation procedure below should be used.

ALTERNATIVE INCUBATION PROCEDURE:
(1) Use frozen sections but do not expose them to fixative solution.
(2) Place the sections in acid preincubation solution for 5 to 30 minutes and drain the excess solution.
(3) Continue with the rest of the regular procedure, steps 4 through 11.

RESULTS:
A heavy reaction should be seen in the red (beta) muscle fibers; this reaction represents mitochondrial ATPase or ATPase of the sarcoplasmic reticulum (or both). Only a minimal reaction should be seen in the white (alpha) muscle fibers.

METHOD FOR SODIUM-POTASSIUM-ACTIVATED ATPase
(Guth and Albers, 1974)

TISSUES:
Guth and Albers recommend perfusion of experimental animals with a polyvinyl pyrrolidone solution. Following perfusion, the desired tissues are immediately removed and quick-frozen. Cryostat sections are used.

PREPARATION OF POLYVINYL PYRROLIDONE PERFUSION SOLUTION:
Prepare the following solution:

Polyvinyl pyrrolidone (M.W. 40,000)	12 percent
$CaCl_2$	$1.8mM$
Histidine	$100mM$

Adjust the pH to 7.25. Use 150 ml for animals weighing between 300 and 500 gm.

PREPARATION OF INCUBATION MEDIUM:
Prepare the following solution:

2-Amino-2-methyl-1-propanol	$70mM$
p-Nitrophenyl phosphate	$5mM$
KCl	$30mM$
$MgCl_2$	$5mM$
Dimethyl sulfoxide	25 percent v/v

Adjust the pH to 9.0 with HCl.

PROCEDURE:

(1) Allow the sections to dry on the slides for 15 to 20 minutes at room temperature.
(2) Incubate for 1 to 3 hours at 37°C.
(3) Place sections directly into a 2 percent solution of cobalt chloride for 5 minutes.
(4) Rinse briefly in distilled water.
(5) Wash in three changes of 2-amino-2-methyl-1-propanol buffer ($70mM$, pH 9.0) for 30 seconds each.
(6) Place in 2 percent ammonium sulfide solution for 3 minutes.
(7) Wash in running tap water for 15 minutes.
(8) Dehydrate through graded alcohols, clear in xylene, and mount with synthetic medium.

CONTROL:

The incubation medium is prepared in the same way as that described previously except that $1mM$ ouabain (final concentration) is added; this incubation medium is substituted in the incubation procedure.

RESULTS:

Sites of Na^+/K^+-activated ATPase are yellowish brown to black. The control sections should have no reaction product.

LEAD METHOD FOR GLUCOSE 6-PHOSPHATASE
(Wachstein and Meisel, 1956)

TISSUES:

Use fresh-frozen cryostat sections.

PREPARATION OF INCUBATION MEDIUM:

Glucose 6-phosphate, K salt, 0.125 percent	4.0 ml
Acetate buffer, 0.1M, pH 6.5	4.0 ml
Lead nitrate, 2 percent	0.6 ml
Distilled water	1.4 ml

All solutions should be made with CO_2-free water. The medium should be filtered just before use.

PROCEDURE:

(1) Incubate sections for 5 to 15 minutes at 37°C.
(2) Wash in distilled water.
(3) Treat sections with 0.5 percent ammonium sulfide for 1 minute.
(4) Wash in distilled water.
(5) Postfix in 10 percent formalin for 2 minutes.
(6) Mount in glycerin jelly.

CONTROLS:

For controls, substitute an incubation medium in which β-glycerophosphate is used instead of glucose 6-phosphate.

RESULTS:

Glucose 6-phosphatase activity is shown by deposits that are yellowish brown to black. Similar deposits in the control section indicate interfering acid phosphatase activity.

REFERENCES

Borgers, M. (1973).   The cytochemical application of new potent inhibitors of alkaline phosphatases. *J. Histochem. Cytochem.* 21:812–824.

Borgers, M., and Thoné, F. (1975).   The inhibition of alkaline phosphatase by L-*p*-bromotetramisole. *Histochemistry* 44:277–280.

Borgers, M., and Thoné, F. (1976).   Further characterization of phosphatase activities using non-specific substrates. *Histochem. J.* 8:301–317.

Burstone, M. S. (1958a). Histochemical comparison of naphthol AS-phosphates for the demonstration of phosphatases. *J. Natl. Cancer Inst.* 20:601–614.

Burstone, M. S. (1958b).   Histochemical demonstration of acid phosphatases with naphthol AS-phosphates. *J. Natl. Cancer Inst.* 21:523–539.

Burstone, M. S. (1958c).   The relationship between fixation and techniques for the histochemical localization of hydrolytic enzymes. *J. Histochem. Cytochem.* 6:322–339.

Burstone, M. S. (1961).   Histochemical demonstration of phosphatases in frozen sections with naphthol AS-phosphates. *J. Histochem. Cytochem.* 9:146–153.

Burstone, M. S. (1962).   *Enzyme Histochemistry and Its Application in the Study of Neoplasms.* Academic Press, New York.

Chiquoine, A. D. (1953).   The distribution of glucose-6-phosphatase in the liver and kidney of the mouse. *J. Histochem. Cytochem.* 1:429–435.

Chiquoine, A. D. (1955).   Further studies on the histochemistry of glucose-6-phosphatase. *J. Histochem. Cytochem.* 3:471–478.

Cox, R. P., and Griffin, M. S. (1965).   Pyrophosphatase activity of mammalian alkaline phosphatase. *Lancet* 2:1018–1019.

Davis, B. J., and Ornstein, L. (1959).   High resolution enzyme localization with a new diazo reagent, "hexazonium pararosaniline." *J. Histochem. Cytochem.* 7:297–298.

DeDuve, C., and Wattiaux, R. (1966). Functions of lysosomes. *Ann. Rev. Physiol.* 28:435–492.

Felicetti, D., and Rath, F.-W. (1975).   Zum Vorkommen und zur Isolierung einer durch Zink stark aktivierbaren sauren Phosphatase im Grosshirn der Ratte. *Acta Histochem.* 53:281–290.

Ganote, C. E., Rosenthal, A. S., Moses, H. L., and Tice, L. W. (1969).   Lead and phosphate as sources of artifact in nucleoside phosphatase histochemistry. *J. Histochem. Cytochem.* 17:641–650.

Gillis, J. M., and Page, S. G. (1967).   Localization of ATPase activity in striated muscle and probable sources of artifact. *J. Cell Sci.* 2:113–118.

Goff, R. A., and Milaire, J. (1974).   A comparative analysis of the histochemical dephosphorylation of ATP and AMP in cryostat and paraffin sections of early chick embryos. *Acta Histochem.* 51:220–254.

Goldberg, D. M. (1973).   5'-Nucleotidase: Recent advances in cell biology, methodology and clinical significance. *Digestion* 8:87–99.

Gomori, G. (1939).   Microtechnical demonstration of phosphatase in tissue sections. *Proc. Soc. Exp. Biol. Med.* 42:23–26.

Gomori, G. (1941).   Distribution of acid phosphatase in the tissues under normal and under pathologic conditions. *Arch. Pathol.* 32:189–199.

Gomori, G. (1950).   An improved histochemical technic for acid phosphatase. *Stain Technol.* 25:81–85.

Gomori, G. (1952).   *Microscopic Histochemistry: Principles and Practice.* University of Chicago Press, Chicago.

Goodlad, G. A. J., and Mills, G. T. (1957).   The acid phosphatases of rat liver. *Biochem. J.* 66:346–354.

Guth, L., and Albers, R. W. (1974).   Histochemical demonstration of (Na^+-K^+)-activated adenosine triphosphatase. *J. Histochem. Cytochem.* 22:320–326.

Guth, L., and Samaha, F. J. (1969).   Quantitative differences between actomyosin ATPase of slow and fast mammalian muscle. *Exp. Neurol.* 25:138–152.

Guth, L., and Samaha, F. J. (1970).   Procedure for the histochemical demonstration of Actomyosin ATPase. *Exp. Neurol.* 28:365–367.

Hardonk, M. J. (1968).   5'-Nucleotidase. I. Distribution of 5'-nucleotidase in tissues of rat and mouse. *Histochemistry* 12:1–17.

Hardonk, M. J., and Koudstaal, J. (1968). 5'-Nucleotidase. II. The significance of 5'-nucleotidase in the metabolism of nucleotides studied by histochemical and biochemical methods. *Histochemistry* 12:18–28.

King, E. J., Wood, E. J., and Delory, G. E. (1945).   Acid phosphatase of the red cells. *Biochem. J.* 39:xxiv–xxv.

Klaushofer, K., and Böck, P. (1974).   Studies on 5'-nucleotidase histochemistry. II. Differences in 5'-nucleotidase activity in stratified squamous epithelia and skin appendages of mouse, rat and guinea pig. *Histochemistry* 40:39–49.

Koenig, C. S., and Vial, J. D. (1973).   A critical study of the histochemical lead method for localization of Mg-ATPase at cell boundaries. *Histochem. J.* 5:503–518.

Kutscher, W., and Wolbergs, H. (1935).   Prostatic phosphatase. *Hoppe Seylers Z. Physiol. Chem.* 236:237–240.

Lin, C.-W., and Fishman, W. H. (1972).   Microsomal and lysosomal acid phosphatase isoenzymes of mouse kidney. Characterization and separation. *J. Histochem. Cytochem.* 20:487–498.

Manheimer, L. H., and Seligman, A. M. (1948).   Improvement in the method for the histochemical demonstration of alkaline phosphatase and its use in a study of normal and neoplastic tissues. *J. Natl. Cancer Inst.* 9:181–199.

Marchesi, V. T., and Palade, G. E. (1967).   The localization of Mg-Na-K-activated adenosine triphosphatase on red cell ghost membranes. *J. Cell Biol.* 35:385–404.

McClurkin, I. T. (1964).   A method for the cytochemical demonstration of sodium-activated adenosine triphosphatase. *J. Histochem. Cytochem.* 12:654–658.

Menten, M. L., Junge, J., and Green, M. H. (1944).   A coupling histochemical azo dye test for alkaline phosphatase in the kidney. *J. Biol. Chem.* 153:471–477.

Moss, D. W. (1969a).   Biochemical studies on phosphohydrolase isoenzymes. *Ann. N. Y. Acad. Sci.* 166:641–652.

Moss, D. W. (1969b).   The influence of metal ions on the orthophosphatase and inorganic pyrophosphatase activities of human alkaline phosphatase. *Biochem. J.* 112:699–701.

Moss, D. W., and Walli, A. K. (1969).   Intermediates in the hydrolysis of ATP by human alkaline phosphatase. *Biochim. Biophys. Acta* 191:476–477.

Nordlie, R. C. (1979).   Multifunctional glucose-6-phosphatase: Cellular biology. *Life Sci.* 24:2397–2404.

Novikoff, A. B. (1967).   Enzyme localizations with Wachstein-Meisel procedures: Real or artifact. *J. Histochem. Cytochem.* 15:353–354.

Otte, P. (1958).   Die Regenerationsunfähigkeit des Gelenkknorpels. *Z. Orthop.* 90:299–303.

Padykula, H. A., and Herman, E. (1955a).   Factors affecting the activity of adenosine triphosphatase and other phosphatases as measured by histochemical techniques. *J. Histochem. Cytochem.* 3:161–169.

Padykula, H. A., and Herman, E. (1955b).   The specificity of the histochemical method for adenosine triphosphatase. *J. Histochem. Cytochem.* 3:170–195.

Pearse, A. G. E. (1968). *Histochemistry: Theoretical and Applied* (3rd ed.), Vol. 1. Little, Brown, Boston.

Rath, F.-W., and Felicetti, D. (1975). Histochemische Darstellung einer durch Zink aktivierten tartratresistenten sauren Phosphatase in experimentell induzierten gliösen Mikrotumoren des Rattengrosshirns. *Acta Histochem.* 53:291–301.

Rosen, S. I. (1974). Histochemical clues to problems of glucose homeostasis. The glucogenic potential of peripheral tissues and glucose-6-phosphatase activity. *Acta Histochem.* 50:1–18.

Rosenbaum, R. M., and Rolon, C. I. (1962). Species variability and the substrate specificity of intercellular acid phosphatases: A comparison of the lead-salt and azo dye methods. *Histochemistry* 3:1–16.

Rosenthal, A. S., Moses, H. L., Beaver, D. L., and Schuffman, S. S. (1966). Lead ion and phosphatase histochemistry. I. Non enzymatic hydrolysis of nucleoside phosphates by lead ion. *J. Histochem. Cytochem.* 14:698–701.

Singh, I., and Tsang, K. Y. (1975). An in vitro production of bone specific alkaline phosphatase. *Exp. Cell Res.* 95:347–358.

Takada, K., Tamura, K., and Mori, M. (1968). Effect of Mg ion on histochemical demonstration of alkaline phosphatase in decalcified hard tissue. *Acta Histochem. Cytochem.* 1:37–42.

Takamatsu, H. (1939). Histochemische Untersuchungen der Phosphatase und deren Verteilung in verschiedenen Organen und Gewebe. *Trans. Soc. Pathol. Jap.* 29:492.

Tormey, J. McD. (1966). Significance of the histochemical demonstration of ATPase in epithelia noted for active transport. *Nature* 210:820–822.

Wachstein, M., and Meisel, E. (1956). On the histochemical demonstration of glucose-6-phosphatase. *J. Histochem. Cytochem.* 4:592.

Wachstein, M., and Meisel, E. (1957). Histochemistry of hepatic phosphatases at a physiologic pH. *Am. J. Clin. Pathol.* 27:13–23.

Wachstein, M., Meisel, E., and Niedzwiedz, A. (1960). Histochemical demonstration of mitochondral adenosine triphosphatase with the lead-adenosine triphosphate technique. *J. Histochem. Cytochem.* 8:387–388.

13

ESTERASES AND LIPASES

BIOCHEMISTRY OF ESTERASES

Esterases are a very diverse group of enzymes, and perhaps the most significant enzymatic property that they have in common is that their natural substrates are esters of carboxylic acids. The specificity of some esterases, which is relatively low, overlaps with that of peptidases and amidases. The normal role of the esterases in metabolism is poorly understood and probably represents a wide variety of functions.

Many esterases have low substrate specificities and are referred to as nonspecific esterases, while others exhibit fairly high substrate specificities or preferences and are named with more specific terms.

Nonspecific esterases have been divided into A-esterases and B-esterases (Aldridge, 1953a and 1953b) on the basis of their response to inhibitors. B-esterases are readily inhibited by diethyl-p-nitrophenyl phosphate (E600) and by diisopropyl fluorophosphate (DFP), whereas A-esterases are not readily inhibited by these organophosphates. A-esterases in fact hydrolyze organophosphates; they also hydrolyze acetates more readily than butyrates whereas the opposite is generally true for B-esterases.

C-esterase, characterized by Bergmann et al. (1957), is not inhibited by DFP and it does not hydrolyze DFP. Furthermore, C-esterase is activated by sulfhydryl reagents (particularly p-chloromercuribenzoate) whereas A- and B-esterases are both inhibited.

No strict distinction can be made between lipases and nonspecific esterases since both hydrolyze glycerol esters of fatty acids. Nonspecific esterases have a preference for esters of short-chain fatty acids whereas lipases prefer esters of long-chain fatty acids. There is no sharp cutoff point for either enzyme and there is some overlap of activity in terms of fatty acid chain length.

Cholinesterases are distinguished from other esterases by the fact that they hydrolyze esters of choline and by the fact that they are not inhibited by eserine. Two types of cholinesterases can be distinguished. One type has a high affinity for its natural substrate, acetylcholine, compared with its affinity for other esters; it has been called *true* or *acetyl*cholinesterase. The other type splits long-chain fatty acid esters of choline more rapidly than acetylcholine and is therefore called *pseudo*cholinesterase. Acetylcholinesterase is inhibited by high concentrations of acetylcholine. It is capable of hydrolyzing the unnatural substrate acetyl-β-methylcholine but not benzoylcholine or butyrylcholine. Pseudocholinesterase, on the other hand, can hydrolyze benzoylcholine and butyrylcholine but not acetyl-β-methylcholine. The use of these substrates has provided a method for distinguishing between the two enzymes.

In recent years it has been possible to separate many more esterases with newer methods. Holmes and Masters (1967), for instance, found 24 different esterases in various guinea pig tissues, not all of which fall into the classification categories described above. The esterases have continuously varying enzymatic properties, rendering the old classification system less meaningful.

Choudhury (1972) studied substrate preferences and inhibition of esterases separated on starch gel electrophoresis. Each esterase showed optimal hydrolytic activity toward an esterase of a particular carbon chain length. Furthermore, the carbon chain length optimum became progressively longer for the esterases farther from the cathode. In fact, those near the anode began to behave like lipases. The enzymes also showed progressively less sensitivity toward inhibition by organophosphatase inhibitors as the distance from the cathode increased. Choudhury therefore proposed that the multiplicity of esterases might be explained with a subunit concept, similar to the classic lactate dehydrogenase system. This kind of explanation has strong appeal, although solid evidence to support it is still lacking.

HISTOCHEMISTRY OF LIPASES

The natural substrates for lipases—fats and oils—are practically insoluble in water. This creates a fundamental problem in demonstrating lipases, because organic solvents are detrimental to enzyme activity. It is of course possible to prepare emulsions of fats and oils, but emulsions cannot readily penetrate into the tissues.

Gomori (1945) overcame the problem of substrate insolubility by using Tweens as substrates. Tweens are esters of long-chain fatty acids and polyalcohols, usually sorbitol, and are readily soluble in aqueous media. They are readily hydrolyzed by lipase, although they are certainly not the natural substrates. In Gomori's method (1945), the liberated fatty acids are trapped with calcium cations to form insoluble calcium soaps. Subsequently, the tissue sections are treated with lead and then with ammonium sulfide to form dark brown precipitates of lead sulfide.

Most workers have found Gomori's original method unsatisfactory, and it has been modified a number of times. Each modification usually improves one aspect of the technique, but generally the basic difficulties remain. If it is desirable to demonstrate lipase with this method, the modification of Bokdawala and George (1964) is recommended. Their version of the method, which visualizes the salts of calcium and fatty acids with alizarin red S, is claimed to be specific and sensitive.

A somewhat more reliable method has been developed by Abe et al. (1964) based on the hydrolysis of naphthol AS-nonanoate. Esterases could of course contribute to the hydrolysis of a lipid substrate and make interpretation difficult. Abe and co-workers therefore attempted to make the method more selective for lipase in two ways: by using taurocholate to activate lipase specifically and by using the nonanoate ester for which lipase has a high affinity. Caution should still be exercised in interpreting results because the contribution of esterases could be significant in cells of high esterase activity such as macrophages, mast cells, and pericytes.

HISTOCHEMISTRY OF NONSPECIFIC ESTERASES
Nonspecific esterases were first demonstrated with naphthyl acetate substrates. Nachlas and Seligman (1949) used both α- and β-naphthyl acetate but preferred the latter because it gave them better color characteristics. Gomori (1950) found that β-naphthol coupled very poorly with diazonium compounds, a finding that undermined confidence in the results of Nachlas and Seligman (1949). Gomori (1950) showed that α-naphthol coupled instantaneously with diazonium compounds, suggesting that α-naphthyl acetate is a more satisfactory esterase substrate. This method gives good results, especially when hexazotized pararosanilin is used as the coupler.

Substituted naphthol AS acetates have been used as substrates with excellent results (Pearse, 1954; Burstone, 1957a, 1957b; and Gössner, 1958). Because of the highly insoluble nature of the final reactions, very sharp localization of enzymes is possible. Burstone's (1962) method using naphthol AS-D-acetate is presented in the following methods section. Naphthol AS-D-acetate has the following structure:

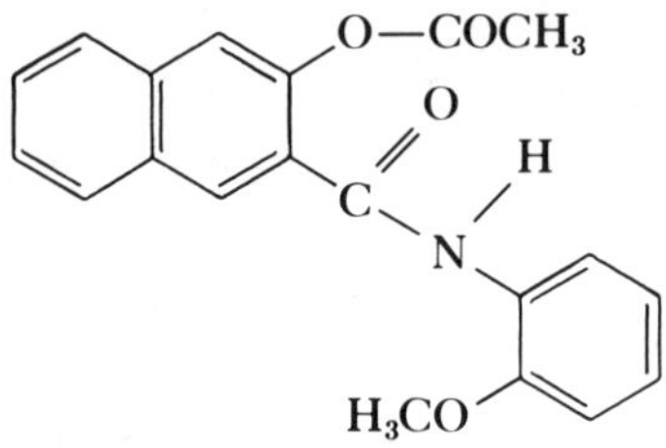

Barrnett and Seligman (1951) and Holt (1952) demonstrated esterase with a new principle. Indoxyl acetate or butyrate, which was used as the substrate, is readily hydrolyzed by most esterases, including cholinesterases. Esterase activity frees indoxyl, which is rapidly oxidized to an indigo pigment:

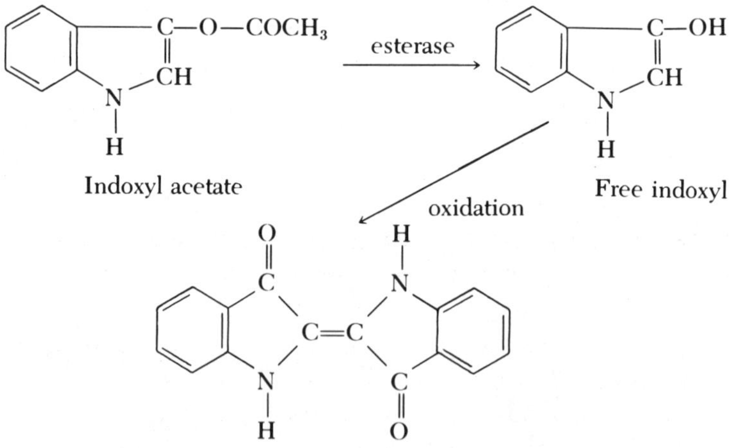

Indoxyl acetate   Free indoxyl

Indigo pigment

The final reaction product is stable and highly insoluble; it has a deep blue color. However, the indigo pigment was deposited in the form of large crystals and there was considerable evidence of diffusion. Pigment crystallization was worse at acid conditions than at alkaline conditions. Originally, the oxidizing agent was simply the dissolved atmospheric oxygen.

Holt (1956) and Holt and Sadler (1958) reported on the use of a number of newly synthesized halogen-substituted indoxyl acetates with the goal of improving the quality of the histochemical reaction products. They settled on the use of 5-bromo-4-chloroindoxyl acetate because it gave a completely nongranular pigment with good substantivity (Holt, 1958). Improvements were also realized by adding equimolar potassium ferricyanide and potassium ferrocyanide (Holt, 1956). The oxidation is effected by the ferricyanide while ferrocyanide prevents overoxidation (to noncolored compounds).

Hanker et al. (1972) developed a method for demonstrating esterases that was based on a novel reaction mechanism. They used 2-thiolacetoxybenzanilide (TAB) as the esterase substrate. The incubation medium also contained cupric sulfate and potassium ferricyanide. By a mechanism that is not entirely understood, cupric ferrocyanide (Hatchett's brown) is precipitated as an indirect result of esterase activity. In a postincubation step, the tissue sections are exposed to diaminobenzidine (DAB), and during this step Hatchett's brown catalyzes the polymerization of DAB (polymerized DAB is a deep brown pigment). Alternatively, the postincubation step can consist of treating the sections with thiocarbohydrazide and subsequently with osmium tetroxide. For light microscopy, the DAB step is preferable.

HISTOCHEMISTRY OF CHOLINESTERASES

Cholinesterases were first demonstrated histochemically by Koelle and Friedenwald (1949). Fresh-frozen sections were incubated in a medium containing acetylthiocholine (substrate) and copper glycinate (capture agent). The enzymatically liberated thiocholine formed white precipitates, presumably of copper mercaptide, at the sites of enzyme activity. The tissue sections were then treated with dilute ammonium sulfide to convert the precipitates to visible brown copper sulfide. This method results in considerable diffusion and needlelike crystallization of the reaction product. Koelle (1951) introduced several modifications to remedy the problems; however, the improvements were marginal, and numerous modifications have been introduced by other workers. This method is reliable enough that it has been very widely used for demonstrating cholinesterases. Koelle's method has now been largely replaced by newer methods that are more reliable and more convenient to use.

Karnovsky and Roots (1964) published a "direct coloring" method for cholinesterases. In this method, acetylthiocholine again serves as the substrate but the mechanism of forming the precipitate is entirely different. The incubation medium also contains cupric ions and ferricyanide. The cupric ions are loosely complexed with citrate to prevent the formation of copper ferricyanide. The liberated thiocholine serves as a catalyst to reduce the ferricyanide to ferrocyanide, which is then precipitated by the cupric ions. The resulting copper ferrocyanide precipitate (Hatchett's brown) has a light brown color. Thus the ammonium sulfide conversion step of Koelle's method, which can be troublesome, is eliminated. Also, formation of needlelike crystals is eliminated and the development of color can be monitored directly.

Hanker et al. (1973) modified the procedure of Karnovsky and Roots somewhat and added a step to intensify the color of the final reaction product. This step consisted either of bridging Hatchett's brown to osmium tetroxide with thiocarbohydrazide or of treating the sections with DAB. In the latter case, Hatchett's brown serves as a catalyst to form a deep brown polymer of DAB. With this amplification step, a shorter incubation period could be used because only a small amount of Hatchett's brown is sufficient.

TWEEN METHOD FOR LIPASE
(Bokdawala and George, 1964)

TISSUES:
Cut fresh-frozen sections directly into 6 percent neutral formalin and fix for 4 hours at 5°C.

INCUBATION MEDIUM:

Tween 85, 5 percent	2 ml
Sodium bicarbonate, $0.2M$	5 ml
$CaCl_2$, 10 percent	2 ml
Distilled water	40 ml

The medium may be stored in the refrigerator with a crystal of thymol but it must be filtered before use.

PROCEDURE:

(1) Mount sections on clean glass slides and allow them to dry.

(2) Apply a thin coat of 1 percent gelatin.

(3) Fix for 30 minutes in cold neutral formalin.

(4) Wash in running water for 30 minutes and rinse in distilled water.

(5) Place in 0.2M borate buffer containing 0.002M ethylenediaminotetraacetate (EDTA) for 1 to 15 minutes at 4°C.

(6) Wash in distilled water.

(7) Incubate for 16 hours at 37°C.

(8) Wash in distilled water.

(9) Treat with 1 percent alizarin red S, pH 6.3 to 6.8, for 1 minute.

(10) Rinse in distilled water.

(11) Mount with glycerin jelly.

CONTROLS:

Treat sections with boiling water for 10 minutes prior to incubation in order to inactivate the enzyme.

RESULTS:

Red-orange deposits indicate sites of lipase activity. Control sections should be completely negative.

NAPHTHOL METHOD FOR LIPASE
(Abe et al., 1964)

TISSUES:

Fix tissues with formol-calcium for 24 hours at 0° to 4°C. Rinse three times with physiological saline and place in gum-sucrose solution for a further 24 hours at 0° to 4°C. Use cryostat sections.

PREPARATION OF INCUBATION MEDIUM:

Tris buffer, 0.4M, pH 7.4	5.0 ml
Sodium taurocholate, 2.5 percent	1.0 ml
Distilled water	3.9 ml
Naphthol AS-nonanoate, 2 percent, in dimethylacetamide	0.1 ml
Fast blue BB	10 mg

Add the naphthol substrate to a mixture of the first three items slowly with stirring. Then dissolve the fast blue BB and filter.

PROCEDURE:
(1) Incubate free-floating sections at 37°C for 40 to 90 minutes.
(2) Wash in distilled water for 10 minutes.
(3) Mount on slides with 90 percent (w/v) polyvinyl pyrrolidone or another suitable aqueous mounting medium.

CONTROLS:
Prepare an incubation medium in which the sodium taurocholate is replaced with an equal amount of water. A parallel esterase procedure should also be performed.

RESULTS:
Deposits of blue dye indicate sites of lipase activity, provided that the parallel esterase procedure is negative at the same sites and that the omission of taurocholate results in diminished staining.

α-NAPHTHYL ACETATE METHOD FOR ESTERASE

TISSUES:
Fix tissue blocks in formol-calcium for 12 to 24 hours. Use frozen sections. Paraffin sections of cold acetone-fixed tissue blocks can also be used, but with a considerable loss of enzyme activity.

PREPARATION OF INCUBATION MEDIUM:
Dissolve 5 mg α-naphthyl acetate in 0.25 ml acetone and slowly add 10 ml 0.1M phosphate buffer, pH 7.4. Shake thoroughly to resolve some of the cloudiness. Add 20 mg fast blue B, shake, and filter. (Fast blue RR or fast red RC may be substituted.)

PROCEDURE:
(1) Incubate frozen sections for 2 to 20 minutes; paraffin sections should be incubated for a longer time.
(2) Rinse in distilled water.
(3) Counterstain nuclei with 1 percent methyl green in acetate buffer, pH 4.0, if desired.

(4) Wash well in tap water.

(5) Mount in glycerin jelly.

RESULTS:

Deposits of blue dye indicate the sites of esterase activity. Nuclei (if counterstained) will be green.

SUBSTITUTED NAPHTHOL METHOD FOR ESTERASES
(Burstone, 1962)

TISSUES:

Fix tissues in cold formol-calcium and use either frozen sections or paraffin sections.

INCUBATION MEDIUM:

Place 5.0 mg naphthol AS-D acetate in a 50-ml container. Add 0.5 ml dimethylformamide or acetone to dissolve the substrate. Now dilute with 25 ml distilled water and 25 ml $0.2M$ tris buffer, pH 7.1. Dissolve 20 to 40 ml fast garnet GBC or fast blue RR and filter into a Coplin jar. Eserine $10^{-5}M$ may be added to inhibit cholinesterase activity.

PROCEDURE:

(1) Deparaffinize sections and bring to water.

(2) Incubate for 5 minutes to 2 hours.

(3) Wash in running water.

(4) Mount with glycerin jelly.

RESULTS:

Esterase activity will be represented by an intense red dye.

INDIGOGENIC METHOD FOR ESTERASE
(Holt, 1958)

TISSUES:

Fix tissues in formol-calcium for 24 to 36 hours at 0° to 2°C. Blot gently and place directly in gum-sucrose solution for 24 hours at 0° to 2°C. Use free-floating cryostat sections.

SUBSTRATE STOCK SOLUTION:
Dissolve 1.5 mg of 5-bromo-4-chloroindoxyl acetate in 0.1 ml absolute ethanol.

OXIDANT STOCK SOLUTION:
Potassium ferricyanide   0.184 gm
Potassium ferrocyanide   0.165 gm
Distilled water   10 ml

Store in a refrigerator and discard after 1 week.

INCUBATION MEDIUM:
Tris buffer, 0.1M, pH 8.5   2.0 ml
Substrate stock solution   0.1 ml
Oxidant stock solution   1.0 ml
Calcium chloride, 1.0M 0.1 ml
Sodium chloride, 2.0M 5.0 ml

PROCEDURE:
(1) Incubate free-floating sections for 5 minutes to 1 hour at 37°C.
(2) Remove sections with glass rods and place into 30 percent ethanol containing 0.1 percent acetic acid.
(3) Transfer to distilled water to flatten out the sections.
(4) Place on slides and mount with glycerin jelly.

RESULTS:
Esterase activity is indicated by deposits of a bluish green indigo pigment.

HATCHETT'S BROWN METHOD FOR NONSPECIFIC ESTERASES
(Hanker et al., 1972)

TISSUES:
Hanker and associates fixed the tissues in a fixative containing equal volumes of (1) 20 percent formalin containing 2 percent calcium chloride (dihydrate) and (2) 0.2M acetate buffer, pH 5.6. Fix the tissue

blocks for 4 days at 0° to 4°C. Place the tissues in gum-sucrose solution for an additional 24 hours at 0° to 4°C. Use cryostat sections.

PREPARATION OF INCUBATION MEDIUM:
Dissolve 2.5 mg of 2-thiolacetoxybenzanilide in 0.1 ml acetone. Then add the following reagents in order, with constant stirring:

Sodium acetate, 0.06M	7.90 ml
Acetic acid, 0.1N	0.25 ml
Trisodium citrate, 0.1M	0.60 ml
Copper sulfate, 0.03M	1.25 ml
Potassium ferricyanide, 0.005M	1.25 ml

The last two reagents should be added drop by drop. Adjust the pH to 5.5 to 5.6 if necessary and filter immediately before use.

PROCEDURE:
(1) Incubate cryostat sections for 15 to 30 minutes at room temperature.
(2) Wash in distilled water three times for 5 minutes each.
(3) Immerse in DAB solution (5 mg 3,3'-diaminobenzidine tetrahydrochloride dissolved in 10 ml 0.05M acetate buffer, pH 5.6) for 20 minutes at room temperature.
(4) Wash in distilled water three times for 5 minutes each.
(5) Osmicate wet tissue sections for 15 minutes in a closed Coplin jar containing 1.5 ml of a 2 percent solution of osmium tetroxide. Heat the jar to 50° to 55°C in a water bath. At the end of the period, cool the jar on ice before opening. (This step must be done in a well-ventilated hood. The goal of osmication is the further intensification of the visibility of the DAB polymer. This step is optional.)
(6) Wash well with distilled water.
(7) Dehydrate through graded alcohols, clear with xylene, and mount with a synthetic resin.

RESULTS:
Sites of esterase activity are indicated by brown deposits of the DAB polymer (or black osmiates, if osmicated).

DIRECT COLORING METHOD FOR CHOLINESTERASE
(Karnovsky and Roots, 1964)

TISSUES:
Fix tissues in cold formol-calcium overnight and transfer to gum-sucrose solution for 24 hours or longer at 0° to 4°C. Use cryostat sections.

INCUBATION MEDIUM:
Dissolve 5 gm acetylthiocholine or butyrylthiocholine iodide in 6.5 ml $0.1M$ acetate or phosphate buffer, pH 6.0. Then add the following items in order while stirring:

Trisodium citrate, $0.1M$	0.5 ml
$CuSO_4$, $0.03M$	1.0 ml
Potassium ferricyanide, $0.005M$	1.0 ml
Distilled water	1.0 ml

These stock solutions can be stored in a refrigerator for several weeks.

PROCEDURE:
(1) Incubate sections at 37°C for up to 2 hours.
(2) Rinse in distilled water.
(3) Counterstain with hematoxylin if desired.
(4) Dehydrate in graded ethanols, clear in xylene, and mount with a synthetic resin.

CONTROLS:
Prepare an incubation medium in which 1.0 ml $0.001M$ eserine sulfate is added instead of the distilled water.

RESULTS:
Deposits of the brownish pigment (Hatchett's brown) are the sites of cholinesterase activity. Eserine in the controls will inhibit cholinesterases and allow only interfering nonspecific esterases to be demonstrated.

HATCHETT'S BROWN METHOD FOR CHOLINESTERASE
(Hanker et al., 1973)

TISSUES:
Fix tissues with formol-calcium for 24 hours at 0° to 4°C. Use cryostat sections.

PREPARATION OF INCUBATION MEDIUM:
Dissolve 6 mg of acetylthiocholine or butyrylthiocholine iodine in 7.9 ml of 0.06N sodium acctate and add the following reagents in order with continuous stirring:

Acetic acid, 0.1N	0.25 ml
Trisodium citrate, 0.1M	0.60 ml
Copper sulfate, 0.03M (added drop by drop)	1.25 ml
Distilled water (or inhibitor solution)	0.50 ml
Potassium ferricyanide, 0.005M (added drop by drop)	1.25 ml
Lead nitrate, 0.015M (optional)	0.10 ml

Adjust the pH to 5.5 to 5.6 if necessary, cool to 0° to 4°C, and filter immediately before use.

In order to demonstrate acetylcholinesterase, use acetylthiocholine iodide as the substrate and replace the distilled water with a 0.024mM solution of tetraisopropylpyrophosphoramide (isoOMPA), an inhibitor of pseudocholinesterase.

In order to demonstrate pseudocholinesterase, use butyrylthiocholine iodide as the substrate and replace the distilled water with a 1.2mM solution of 1,5-bis(4-allyldimethylammoniumphenyl)pentan-3-one dibromide (BW 284 C 51), an inhibitor of acetylcholinesterase.

PROCEDURE:
(1) Incubate cryostat sections for 5 to 30 minutes at 0° to 4°C.
(2) Rinse quickly in cold distilled water.
(3) Immerse in DAB solution (5 mg 3,3'-diaminobenzidine·4HCl dissolved in 10 ml 0.05M acetate buffer, pH 5.6) for 20 minutes at 0° to 4°C.

(4) Rinse in four changes of distilled water for 5 minutes each at 0° to 4°C.

(5) Osmicate wet tissues for 15 minutes in a closed Coplin jar containing 1.5 ml of a 2 percent osmium tetroxide solution. Heat the jar to 50° to 55°C with a water bath. At the end of the period, cool the jar on ice before opening. (This step must be done in a well ventilated hood. The goal of osmication is the further intensification of the visibility of the DAB polymer. This step is optional.)

(6) Wash well with distilled water.

(7) Dehydrate through graded alcohols, clear with xylene, and mount with a synthetic resin.

RESULTS:

Sites of cholinesterase activity are indicated by brown deposits of the DAB polymer (or black osmiates, if osmicated).

REFERENCES

Abe, M., Kramer, S. P., and Seligman, A. M. (1964). The histochemical demonstration of pancreatic-like lipase and comparison with the distribution of esterase. *J. Histochem. Cytochem.* 12:364–383.

Aldridge, W. N. (1953a). Serum esterases. I. Two types of esterase (A and B) hydrolysing p-nitrophenyl acetate, proprionate and butyrate, and a method for their determination. *Biochem. J.* 53:110–117.

Aldridge, W. N. (1953b). Serum esterases. II. An enzyme hydrolysing diethyl p-nitrophenyl phosphate (E600) and its identity with the A-esterase of mammalian sera. *Biochem. J.* 53:117–124.

Barnett, R. J., and Seligman, A. M. (1951). Histochemical demonstration of esterases by production of indigo. *Science* 114:579–582.

Bergmann, F., Segal, R., and Rimon, S. (1957). A new type of esterase in hog-kidney extract. *Biochem. J.* 67:481–486.

Bokdawala, F. D., and George, J. C. (1964). Histochemical demonstration of muscle lipase. *J. Histochem. Cytochem.* 12:768–771.

Burstone, M. S. (1957a). The cytochemical localization of esterase. *J. Natl. Cancer Inst.* 18:167–173.

Burstone, M. S. (1957b). Esterase activity of developing bones and teeth. *Arch. Pathol.* 63:164–167.

Burstone, M. S. (1962). *Enzyme Histochemistry and Its Application to Neoplasms.* Academic Press, New York.

Choudhury, S. R. (1972). The nature of nonspecific esterases: A subunit concept. *J. Histochem. Cytochem.* 20:507–517.

Gomori, G. (1945). The microchemical demonstration of sites of lipase activity. *Proc. Soc. Exp. Biol. Med.* 58:362–364.

Gomori, G. (1950). Sources of error in enzymatic histochemistry. *J. Lab. Clin. Med.* 35:802–809.

Gössner, W. (1958). Histochemischer Nachweis hydrolytischer Enzyme mit Hilfe der Azofarbstoffmethode. Untersuchungen zur Methodik und vergleichenden Histotopik der Esterasen und Phosphatasen bei Wirbeltieren. *Histochemistry* 1:48–96.

Hanker, J. S., Yates, P. E., Clapp, D. H., and Anderson, W. A. (1972). New methods for the demonstration of lysosomal hydrolases by the formation of osmium blacks. *Histochemistry* 30:204–214.

Hanker, J. S., Thornburg, L. P., Yates, P. E., and Moore, H. G. (1973). The demonstration of cholinesterases by the formation of osmium blacks at the sites of Hatchett's Brown. *Histochemistry* 37:223–242.

Holmes, R. S., and Masters, C. J. (1967). The developmental multiplicity and isoenzyme status of cavian esterases. *Biochem. Biophys. Acta* 132:379–399.

Holt, S. J. (1952). A new principle for the histochemical localization of hydrolytic enzymes. *Nature* 169:271–273.

Holt, S. J. (1956). The value of fundamental studies of staining reactions in enzyme histochemistry, with reference to indoxyl methods for esterases. *J. Histochem. Cytochem.* 4:541–552.

Holt, S. J. (1958). Indigogenic Staining Methods for Esterases. In J. F. Danielli (Ed.), *General Cytochemical Methods,* Vol. 1. Academic Press, New York.

Holt, S. J., and Sadler, P. W. (1958). Studies in enzyme cytochemistry. II. Synthesis of indigogenic substrates for esterases. *Proc. R. Soc. Lond. [Biol.]* 148:481–494.

Karnovsky, M. J., and Roots, L. (1964). A "direct coloring" thiocholine method for cholinesterases. *J. Histochem. Cytochem.* 12:219–221.

Koelle, G. B. (1951). The elimination of enzymatic diffusion artifacts in the histochemical localization of cholinesterases and a survey of their cellular distributions. *J. Pharmacol. Exp. Ther.* 103:153–171.

Koelle, G. B., and Friedenwald, J. S. (1949). A histochemical method for localizing cholinesterase activity. *Proc. Soc. Exp. Biol. Med.* 70:617–622.

Nachlas, M. M., and Seligman, A. M. (1949). The histochemical demonstration of esterase. *J. Natl. Cancer Inst.* 9:415–425.

Pearse, A. G. E. (1954). Azo-dye methods in histochemistry. *Int. Rev. Cytol.* 3:329–358.

14

MISCELLANEOUS HYDROLASES

The miscellaneous hydrolases that are of histochemical interest can be divided into two categories on the basis of the method of histochemical demonstration. The first group of hydrolases can be demonstrated with the more common method of offering a low-molecular-weight substrate to the enzyme and trapping one of the reaction products. The second group will hydrolyze only high-molecular-weight substances and can be demonstrated only with substrate film methods. Peptidases, glycosidases, and sulfatases are included in the first category, while proteases, nucleases, hyaluronidase, and amylase are included in the second group because they can be demonstrated with the substrate film techniques.

ENZYMES DEMONSTRATED WITH CONVENTIONAL SUBSTRATE METHODS

Peptidases

Extracts from many different tissues show enzymatic activity toward peptides of many different amino acids. It has been questioned whether this activity represents one enzyme with varying specificities toward the peptides, or whether multiple enzymes with limited specificities are involved. It was shown many years ago that there is an intestinal enzyme that specifically hydrolyzes leucine aminopeptides (Linderstrøm-Lang, 1929). Later an enzyme was isolated and purified that had high activity toward leucine aminopeptides and very little activity toward peptides of other amino acids (Smith and Bergmann, 1944; Spackman et al., 1955); this firmly established the existence of leucine aminopeptidase. Hirschman and Hirschman (1977) tried unsuccessfully to separate the various peptidase activities from tissue extracts with disc electrophoresis, DEAE cellulose chromatography, and gel filtration, suggesting that only one enzyme was involved. However, they obtained different profiles of activity when the extracts were tested against a number of different amino acid naphthylamides, confirming earlier evidence of Nachlas et al. (1962). This seems to implicate multiple enzymes with limited specificities and similar physiochemical properties.

Aminopeptidase activity was first demonstrated histochemically by Gomori (1954), who used alanyl-β-naphthylamide as the substrate. Burstone and Folk (1956) published a method for demonstrating leucine aminopeptidase in freeze-dried sections using the leucyl derivative of β-naphthylamide, and Nachlas et al. (1957) used the same substrate to demonstrate the enzyme in frozen sections. With this substrate, a strongly colored final reaction product was obtained. However, the rate of coupling was too slow, resulting in diffusion of the final reaction product.

Nachlas et al. (1960) reported on the synthesis of a new compound, leucyl-4-methoxy-β-naphthylamide, which gave greatly improved results. The rate of coupling of fast blue B with 4-methoxy-β-naphthylamine is said to be 40 times greater than with the unsubstituted β-naphthylamine. Nachlas et al. (1960) also included a cupric ion chelation step, which makes the pigment resistant to organic reagents. This remains the method of choice for demonstrating leucine aminopeptidase.

β-Glucuronidase

The enzyme β-glucuronidase is widely distributed in animal tissues, and the preputial gland of the female rat is an especially rich source. The known sources, along with the amounts of activity, are tabulated by Levvy and Marsh (1959). The enzyme catalyzes the hydrolysis of glucuronate-containing oligosaccharides, which are degradation products of hyaluronic acid and chondroitin sulfate. The enzyme therefore probably plays an important role in the metabolism of glycosaminoglycans. Glucuronides of steroids and bilirubin are also hydrolyzed by β-glucuronidase. This reaction is reversible, as are many other hydrolase reactions, and it was once thought that the physiological significance of the enzyme was in the synthesis of glucuronide conjugates (Fishman, 1940).

Early studies showed that the pH optimum of β-glucuronidase is in the acid range. It was therefore natural to relegate the enzyme to the group of lysosomal enzymes. However, cell fractionation studies of DeDuve et al. (1955) showed that a considerable portion of the enzyme activity also occurs in the soluble fraction. Strong histochemical evidence presented by Fishman et al. (1967) showed that β-glucuronidase activity occurred in lysosomes as well as in the endoplasmic reticulum. DeLellis and Fishman (1968) further showed that the nonlysosomal fraction of the thyroid gland was under hormonal control.

Biochemical studies of Mills et al. (1953) revealed the presence of three electrophoretically separable fractions of β-glucuronidase in bovine spleen. The three fractions had pH optima of 3.4, 4.5, and 5.2. The fraction with a pH optimum of 3.4 was completely inhibited by a $10^{-4}M$ concentration of D-saccharate. Less potent inhibitors were D-mucate, oxalate, citrate, and galacturonate, in decreasing order of potency. The fractions with pH 4.5 and 5.2 optima were successively less sensitive to D-saccharate inhibition.

Friedenwald and Becker (1948) published two methods for demonstrating this enzyme. In the first method, a naphthyl glucuronide compound was used as a substrate, and no coupling was required. In the second method, 8-hydroxyquinoline glucuronide served as the substrate in the presence of ferric ions. It was thought that the liberated 8-hydroxyquinoline was precipitated by the ferric ions. The ferric precipitate was then converted to Prussian blue by the usual method.

Neither of the methods evolved by Friedenwald and Becker was considered satisfactory by subsequent workers. This led to the development of many modifications as well as some completely new methods. One of the more successful methods was published by Rath and Otto (1966), who also used 8-hydroxyquinoline glucuronide and simultaneously coupled the liberated 8-hydroxyquinoline with fast black K salt. Bulmer (1966) also obtained good results with a similar method (using different coupling salts), but he felt that even better results could be obtained with naphthol compounds.

The method of Hayashi et al. (1964) is commonly used for demonstrating β-glucuronidase; this method uses naphthol AS-BI-β-glucuronide as the substrate and hexazotized pararosanilin as the simultaneous coupling agent. Enzyme localization, which is more precise than with the older methods, is said to be lysosomal in distribution.

The indigogenic methods of Pearson et al. (1967) also give excellent results. There are two main advantages of the indigogenic methods (1) no coupling agent is required (the freed indoxyl compound is auto-oxidizable) and (2) the final reaction product has high substantivity.

N-Acetyl-β-Glucosaminidase

Another important lysosomal glycosidase is N-acetyl-β-glycosaminidase, which is widely distributed in animal tissues and is especially prevalent in tissues with a high turnover of polysaccharides such as salivary glands and mucous epithelia.

Biochemical studies of Robinson and Stirling (1968) showed two distinct forms of the enzyme in the spleen. Form A is an acidic protein and is the less stable of the two. Form B is a basic protein and is fairly stable. Cell fractionation showed that both forms occurred in the lysosomal fraction but form B also occurred in the supernatant. Price and Dance (1967) also found a lysosomal and a microsomal fraction in the rat kidney, although the former was more abundant.

Purified N-acetyl-β-glucosaminidase also has N-acetyl-β-galactosaminidase activity, although the V_{max} (see p. 196) for the galactose epimer is considerably lower than that for the glucose epimer (Verpoorte, 1972); the lowered V_{max} holds for forms A and B of the enzyme. Histochemical evidence has also shown that the same enzyme hydrolyzes both substrates (Gossrau, 1972) but is less effective toward the galactose epimer.

N-Acetyl-β-glucosaminidase was demonstrated by Pugh and Walker (1961) with a method using α-naphthyl-N-acetyl-β-glucosaminide as the substrate and fast garnet GBC as the coupling salt. The results were not very satisfactory, primarily because the substantivity of α-naphthol was too low. Later Hayashi (1965) published two methods for demonstrating the enzyme, both using naphthol AS-BI-N-acetyl-β-glucosaminide as the substrate. In the first method, hexazotized pararosanilin was used as the coupler, and in the second method fast garnet GBC was used. The second method is somewhat more convenient, but the first gives superior results and is the method of choice.

β-Galactosidase

β-Galactosidase activity, which is found in many animal tissues, is the result of several types of β-galactosidases. Most of the activity is due to a lysosomal enzyme with a pH optimum somewhere between 3 and 4.5, depending on the organ and species of origin. The enzyme hydrolyzes various β-galactosides but is completely ineffective toward β-glucosides, and it is inhibited by p-chloromercuribenzoate and galactonolactone but not by gluconolactone. Its physiological function is thought to involve the catabolism of certain polysaccharides and glycolipids. β-Galactosidase is generally absent in Hurler's syndrome (Ho and O'Brien, 1969) and in generalized gangliosidosis (van Hoof and Hers, 1968).

Another β-galactosidase, found in intestine and kidney, has considerably different properties (Kraml et al., 1969; Swaminathan and Radhakrishnan, 1969; Semenza et al., 1965). It is not a lysosomal enzyme but is firmly bound to the brush borders of intestinal epithelium and the proximal tubules of the kidney, and it has a pH optimum of 5.5 to 6.0. This form of β-galactosidase is inhibited by gluconolactone as well as by galactonolactone; it is sometimes called lactase because its physiological function, at least in the intestine, is to split ingested lactose.

During tissue fractionation studies of rat liver, Price and Dance (1967) found a soluble β-galactosidase in addition to the one associated with lysosomes; it has a pH optimum of 5.5 to 6.5. They presented evidence that this enzyme also has β-glucosidase activity.

Several serious but common problems were encountered in developing histochemical methods for demonstrating β-galactosidase. First, the enzyme was readily inhibited by diazonium salts, which practically precluded a simultaneous coupling procedure. Second, the enzyme or at least a fraction of the enzyme, is very soluble and can therefore readily diffuse out of the tissue section during incubation. A postcoupling method was developed by Cohen et al. (1952) and an improved version was later developed in the same laboratory (Rutenburg et al., 1958). Diffusion of the enzyme remained a problem, particularly in sites of high enzyme concentration.

An indigogenic method was developed by Pearson et al. (1963) using 5-bromo-4-chloroindoxyl-β-D-galactoside as the substrate. The enzymatically liberated 5-bromo-4-chloroindoxyl readily oxidizes to form a very fine granular precipitate that is very substantive. Lojda (1970) evaluated and improved the method by adding an oxidation catalyst. Lake (1974) increased the sensitivity of the method by simply adding 0.1M NaCl. According to Lake, the color yield was increased because NaCl activates the enzyme and decreases its solubility.

Sulfatase

Throughout the plant and animal kingdoms there exist a number of sulfatases, including arylsulfatase, alkylsulfatase, cholinesulfatase, glycosulfatase, steroid sulfatase, and myrosulfatase. In mammalian tissue, arylsulfatase is the only one of real interest histochemically. The other either are absent from mammalian tissues or are not demonstrable histochemically.

Arylsulfatases are divided into type I and type II. Type I is a microsomal enzyme with a relatively high pH optimum, and it is readily inhibited by cyanide ions but is not affected by sulfate or phosphate ions. The enzyme is a lipoprotein and is extremely insoluble, which has hindered biochemical characterization. Its physiological function is probably to hydrolyze steroid sulfates.

Type II arylsulfatase is further subdivided into arylsulfatase A and arylsulfatase B. Both are lysosomal enzymes and have pH optima of 5.0 and 5.5, and both are inhibited by sulfate and phosphate ions but unaffected by cyanide ions.

Arylsulfatases have recently been reviewed with special reference to genetic storage diseases (Farooqui and Mandel, 1977; Roy, 1976).

One approach to the demonstration of arylsulfatases was based on the capture of the sulfate ions liberated by the enzyme. Goldfischer (1965) used lead for this purpose with a Gomori-type medium, whereas Hopsu-Havu et al. (1967) used barium. Neither of these methods is favored today.

Rutenburg et al. (1952) published a postcoupling method for demonstrating sulfatase activity using 6-benzoyl-2-naphthyl sulfate as the substrate. It was not considered a reliable method by subsequent workers.

A simultaneous coupling method was attempted by Gössner (1958) using naphthol AS-sulfate as the substrate and fast red TR as the coupling salt. This method resulted in a scant and coarse-grained precipitate and was considered unsatisfactory by Gössner. Wooshmann and Hartrodt (1964) developed this method further by using substituted naphthol AS compounds, with greatly improved results. Wächtler and Pearse (1966) used this method for studying sulfatases in amphibian pituitary glands. However, they replaced fast red TR with hexazotized pararosanilin, which according to Pearse (1972) gave "greatly improved results." Pearse indicated that, unfortunately, similar results could not be obtained in mammalian tissues. The failure was blamed on the impermeability of the mammalian lysosomal membranes to the sulfates of substituted naphthol AS compounds. My own experience has shown that naphthol AS-BI-sulfate is readily precipitated by fast red TR, hexazotized pararosanilin, and other stabilized coupling compounds, and that it is difficult to obtain positive results.

Hanker et al. (1975) developed a new method for demonstrating arylsulfatase with an entirely different chemical basis of precipitate formation. The incubation medium contains p-nitrocatechol sulfate as the enzyme substrate, as well as cupric sulfate and potassium ferricyanide. The enzyme liberates p-nitrocatechol, which in turn causes the precipitation of cupric ferrocyanide (Hatchett's brown). The color of the reaction product is fairly weak, but is can be intensified by treating the sections with diaminobenzidine. Hatchett's brown serves as a catalyst for polymerizing diaminobenzidine, forming a highly visible deep brown pigment. This is probably the most satisfactory method for demonstrating arylsulfatase.

ENZYMES DEMONSTRATED WITH
SUBSTRATE FILM METHODS

The methods involving the use of films of high-molecular-weight enzyme substrates are useful for studying enzymes that are unable to hydrolyze low-molecular-weight substrates. These techniques require the enzymes to be somewhat soluble (in the more conventional methods, this is undesirable). Tissue sections, usually fresh-frozen sections, are placed on a thin film of the substrate. The enzyme will diffuse from the section into the film, hydrolyzing the substrate at the loci of the enzyme. Following an appropriate period of incubation, the substrate film is stained to visualize the remaining substrate. Unstained "holes" correspond to the sites where the enzyme is located in the tissue. An adjacent tissue section is also stained (or the same section if it can be salvaged) and the morphology is related to the substrate film. The RNase method (Fig. 14-1) exemplifies these methods.

Precise localization is not possible with these techniques, although they allow the somewhat crude histochemical localization of enzymes that would otherwise be impossible.

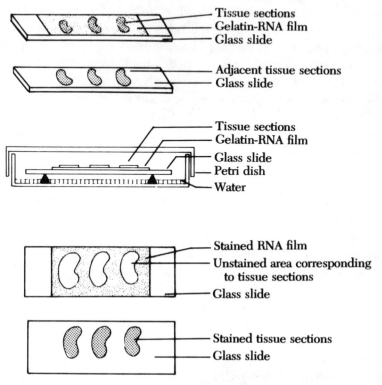

Figure 14-1. The substrate film method for demonstrating RNase. (Redrawn and modified from Roger Daoust, Modified procedure for the histochemical localization of ribonuclease activity by the substrate film method, *J. Histochem. Cytochem.* 14:254, 1966. © 1966 The Williams & Wilkins Co., Baltimore.)

The pioneering work with substrate film techniques was carried out by Daoust (1957), who was interested in demonstrating nucleases. The methods developed subsequently have generally been based on the Daoust methods, although a few novel approaches have been developed as well.

Deoxyribonuclease (DNase)

DNase was the first enzyme to be demonstrated with the substrate film technique by a method developed by Daoust (1957). Preparation of a suitable film of substrate is of crucial importance. Daoust incorporated 0.2 percent DNA into a 5 percent solution of gelatin and spread a thin film of it over a clean glass slide. The film was fixed in formalin to stabilize the gelatin. Daoust and Amano (1961) published a somewhat refined and improved version of the method. Daoust (1968) reviewed the method, pointing out artifacts that may occur and pitfalls to watch for.

Ribonuclease (RNase)

Daoust and Amano (1960) presented a method for demonstrating RNase that is very similar to the method for DNase. All the steps of the method were carefully reevaluated and a revised method (Daoust, 1966) was recommended that was more convenient to carry out and gave better resolution. The Daoust (1968) paper should be consulted by those who wish to use the method.

Amylase

Szemplinska et al. (1962) developed a method for demonstrating tissue amylase. The substrate film was prepared by incorporating starch into a gelatin film, and the tissue sections were first spread on top of a pure gelatin film and then inverted on top of a starch-gelatin film. Following an appropriate incubation period, the starch-gelatin film was stained with iodine.

A second method was published by Tremblay (1963) independently and evidently without knowledge of the first method. In this method, a film of starch only (no gelatin) was prepared, and fresh-frozen sections were placed directly on the substrate film. Following incubation, the starch was visualized by the periodic acid–Schiff (PAS) method. Since then a number of additional methods have been published for the detection of amylase. Smith and Frommer (1973) evaluated all the methods and variations for amylase and suggested a method that incorporates all the best features of each of the former methods. This is the method of choice and is given in the following methods section.

Hyaluronidase

McCombs and White (1968) modified the Daoust method so that tissue hyaluronidase could be demonstrated. The substrate film was prepared by incorporating hyaluronic acid into a gelatin film. Frozen sections of bovine testis, known to contain high concentrations of hyaluronidase, were used to test the system. Following the incubation, the substrate films were stained with toluidin blue without removing the tissue sections. The undigested hyaluronic acid in the film stained metachromatically (purple) but areas corresponding to sites of hyaluronidase were not stained. Toluidin blue also stained the cell nuclei, which helped to resolve the tissue morphology. McCombs and White (1968) were able to localize hyaluronidase to the area of secondary spermatocytes, spermatids, and sperm. The method seems to have given them good results, although it has not been used extensively by other investigators.

Proteases

An interesting variant of the original Daoust methods is a method using blackened photographic film (with a gelatin base) to demonstrate tissue proteases. The method was developed by Adams et al. (1958) and refined by Adams and Tuqan (1961). Glass photographic plates were prepared by exposing them to daylight and developing them in the usual way, thus producing gelatin films thoroughly impregnated with fine silver grains. Formalin-fixed frozen sections were placed in contact with the film. Areas in the film corresponding to sites of tissue proteases were cleared of their silver granules by the digestion of the gelatin matrix.

A number of variations on this method have been developed using film impregnated with India ink (Owers and Blandau, 1971), film impregnated with an azo dye (Cunningham, 1967), and reversible daylight color film (Fratello, 1968). The third of these methods gave particularly good resolution because of the thinness of the film. An added advantage was that the shade of the film color indicated the intensity of protease activity.

Denker (1974) published a method using gelatin films that had not been impregnated with dyes or particles in order to avoid possible enzyme inhibition by the impregnating substance. Postincubation staining was done with toluidin blue. (Any protein-specific stain could have been used.) The method appears to give excellent results, although the preparation of the substrate films is somewhat complicated. Both the Adams and Tuqan (1961) and Denker (1974) methods are given in the following methods section.

METHOD FOR LEUCINE AMINOPEPTIDASE
(Nachlas et al., 1960)

TISSUES:
Use fresh-frozen cryostat sections. Tissues with high enzyme activity may be fixed with formol-calcium at 4°C.

INCUBATION MEDIUM:

L-leucyl-4-methoxynaphthylamide·HCl (4 mg/ml)	1.0 ml
Acetate buffer, 0.1M, pH 6.5	5.0 ml
NaCl, 0.85 percent	3.5 ml
Potassium cyanide, 0.02M, 1.3 mg/ml	0.5 ml
Fast blue B	5.0 mg

At 4°C this solution is stable for several months.

PROCEDURE:
(1) Incubate sections for 5 minutes to 2 hours.
(2) Rinse in saline for 2 to 3 minutes.
(3) Place in 0.1M cupric sulfate for 2 or 3 minutes.
(4) Dehydrate through graded alcohols, clear in xylene, and mount with synthetic mountant.

RESULTS:
Deposits of the dye indicate the sites of enzyme activity.

METHOD FOR β-GLUCURONIDASE
(Hayashi et al., 1964)

TISSUES:
Fix tissues in formol-calcium for 24 hours at 4°C. Transfer the tissues to gum-sucrose solution for an additional 24 hours or longer at 4°C. Use cryostat sections.

PREPARATION OF STOCK SOLUTION A:
Dissolve 1 gm pararosanilin·HCl in 25 ml 2N HCl by stirring and warming. Cool, filter, and store at room temperature.

PREPARATION OF STOCK SOLUTION B:
Prepare a 4 percent solution of sodium nitrite in distilled water. Store in refrigerator and prepare fresh each week.

PREPARATION OF HEXAZOTIZED PARAROSANILIN:
Combine 0.3 ml stock solution A with 0.3 ml stock solution B and allow it to stand for 2 or 3 minutes.

PREPARATION OF STOCK SUBSTRATE SOLUTION:
Prepare a $0.5mM$ stock substrate solution by dissolving 28 mg naphthol AS-BI-glucuronide (free acid) in 1.2 ml $0.05M$ sodium bicarbonate (420 mg $NaHCO_3$ in 100 ml distilled water). Add $0.2N$ acetate buffer, pH 5, to make 100 ml. This solution may be stored at room temperature for several weeks.

PREPARATION OF INCUBATION MEDIUM:
To 0.6 ml of the hexazotized pararosanilin solution, add 10 ml of the stock substrate solution. Adjust the pH to 5.2 with $1N$ NaOH using a pH meter. Add distilled water to make a final volume of 20 ml. Filter through a No. 1 Whatman filter paper. The mixture should be clear and slightly yellow.

PROCEDURE:
(1) Incubate free-floating frozen sections at 37°C for 5 to 30 minutes.
(2) Rinse well in distilled water.
(3) Mount on slide.
(4) Counterstain with 1 percent methyl green in $0.1M$ veronal acetate buffer, pH 4.0, if desired.
(5) Dehydrate in graded alcohols, clear in xylene, and mount with a synthetic resin.

RESULTS:
Deposits of bright red dye indicate sites of enzyme activity. Nuclei (if counterstained) are green.

INDIGOGENIC METHOD FOR β-GALACTOSIDASE
(Lake, 1974)

TISSUES:
Use fresh-frozen cryostat sections.

PREPARATION OF INCUBATION MEDIUM:
Dissolve 3 mg of 5-bromo-4-chloroindoxyl-β-D-galactoside in two drops of methoxyethanol. Add the following components:

Phosphate-citrate (McIlvaine's) buffer, pH 4.0	7.0 ml
Potassium ferrocyanide (50mM)	0.5 ml
Potassium ferricyanide (50mM)	0.5 ml
Sodium chloride (final concentration, 0.1M)	47.5 mg

PROCEDURE:
(1) Incubate sections for up to 17 hours at 37°C.
(2) Wash in tap water.
(3) Counterstain in 1 percent neutral red for 3 minutes (this step is optional).
(4) Dehydrate through graded alcohols, clear in xylene, and mount with a synthetic resin.

RESULTS:
Deposits of the blue indigo pigment indicate areas of β-galactosidase activity and nuclei (if counterstained) are red.

METHOD FOR N-ACETYL-β-GLUCOSAMINIDASE
(Hayashi, 1965)

TISSUES:
Fix tissues in formol-calcium for 24 hours at 4°C. Transfer the tissues to gum-sucrose solution for additional 24 hours or longer at 4°C. Use cryostat sections.

PREPARATION OF STOCK SOLUTION A:
Dissolve 1 gm pararosanilin HCl in 25 ml 2N HCl by stirring and warming. Cool, filter, and store at room temperature.

PREPARATION OF STOCK SOLUTION B:
Prepare a 4 percent solution of sodium nitrite in distilled water. Store in a refrigerator, and prepare fresh each week.

PREPARATION OF HEXAZOTIZED PARAROSANILIN:
Combine 0.3 ml stock solution A with 0.3 ml stock solution B and allow it to stand for 2 or 3 minutes.

PREPARATION OF INCUBATION MEDIUM:
Dissolve 3 mg naphthol AS-BI-N-acetyl-β-glucosaminide in 0.5 ml ethylene glycol monomethyl ether. Add 5.0 ml 0.1M citrate buffer, pH 5.2, and immediately add 0.6 ml hexazotized pararosanilin. Adjust the pH to 5.2 with 1N NaOH and bring the volume to 10 ml with distilled water. Filter the medium immediately before incubation.

PROCEDURE:
(1) Treat free-floating cryostat sections with ethanol at 0°C for 5 minutes.
(2) Rinse briefly with distilled water.
(3) Incubate at 37°C for 15 to 30 minutes.
(4) Rinse in 2 changes of distilled water.
(5) Place sections in 30 percent ethanol briefly and then float them on water.
(6) Mount on gelatinized slides, blot, and place in a Coplin jar containing a 1-cm layer of full-strength formalin for 5 minutes.
(7) Rinse in distilled water.
(8) Counterstain in 1 percent methyl green in 0.1M veronal acetate buffer, pH 4.0 (this step is optional).
(9) Dehydrate in graded alcohols, clear in xylene, and mount with a synthetic medium.

RESULTS:
Bright red deposits of dye indicate the sites of enzyme activity. If counterstained, nuclei are green.

HATCHETT'S BROWN METHOD FOR ARYLSULFATASES
(Hanker et al., 1975)

TISSUES:

Fix tissues in neutral buffered formalin for 24 hours and store in gum-sucrose solution for 3 days at 0° to 4°C. Use cryostat sections.

PREPARATION OF INCUBATION MEDIUM:

Dissolve 20 mg 4-nitro-1,2-benzenediol mono(hydrogen sulfate) (nitrocatechol sulfate) in 7.9 ml of 0.06N sodium acetate. Then add the following reagents in order with continuous stirring:

Acetic acid, 0.1N	0.25 ml
Trisodium citrate, 0.1M	0.60 ml
Copper sulfate, 0.03M	1.25 ml
Potassium ferricyanide, 0.005M	1.25 ml

The last two reagents should be added drop by drop. Adjust the pH to 5.5 to 5.6 and filter immediately before use.

PROCEDURE:

(1) Incubate cryostat sections for 15 to 90 minutes at room temperature.
(2) Rinse well in distilled water.
(3) Immerse in diaminobenzidine solution (5 mg 3,3′-diaminobenzidine tetrahydrochloride in 10 ml 0.05M acetate buffer, pH 5.6) for 20 minutes at 0° to 4°C.
(4) Wash in four changes of distilled water for 5 minutes each.
(5) Osmicate wet tissue sections for 15 minutes in a closed Coplin jar containing 1.5 ml of a 2 percent solution of osmium tetroxide. Heat the jar to 50° to 55°C in a water bath. At the end of the period, cool the jar on ice before opening. (This step must be done in a well ventilated hood. The goal of osmication is the further intensification of the visibility of the diaminobenzidine polymer. This step is optional.)
(6) Wash well with distilled water.
(7) Dehydrate through graded alcohols, clear with xylene, and mount with a synthetic resin.

RESULTS:

Sites of arylsulfatase activity are indicated by brown deposits of the diaminobenzidine polymer (or black osmiates, if osmicated).

SUBSTRATE FILM METHOD FOR DNase
(Daoust and Amano, 1961)

PREPARATION OF GELATIN-DNA FILMS:

Prepare the gelatin-DNA mixture by combining equal volumes of a 5 percent solution of gelatin and a 7 percent solution of DNA. Place two or three drops of the warmed mixture on a glass slide and spread rapidly with the tip of the pipette over an area about 2.5 by 4 cm. Allow the excess to drain by holding the slide vertically on a piece of filter paper. Wipe the solution off the lower end of the slide with the filter paper and allow the slide to dry at room temperature in a vertical position. Fix the film overnight in neutral 50 percent formalin (20 percent formaldehyde) at room temperature. Wash in three successive baths of distilled water for 5 minutes each. Allow the film to dry.

PREPARATION OF GELATIN-GLYCEROL SLIDES:

Prepare a mixture containing 7 percent gelatin and 40 percent glycerol (final concentrations). Place 0.5 ml of this liquefied mixture on a slide and spread with the tip of the pipette over an area 2.5 by 4 cm. Leave the slide at room temperature for 2 to 3 minutes to allow it to gel.

MOUNTING OF TISSUE SECTIONS:

Cryostat sections are cut at 15 μ and placed on the gelatin-glycerol slide. The time that the slide is exposed to the cryostat chamber should be minimized to prevent the film from freezing. Place the slide with the tissue section on a level surface at 40°C until the gelatin-glycerol liquefies and the section spreads on its surface. Transfer to a level surface at room temperature to allow it to gel (about 5 to 10 minutes).

EXPOSURE OF TISSUE SECTION TO FILM SUBSTRATE:

Place a drop of the gelatin-glycerol mixture on the slide containing the gelatin-DNA film. Put the two slides face to face so that the tissue section is in contact with the substrate film. Place them on a table at room temperature for 5 to 40 minutes. The tissue section should be above the gelatin-DNA film.

STAINING OF GELATIN-DNA FILM:
After exposure, the substrate film is stained by dipping the slide into a 0.2 percent solution of toluidin blue for 5 minutes. Wash the film with two baths of distilled water for 5 minutes each. Dry at room temperature and mount with Canada balsam.

TREATMENT OF GELATIN-GLYCEROL SLIDE WITH TISSUE SECTION:
After exposure, dip the slide into a 1:3 mixture of acetic acid and alcohol. (This prevents staining of the gelatin.) Stain with a 0.1 percent solution of toluidin blue for 1 to 2 minutes. Wash in distilled water to remove excess stain and dry at room temperature. Mount with Canada balsam.

RESULTS:
The substrate film will be stained except at sites corresponding to tissue DNase.

CONTROLS:
Films made with 2.5 or 5 percent gelatin (omitting DNA) should not stain with toluidin blue.

Fixed gelatin-DNA films exposed to a solution of DNase for 8 to 24 hours should be negative when stained with toluidin blue. (Prepare the DNase solution by dissolving 0.1 mg crystalline DNase for each milliliter of phosphate-citrate buffer, pH 6.5.)

To control for possible spontaneous loss of DNA from the film, a slide with the film can be stored in water or phosphate-citrate buffer for 24 hours, followed by toluidin blue staining.

Heat-inactivated sections (90°C for 10 minutes) should have no effect on the substrate film.

SUBSTRATE FILM METHOD FOR RNase
(Daoust, 1966)

PREPARATION OF SUBSTRATE FILM:
Make a 10 percent solution of sodium ribonucleate. Chill the solution to 5°C and add 2 volumes of absolute alcohol, also chilled to 5°C.

Collect the precipitated RNA by centrifugation, dry in vacuo, and grind in a mortar and pestle. Make a 5 percent solution of gelatin (commercial sheet form) by gently heating it in an oven to 60°C. Combine equal volumes of a 10 percent solution of purified RNA and the 5 percent gelatin solution. Filter it at 60°C. Place 3 to 5 drops of the gelatin-RNA mixture on a glass slide and spread rapidly with the tip of the pipette over a 2.5 by 5 cm area, and allow the excess to drain by holding the slide vertically on a piece of filter paper. Wipe off the solution covering the lower end of the slide with filter paper. Allow the film to dry at room temperature in a vertical position. Fix the film overnight at 5°C with 50 percent formalin (20 percent formaldehyde) neutralized with an excess of calcium carbonate. Wash films in three baths of distilled water for 5 minutes each and allow them to dry at room temperature.

EXPOSURE OF TISSUE SECTIONS TO SUBSTRATE FILM:
Cut cryostat sections at 10 to 15 μ. The slide with the substrate film is introduced into the cryostat chamber just long enough to pick up the tissue section. Adjacent cryostat sections are picked up by a clean glass slide. Warm the substrate film slide with the section by touching the bottom of it with the palm of the hand. Sections will thaw and adhere to the film. The slide is then placed in a Petri dish above a layer of water to saturate the atmosphere with water vapor. Leave (incubate) for 1 minute to 2 hours at room temperature. Following incubation, the tissue sections can be floated off the film with water. Again place the film in 50 percent formalin for 1 hour to stop enzyme action. Wash the film in three baths of distilled water, 5 minutes each, and allow it to dry.

PREPARATION OF SLIDES WITH ADJACENT SECTIONS:
Slides with adjacent sections are removed from the cryostat chamber and warmed with the palm of the hand to promote thawing and adhesion to the slide. Dry for 3 to 5 minutes. Fix in Carnoy's fluid for 1 hour. Wash in three baths of distilled water for 5 minutes each. Stain the sections with 0.1 percent toluidin blue for 2 minutes. Wash in distilled water, dry, and mount.

STAINING OF SUBSTRATE FILM:
Stain the slides with the substrate film in 0.2 percent toluidin blue for 5 minutes. Wash in 3 baths of distilled water for 5 minutes each. Dry at room temperature. Mount with Canada balsam.

RESULTS:
The substrate film will be stained except at sites corresponding to tissue RNase.

SUBSTRATE-FILM METHOD FOR AMYLASE
(Smith and Frommer, 1973)

TISSUES:
Use fresh-frozen cryostat sections, 8 μ thick.

PREPARATION OF SUBSTRATE FILM:
Dissolve 5 gm hydrolyzed starch in 100 ml borate buffer (0.02M boric acid and 0.01M sodium hydroxide). In order to demonstrate weak activities of amylase, it may be necessary to reduce the starch concentration to 3 percent. Heat the suspension with continuous stirring; temperatures approaching 100°C will be required. Thoroughly degas the solution. Allow the starch solution to reach 70°C and the slides to reach 22°C. Dip the slides into the starch solution for 1 minute. Set the slides on end at 22°C. The slides may be used within 2 to 4 hours, and they remain usable for 7 to 10 days.

PROCEDURE:
(1) Place sections on starch substrate film slide and incubate in a moist chamber for up to 45 minutes.
(2) Fix starch film and tissue section for 1 hour in a mixture of methanol, acetic acid, and distilled water (in a ratio of 5:1:5).
(3) Rinse in running tap water for 1 minute.
(4) Stain in Lugol's iodine solution for 1 minute *or* stain with the periodic acid–Schiff technique (see p. 104).
(5) Wipe starch film off unused side of the slide.

RESULTS:

Areas on the starch film corresponding to sites of amylase in the tissue section will be unstained.

SUBSTRATE FILM METHOD FOR HYALURONIDASE
(McCombs and White, 1968)

TISSUES:

Use fresh-frozen cryostat sections 10 μ thick.

PREPARATION OF SUBSTRATE FILM:

Dissolve 5 mg of hyaluronic acid, sodium salt, in 0.8 ml distilled water at room temperature. Add the hyaluronate solution to 0.4 ml of a 12 percent aqueous gelatin solution, maintained at 80°C in a water bath. Agitate the mixture for 1 to 2 minutes to ensure thorough mixing. While the solution is still warm, place one drop on the center of each slide. (Acid-cleaned slides should be used.) The drop of hyaluronate-gelatin solution will spread to a diameter of about 0.5 to 1.0 cm. Allow the slides to dry in a horizontal position for 24 hours in a dust-free place. Fix the slides in full-strength formalin (37 percent formaldehyde) for 1 hour. Wash thoroughly for 1 hour in tap water, rinse in distilled water, and allow them to dry for several hours before using. Slides may be stored indefinitely in a dust-free slide box at 20° to 25°C.

PREPARATION OF COATING MIXTURE:

Acidify 60 ml of distilled water to pH 4.0 with dilute HCl, add 40 ml glycerol, and then suspend 7 gm gelatin in that solution. Heat the mixture in a boiling water bath until the gelatin is dissolved.

INCUBATION PROCEDURE:

Place the frozen sections onto the slides, making sure that the tissue does not entirely cover each substrate film spot. Warm the coating solution just enough to liquefy it. Apply sufficient coating solution to cover all of the tissue section and the substrate spot. An incubation time of 5 minutes at room temperature is adequate for bovine testis sections. At the end of the incubation period, remove the coating solution by agitating the section very gently for 30 to 60 seconds in a large staining dish full of lukewarm water. All of the coating solution must be removed or staining will not be adequate.

FIXATION, STAINING, AND MOUNTING:
Fix the slides for 10 minutes in full-strength formalin (37 percent formaldehyde) and rinse in distilled water (several dips). Stain for 3 minutes in 0.05 percent toluidin blue in phosphate buffer, pH 4.0. Rinse with several dips in two changes of distilled water. Destain uniformly with five 1-second dips in absolute ethanol. Immerse immediately in xylene for several minutes and mount with a synthetic resin.

RESULTS:
The substrate film will be stained blue except at sites corresponding to tissue hyaluronidase.

GELATIN SILVER FILM SUBSTRATE METHOD FOR PROTEASE
(Adams and Tuqan, 1961)

TISSUES:
Fix tissues in 10 percent formalin-saline at 4°C for 24 to 72 hours. Use cryostat sections.

PREPARATION OF SUBSTRATE FILM:
Expose panchromatic photographic plates to full daylight for 10 to 15 minutes. Develop and fix the plates in the usual way and dry. Cut the plate into the size of conventional microscope slides with the aid of a glass-cutter.

PROCEDURE:
(1) Place sections on the substrate film and allow it to become just dry.
(2) Moisten the section and an area surrounding it with 0.15M phosphate buffer, pH 7.6, with 0.15M acetate buffer, pH 5.0, or with another buffer, depending on the pH optimum of the protease of interest. It is important that no free fluid should be left over the section. The buffer should be entirely absorbed in the film.

(3) Incubate the preparation in a moisture-saturated atmosphere at 37°C for 30 to 60 minutes. Inspect the preparation every 5 to 10 minutes and apply additional buffer if necessary.

(4) Following incubation, dry without washing, dehydrate in alcohol, clear in xylene, and mount with a synthetic mounting medium. Dehydration and clearing should be done within 2 minutes to avoid a finely grazed appearance of the gelatin. Do not mount with glycerin jelly, since this will dissolve the gelatin silver film.

RESULTS:
Clear areas in the blackened film indicate sites of protease activity.

SUBSTRATE FILM METHOD FOR PROTEASE
(Denker, 1974)

TISSUES:
Use fresh-frozen cryostat sections.

PREPARATION OF SUBSTRATE FILM:
Use slides cleaned in acid alcohol. Spread 0.10 to 0.15 ml of an aqueous gelatin solution (50°C) on an area 2.5 by 5.0 cm. Allow it to solidify in a horizontal position and let it dry completely at room temperature (22° to 23°C). Store overnight in a wet chamber (a covered Coplin jar with distilled water covering the bottom) at 4°C. Transfer directly from the refrigerator to an alcoholic formalin fixative (10 ml commercial formalin plus 90 ml 50 percent ethanol) and fix at room temperature for 1 hour. Wash for 20 minutes in running tap water (not more than 20°C), rinse in distilled water and air-dry at room temperature. Swell the gelatin in 7 percent ammonium hydroxide at room temperature until it detaches from the slide (about 3 minutes). It is helpful to use a Coplin jar and to adjust the level of the ammonium hydroxide solution so that a small portion of the gelatin film stays out of the solution and remains attached to the slide. Gently remove the slide (with the gelatin film attached at one end) and allow it to air-dry at room temperature. Store slides with films in covered staining jars at −25°C.

PROCEDURE:

(1) Remove substrate films from the freezer. Keep the jars covered until they have equilibrated with room temperature.

(2) Apply frozen sections to the film and let dry for 1 minute at room temperature.

(3) Incubate in a moisture chamber (Petri dish) containing a small amount of distilled water or buffer of choice and seal with petrolatum. Incubate for up to 4 hours.

(4) Stop the reaction by fixing with 30 percent formalin (30 ml commercial formalin and 70 ml distilled water) for 15 minutes at 4°C.

(5) Wash in running tap water for 3 minutes and dip in distilled water.

(6) Stain in 0.2 percent toluidin blue in borate buffer, pH 10.0, for 30 minutes at 4°C. Alternatively, stain with any specific protein stain.

(7) Wash briefly in distilled water (or in dilute ammonium hydroxide if too much stain is lost with water).

(8) Air-dry the slides.

(9) Dehydrate starting with 80 percent ethanol, clear in xylene, and mount with a synthetic resin.

RESULTS:

Sites of enzyme activity are indicated by unstained areas ("holes") in the stained gelatin film.

REFERENCES

Adams, C. W. M., Fernand, V. S. V., and Schnieden, H. (1958). Histochemistry of a condition resembling Kwashiorkor produced in rodents by a low protein-high carbohydrate diet (cassava). *Br. J. Exp. Pathol.* 39:393–404.

Adams, C. W. M., and Tuqan, N. A. (1961). The histochemical demonstration of protease by a gelatin-silver film substrate. *J. Histochem. Cytochem.* 9:469–472.

Bulmer, D. (1966). Simultaneous coupling azo dye methods for β-glucuronidase. *J. R. Microsc. Soc.* 86:385–395.

Burstone, M. S., and Folk, J. E. (1956). Histochemical demonstration of aminopeptidase. *J. Histochem. Cytochem.* 4:217–226.

Cohen, R. B., Tsou, K.-C., Rutenburg, S. H., and Seligman, A. M. (1952). The colorimetric estimation and histochemical demonstration of β-D-galactosidase. *J. Biol. Chem.* 195:239–249.

Cunningham, L. (1967). Histochemical observations of the enzymatic hydrolysis of gelatin films. *J. Histochem. Cytochem.* 15:292–298.

Daoust, R. (1957). Localization of deoxyribonuclease in tissue sections: A new approach to the histochemistry of enzymes. *Exp. Cell Res.* 12:203–211.

Daoust, R. (1966). Modified procedure for the histochemical localization of ribonuclease activity by the substrate film method. *J. Histochem. Cytochem.* 14:254–259.

Daoust, R. (1968). The localization of enzyme activities by substrate film methods—evaluation and perspectives. *J. Histochem. Cytochem.* 16: 540–545.

Daoust, R., and Amano, H. (1960). The localization of ribonuclease activity in tissue sections. *J. Histochem. Cytochem.* 8:131–134.

Daoust, R., and Amano, H. (1961). Improved DNA film method for localizing DNAase activity in tissue sections. *Exp. Cell Res.* 24:559–564.

DeDuve, C., Pressman, B. C., Gianetto, R., Wattiaux, R., and Appelmans, F. (1955). Tissue fractionation studies. VI. Intracellular distribution patterns of enzymes in rat liver tissue. *Biochem. J.* 60:604–617.

DeLellis, R. A., and Fishman, W. H. (1968). The dual localization of β-glucuronidase in rat thyroid. *Histochemistry* 13:4–6.

Denker, H. W. (1974). Protease substrate film test. *Histochemistry* 38:331–338.

Farooqui, A. A., and Mandel, P. (1977). The clinical aspects of arylsulfatases. *Clin. Chem. Acta* 74:93–100.

Fishman, W. H. (1940). Studies on β-glucuronidase. III. The increase in β-glucuronidase activity of mammalian tissues induced by feeding glucuronidogenic substances. *J. Biol. Chem.* 136:229–236.

Fishman, W. H., Goldman, S. S., and DeLellis, R. (1967). Dual localization of β-glucuronidase in endoplasmic reticulum and in lysosomes. *Nature* 213:457–460.

Fratello, B. (1968). Enhanced interpretation of tissue protease activity by use of photographic color film as a substrate. *Stain Technol.* 43:125–128.

Friedenwald, J. S., and Becker, B. (1948). The histochemical localization of glucuronidase. *J. Cell Comp. Physiol.* 31:303–309.

Goldfischer, S. (1965). The cytochemical demonstration of lysosomal aryl sulfatase activity by light and electron microscopy. *J. Histochem. Cytochem.* 13:520–523.

Gomori, G. (1954). Chromogenic substrates for aminopeptidases. *Proc. Soc. Exp. Biol. Med.* 87:559–561.

Gössner, W. (1958). Histochemischer Nachweis hydrolytischer Enzyme mit Hilfe der Azofarbstoffmethode. Untersuchungen zur Methodik und vergleichenden Histotopik der Esterasen und Phosphatasen bei Wirbeltieren. *Histochemistry* 1:48–96.

Gossrau, R. (1972). On the histochemical demonstration of N-acetyl-β-galactosaminidase. *Histochemistry* 29:315–324.

Hanker, J. S., Thornburg, L. P., Yates, P. E., and Romanovicz, D. K. (1975). The demonstration of arylsulfatases with 4-nitro-1,2-benzenediol mono(hydrogen sulfate) by the formation of osmium *blacks* at the sites of copper capture. *Histochemistry* 41:207–225.

Hayashi, M. (1965). Histochemical demonstration of N-acetyl-β-glucosaminidase employing naphthol AS-BI-N-acetyl-β-glucosaminide as substrate. *J. Histochem. Cytochem.* 13:355–360.

Hayashi, M., Nakajima, Y., and Fishman, W. H. (1964). The cytologic demonstration of β-glucuronidase employing naphthol AS-BI-glucuronide and hexazonium pararosanilin: A preliminary report. *J. Histochem. Cytochem.* 12:293–297.

Hirschman, A., and Hirschman, M. (1977).   Naphthylamindase (arylaminopeptidase) activities in rat cartilages and bone. *Enzyme* 22:2–12.

Ho, M. W., and O'Brien, J. S. (1969).   Hurler's syndrome: Deficiency of a specific beta-galactosidase isoenzyme. *Science* 165:611–613.

Hopsu-Havu, V. K., Arstila, A. U., Helminen, H. J., Kalimo, H. O., and Glenner, G. G. (1967).   Improvements in the method for the electron microscopic localization of arylsulphatase activity. *Histochemistry* 8:54–64.

Kraml, J., Koldovsky, O., Heringova, A., Jirsova, V., Kacl, K., Ledvina, M., and Pelichova, H. (1969).   Characteristics of β-galactosidases in the mucosa of the small intestine of infant rats: Physicochemical properties. *Biochem. J.* 114:621–627.

Lake, B. D. (1974).   An improved method for the detection of β-galactosidase activity and its application to G_{M1}-gangliosidosis and mucopolysaccharidosis. *Histochem. J.* 6:211–218.

Levvy, G. A., and Marsh, C. A. (1959).   Preparation and properties of β-glucuronidase. *Adv. Carbohyd. Chem.* 14:381–428.

Linderstrøm-Lang, K. (1929).   Über Darmerepsin. *Hoppe Seylers Z. Physiol. Chem.* 182:151–174.

Lojda, Z. (1970).   Indigogenic methods for glycosidases: II. An improved method for β-D-galactosidase and its application to localization studies of the enzymes in the intestine and in other tissues. *Histochemistry* 23:266–288.

McCombs, H. L., and White, H. J. (1968).   Histochemistry of hyaluronidase: Tissue localization of hyaluronidase in the testis by a new substrate film technic. *Am. J. Clin. Pathol.* 49:68–74.

Mills, G. T., Paul, J., and Smith, E. E. B. (1953).   Studies on β-glucuronidase. 2. The preparation and properties of three ox-spleen β-glucuronidase fractions. *Biochem. J.* 53:232–245.

Nachlas, M. M., Crawford, D. T., and Seligman, A. M. (1957).   The histochemical demonstration of leucine aminopeptidase. *J. Histochem. Cytochem.* 5:264–278.

Nachlas, M. M., Monis, B., Rosenblatt, D., and Seligman, A. M. (1960).   Improvement in the histochemical localization of leucine aminopeptidase with a new substrate, L-leucyl-4-methoxy-2-naphthylamide. *J. Biophys. Biochem. Cytol.* 7:261–264.

Nachlas, M. M., Goldstein, T. P., and Seligman, A. M. (1962).   An evaluation of aminopeptidase specificity with seven chromogenic substrates. *Arch. Biochem. Biophys.* 97:223–231.

Owers, N. O., and Blandau, R. J. (1971).   Proteolytic Activity of the Rat and Guinea Pig Blastocyst *in Vitro*. In R. J. Blandau (Ed.), *The Biology of the Blastocyst*. The University of Chicago Press, Chicago.

Pearse, A. G. E. (1972).   *Histochemistry: Theoretical and Applied* (3rd ed.), Vol. 2. Little, Brown, Boston.

Pearson, B., Wolf, P. L., and Vazquez, J. (1963).   A comparative study of a series of new indolyl compounds to localize β-galactosidase in tissue. *Lab. Invest.* 12:1249–1259.

Pearson, B., Standen, A. C., and Esterly, J. R. (1967).   Histochemical β-glucuronidase distribution in mammalian tissue as detected by 5-bromo-4-chloroindol-3-yl-β-D-glucopyruroniside. *Lab. Invest.* 17:217–224.

Price, R. G., and Dance, N. (1967). The cellular distribution of some rat-kidney glycosidases. *Biochem. J.* 105:877–883.

Pugh, D., and Walker, P. G. (1961). The localization of N-acetyl-β-glucos-aminidase in tissues. *J. Histochem. Cytochem.* 9:242–250.

Rath, F. W., and Otto, L. (1966). Zur histochemischer Darstellung der β-Glucuronidase. *Acta Histochem.* 25:355–362.

Robinson, D., and Stirling, J. L. (1968). N-acetyl-β-glucosaminidase in human spleen. *Biochem. J.* 107:321–327.

Roy, A. B. (1976). Sulphatases, lysosomes and disease. *Aust. J. Exp. Biol. Med. Sci.* 54:111–135.

Rutenburg, A. M., Cohen, R. B., and Seligman, A. M. (1952). Histochemical demonstration of aryl sulfatase. *Science* 116:539–543.

Rutenburg, A. M., Rutenburg, S. H., Monis, B., Teague, R., and Seligman, A. M. (1958). Histochemical demonstration of β-D-galactosidase in the rat. *J. Histochem. Cytochem.* 6:122–129.

Semenza, G., Auricchio, S., and Rubino, A. (1965). Multiplicity of human intestinal disaccharidases. I. Chromatographic separation of maltases and of two lactoses. *Biochim. Biophys. Acta* 96:487–497.

Smith, E. L., and Bergmann, M. (1944). The peptidases of intestinal mucosa. *J. Biol. Chem.* 153:627–651.

Smith, R. J., and Frommer, J. (1973). The starch substrate film method for the localization of amylase activity. *J. Histochem. Cytochem.* 21:189–190.

Spackman, D. H., Smith, E. L., and Brown, D. M. (1955). Leucine aminopeptidase. IV. Isolation and properties of the enzyme from swine kidney. *J. Biol. Chem.* 212:255–269.

Swaminathan, N., and Radhakrishnan, A. N. (1969). Purification and properties of lactase from monkey kidney. *Biochim. Biophys. Acta* 191:322–328.

Szemplinska, H., Sierakowska, H., and Shugar, D. (1962). Histochemical localization of hyaluronidase and amylase by the film-substrate technique. *Acta Biochim. Pol.* 9:239–244. (In English.)

Tremblay, G. (1963). The localization of amylase activity in tissue sections by a starch film method. *J. Histochem. Cytochem.* 11:202–206.

van Hoof, F., and Hers, H. G. (1968). The abnormalities of lysosomal en-zymes in mucopolysaccharidoses. *Eur. J. Biochem.* 7:34–44.

Verpoorte, J. A. (1972). Purification of two β-N-acetyl-D-glucosaminidases from beef spleen. *J. Biol. Chem.* 247:4787–4793.

Wähtler, K., and Pearse, A. G. E. (1966). The histochemical demonstration of five lysosomal enzymes in the pars distalis of the amphibian pituitary. *Z. Zellforsch.* 69:326–333.

Wooshmann, H., and Hartrodt, W. (1964). Der Nachweis einer phos-phatempfindlichen Sulfatase mit Naphthol-AS-Sulfaten. *Histochemistry* 4:336–344.

15

OXIDOREDUCTASES

NATURE OF OXIDOREDUCTASE REACTIONS

Oxidoreductases are a large group of enzymes that catalyze the oxidation of organic molecules. This process generally involves the removal of hydrogen atoms, or *dehydrogenation,* of the substrate. What happens to the removed hydrogens is of great importance, because they are not simply liberated as hydrogen gas; rather, the hydrogen atoms are transferred to another molecule, a receptor molecule. As the substrate is oxidized, the receptor molecule is simultaneously reduced. Thus, the overall chemical reaction is an oxidation-reduction reaction:

$$SH_2 + A \rightarrow S + AH_2$$

where S is the substrate and A is the acceptor.

Oxidases are oxidoreductases that can utilize molecular oxygen directly as the hydrogen acceptor and water is formed:

$$2SH_2 + O_2 \rightarrow 2S + 2H_2O$$

Dehydrogenases, however, must transfer the hydrogen atoms to another organic molecule, generally thought of as the coenzyme of the dehydrogenase. From the coenzyme, the hydrogen is passed on from one acceptor to another in a highly ordered sequence, called the electron transport system (Fig. 15-1). Finally, at the end of the system, oxygen combines with two of the hydrogen atoms to form water. Actually, it is thought that only the electron is actively transported, while the proton simply "floats along."

Biological oxidation reactions are tremendously important because the hydrogen atoms that are removed from the substrate represent potential energy, and can be compared with water behind a dam. Just as water behind a dam can be harnessed with a water wheel or a turbine, so the potential energy of electrons can be harnessed. In the cell the electron transport system and the oxidative phosphorylation system can be compared with the waterwheel or the turbine. As the electron is passed sequentially down the electron transport system, it gradually yields its energy, which is converted to adenosine triphos-

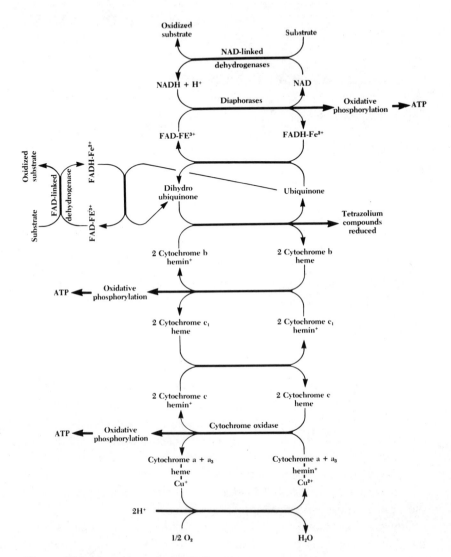

Figure 15-1. The electron transport system.

phate (ATP) by oxidative phosphorylation. Obviously, only the reaction of the dehydrogenases results in useful energy. In the case of oxidases, the potential energy is wasted.

HISTOCHEMISTRY OF DEHYDROGENASES

The development of dehydrogenase histochemistry is strongly linked to the development of tetrazolium compounds, which are heterocyclic ring compounds containing four nitrogen atoms and one carbon atom:

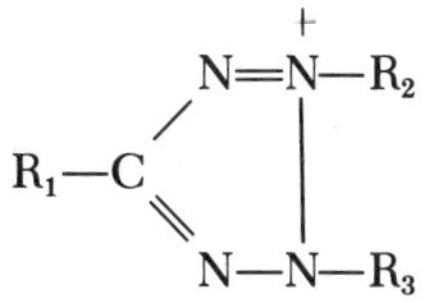

$$R_1-C \underset{N-N-R_3}{\overset{N=N-R_2}{<}}$$

The ring can be reduced with the addition of a hydrogen atom to form a formazan compound:

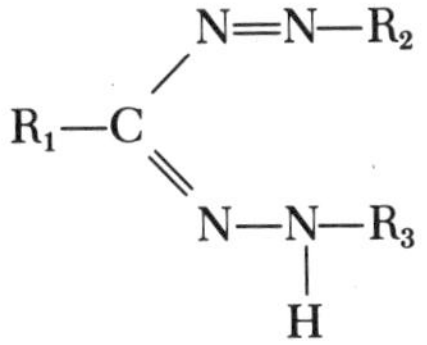

$$R_1-C \underset{\underset{H}{|}}{\overset{N=N-R_2}{<}} N-N-R_3$$

Tetrazolium compounds may intercept the electrons at certain points along the electron transport system, a process that forms the basis of dehydrogenase histochemistry.

Triphenyl tetrazolium chloride (TTC) was the first tetrazolium compound used to demonstrate dehydrogenases. The structure of TTC is as follows:

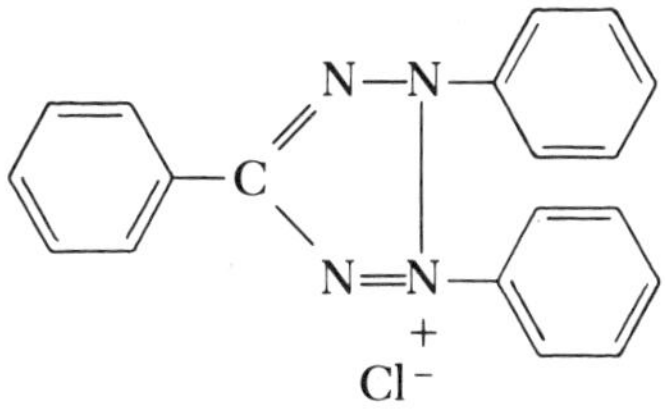

The histochemical results were not satisfactory because the reaction product had a weak red color and was not sufficiently insoluble; in addition, long incubation periods were required because TTC was sluggish as an electron acceptor. The potential of the tetrazolium compounds was nevertheless obvious, and more satisfactory tetrazolium compounds were sought.

Two different but closely related ditetrazolium compounds were synthesized for this purpose: blue tetrazolium (BT) and nitro tetrazolium (NT). BT, which was synthesized and used by Rutenburg et al. (1950), produced a formazan with greatly improved pigment qualities; it also accepted electrons somewhat more readily than TTC. NT was used by Padykula (1952) and Shelton and Schneider (1952); it accepted electrons still more readily but had inferior pigment qualities. A monotetrazolium compound, 2-*p*-iodophenyl-3-*p*-nitrophenyl-5-phenyl tetrazolium chloride (INT), introduced by Atkinson et al. (1950), accepted electrons very readily but again the formazan was unsuitable. Finally, Nachlas et al. (1957) attempted to synthesize a new compound that would combine the good pigment qualities of BT and that would be reduced as easily as NT. Their efforts resulted in the development of nitro blue tetrazolium (nitro BT or NBT):

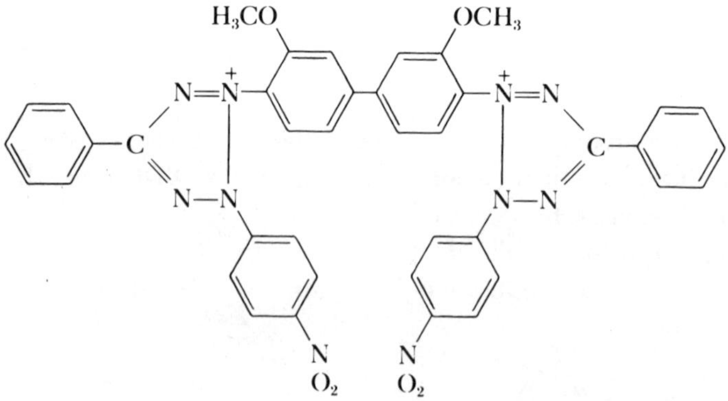

This tetrazolium readily accepts electrons, has good pigment qualities, is not lipid-soluble, and has good protein substantivity. It is now almost universally used for the light microscopic demonstration of dehydrogenases.

Two other noteworthy tetrazolium compounds should be mentioned. A monoformazan, 3-(4,5-dimethylthiazol-2-yl)-2,5-diphenyl-

2H-tetrazolium bromide (MTT), developed by Pearse (1957), is even more readily reduced than INT; however, it is lipid-soluble and has to be chelated with cobalt to prevent crystallization. MTT, like the other older tetrazolium compounds, has given way to nitro BT. Tetranitro blue tetrazolium (TNBT), which was introduced along with nitro BT by Nachlas et al. (1957), has found some favor among histochemists. TNBT differs from nitro BT only in that two more nitro groups have been added; it is reduced somewhat more readily than nitro BT but generally gives similar results.

The ditetrazolium compounds can lead to problems in microdensitometric quantification of enzyme reaction products. The main problem lies in the fact that only one-half of some of the molecules are reduced, and the resulting "half formazans" will have absorption characteristics quite different from the fully reduced formazans. Even the fully reduced formazans show multiple absorption peaks rather than a single one. Another problem with the ditetrazolium compounds is the presence of other tetrazolium compounds as contaminants (Altman, 1976a) with unsuitable characteristics. For quantitative studies, Altman (1976b) adopted a new monotetrazolium compound called 2-(2-benzothiazolyl)-3-(4-phthalhydrazidyl)-5-styryl-tetrazolium chloride (BPST), which has the following structure:

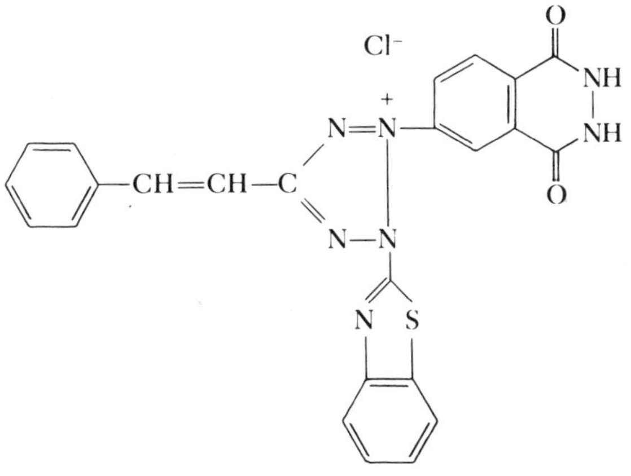

Its formazan has a single, well-defined peak that is well suited for microdensitometric quantification of dehydrogenase reactions. BPST activity can be expressed in terms of absolute units.

Solubility of Enzymes

Some of the dehydrogenases are tightly bound to the mitochondria, particularly succinate dehydrogenase, β-hydroxybutyrate dehydrogenase, and glutamate dehydrogenase, while others are loosely bound or are almost completely soluble in the cytoplasm. The unbound dehydrogenases may diffuse from their normal sites to other areas of the tissue or into the incubation medium. It has been shown, for instance, that there may be 100 percent diffusion of some of the most soluble dehydrogenases (Kalina and Gahan, 1965; Altman and Chayen, 1966). If the enzyme diffuses to another site, obviously the localization will be inaccurate. If the enzyme diffuses into the incubation medium, naturally the coenzyme will be reduced in that medium. The reduced coenzyme may also diffuse back into the tissue where it may be reoxidized by diaphorases, resulting in the deposition of formazan. The distribution of this formazan may therefore be totally unrelated to the distribution of the enzyme of interest.

Several approaches have been taken to solve the problem of dehydrogenase diffusibility. The first approach is to fix the tissue blocks or the tissue sections. However, most dehydrogenases are extremely sensitive to aldehyde fixation, and they are seriously inhibited after only a few minutes of fixation in cold formalin. Nevertheless, some of the dehydrogenases are able to withstand minimal fixation, in which case the problem of diffusion is considerably improved. Lactate dehydrogenase is a soluble enzyme that can withstand considerable fixation, which greatly reduces its diffusion.

Another approach to the problem of dehydrogenase diffusion is to add to the incubation medium an inert polymer that "entangles" large (enzyme) molecules and inhibits their movement. The most commonly used polymers are polyvinyl pyrrolidone (PVP) and polyvinyl alcohol (PVA). Altman and Chayen (1965) found that diffusion of nitrogenous material from tissue sections was greatly delayed by the addition of PVA. PVP and PVA are commercial products that are only secondarily adapted for histochemical purposes. The manufacturers seem oblivious of the needs of histochemists, which is sometimes a source of annoyance (Altman, 1977).

Gelatinous media have been prepared in order to control enzyme diffusion (Fahimi and Amarasingham, 1964; Kalina and Gahan, 1968).

The diffusion of enzymes is greatly inhibited because of their large molecular size, while substrates, cofactors, and other low-molecular-weight substances can diffuse through the gelatin virtually unhindered. A practical problem arises, however, because a certain amount of heat is required to apply the medium, and that heat may be detrimental to enzyme activity. It may also be difficult to remove the gelatinous medium completely at the end of the incubation period. Furthermore, some workers using gelatinous media have seen no improvement in the results.

Diaphorases

Even though there are many dehydrogenases, all of them (with only a few exceptions) use one of two different coenzymes, nicotinamide-adenine dinucleotide (NAD) or nicotinamide-adenine dinucleotide phosphate (NADP). The coenzymes serve as the first of a series of electron carriers; in accepting an electron, the coenzyme becomes reduced to NADH or NADPH. Another enzyme, diaphorase, is required to restore the coenzyme to its oxidized state. While many dehydrogenases exist, there are only two diaphorases, one for reoxidizing NADH and the other for reoxidizing NADPH.

Farber et al. (1956), using neotetrazolium as the electron acceptor, attempted to demonstrate five different specific dehydrogenases, two that were NAD-linked and three that were NADP-linked. They did not find the five different distribution patterns that might have been expected. Instead, the distribution patterns were the same for both NAD-linked dehydrogenases. The distribution patterns for the three NADP-linked dehydrogenases were also the same, but different from the pattern given by the NAD-linked enzymes. This observation was interpreted to mean that the respective diaphorases were being demonstrated rather than the individual dehydrogenases.

Diaphorases can, of course, be demonstrated intentionally. The incubation medium is similar to that used for insoluble dehydrogenases except that the dehydrogenase substrate is omitted and either NADH or NADPH (reduced form) is added. The reduced coenzyme is in fact the substrate for the diaphorase. It is frequently desirable to study the distribution of diaphorases in order to compare them with dehydrogenases.

Intermediate Electron Carriers

In the early development of dehydrogenase histochemistry, it was frequently the practice to add phenazine methosulfate (PMS) or menadione to the incubation medium to enhance formazan production. Presumably these compounds served as intermediate electron carriers from some point in the electron transport system to the tetrazolium compound. Later, workers found no benefit from these compounds and even noted inhibition. There is now general agreement that in demonstrating bound dehydrogenases with nitro BT no benefit is derived from the use of intermediate electron carriers. It is common practice, however, to use PMS in demonstrating the soluble dehydrogenases. Menadione is beneficial in demonstrating flavin adenine dinucleotide (FAD)–linked dehydrogenases, particularly for soluble α-glycerophosphate dehydrogenase.

pH of Incubation Medium for Dehydrogenases

The pH optima of dehydrogenases frequently lie considerably above neutral, and generally it is desirable to carry out an enzyme reaction at or near the enzyme's pH optimum. There are three reasons, however, why the pH of the incubation medium for dehydrogenases must be kept close to neutral. First, at an alkaline pH, the mitochondria tend to swell, which may alter the solubility of the enzyme and the velocity of the enzyme reaction. Second, NAD becomes progressively less stable in alkaline conditions but is stable at acid conditions. The third reason is the problem created by the phenomenon of "nothing dehydrogenase."

"Nothing Dehydrogenase"

At a pH considerably more than 7.0, an interesting but troublesome phenomenon occurs—the formazan may be deposited in the absence of a dehydrogenase substrate, a phenomenon that is therefore called "nothing dehydrogenase." The phenomenon increases progressively as the pH is raised, and it may be a major source of error if the pH of the incubation medium is too high. "Nothing dehydrogenase" activity varies considerably with different tissues and is generally more prominent with NAD than with NADP. The distribution pattern is similar to that of the respective diaphorases.

The most likely explanation of "nothing dehydrogenase" is that protein sulfhydryl groups are able to reduce NAD (and to a lesser extent NADP) at a pH of around 8 (or more). The reduced coenzymes can then serve as a substrate for diaphorase, and it is actually the diaphorase that is being observed. This phenomenon can be prevented with sulfhydryl blocking agents; p-hydroxymercuribenzoate has been used for this purpose (Manns, 1972), but it can be used only with nonsulfhydryl enzymes.

BIOCHEMISTRY AND HISTOCHEMISTRY OF INDIVIDUAL OXIDOREDUCTASES

DOPA-Oxidase

It was observed a long time ago (Block, 1917) that melanocytes of human skin become visible when they are incubated with dihydroxyphenylalanine (DOPA). It is now known that this observation results from a reaction catalyzed by an enzyme. Under certain conditions, tyrosine may serve as the substrate. It seemed as if the same enzyme might catalyze both reactions, although this was difficult to reconcile with the general observation that one enzyme catalyzes one reaction. However, it is now generally accepted that only one enzyme is involved because (1) tyrosine and DOPA-oxidase activities are always physically associated with each other, (2) inhibitors always have similar effects on both activities, and (3) both activities are proportional to the content of copper. A lag time is observed when tyrosine is used as the substrate but not when DOPA is used, suggesting that the enzyme catalyzes two consecutive steps.

Demonstration of DOPA-oxidase is relatively easy, although several factors can complicate interpretation of the results. Preexisting melanin should be evaluated on unincubated sections. The reaction has been repeatedly observed in erythrocytes and granular leukocytes. These cells of course do not contain the melanin-forming enzyme, but peroxidases (the heme from red blood cells) are strongly suspected of causing the reaction.

Theoretical considerations suggest that diffusion of the final reaction product could well be a problem. The enzyme catalyzes only the first of a series of chemical reactions leading to the formation of melanin. Several of the intermediate products would be expected to be somewhat soluble; unless the spontaneous reactions are very rapid, diffusion would be expected. In actual practice, diffusion is a problem only if unfixed tissues are used. Since the enzyme withstands considerable formalin fixation, it is preferable to fix the tissues.

Monoamine Oxidase

Monoamine oxidase, which is found particularly in the brain and liver, is a bound enzyme that is associated with the outer membrane of the mitochondria. A wide variety of physiological monoamines, such as dopamine, epinephrine, norepinephrine, tryptamine, and others, are oxidized to their corresponding aldehydes with the liberation of ammonia and hydrogen peroxide.

Two forms or isozymes of monoamine oxidase that have been recognized in brain tissue are designated A-forms and B-forms. A-form is very effectively inhibited by clorgyline—N-methyl-N-propargyl-3-(2,4-dichlorophenoxy) propylamine—whereas B-form is nearly insensitive to clorgyline. A-form specifically deaminates benzylamine and β-phenylethylamine. Both forms can deaminate tryptamine and dopamine.

Several approaches have been pursued to demonstrate this enzyme histochemically. Oster and Schlossman (1942) first attempted to couple the aldehyde reaction product with Schiff's reagent; this approach was impractical because the final reaction product was not sufficiently insoluble. The first successful attempt to demonstrate the enzyme involved a somewhat novel approach by Koelle and Valk (1954). Prior to incubation, the tissue sections were treated with hydrazine. The incubation medium contained tryptamine and 3-hydroxy-2-naphthoic acid hydrazide, and the aldehyde product was precipitated as its hydrazone. Postcoupling with fast blue B formed a blue pigment at the sites of enzyme activity.

A method using tetrazolium salts was developed by Glenner et al. (1957). This method gives excellent results, is much more straightforward than that of Koelle and Valk, and is the method of choice for demonstrating monoamine oxidase. It is not entirely clear how tetrazolium is reduced by the enzyme. The technique utilizes tryptamine as a substrate; no reaction product is formed if tyramine is used, which is puzzling, since it is well known that tyramine is oxidized at least as well as tryptamine.

Peroxidase

Peroxidases are relatively rare in animal tissues but are very common in plant tissues; they are found in microbodies of the liver and kidneys and in the granules of neutrophils. Horseradish and turnips contain high concentrations of peroxidase and the horseradish enzyme is now widely used as a marker enzyme in immunohistochemical procedures.

Hydrogen peroxide is degraded by peroxidases as well as by catalase. While catalase utilizes two molecules of peroxide for each event, peroxidase utilizes one molecule of peroxide and one of an organic compound:

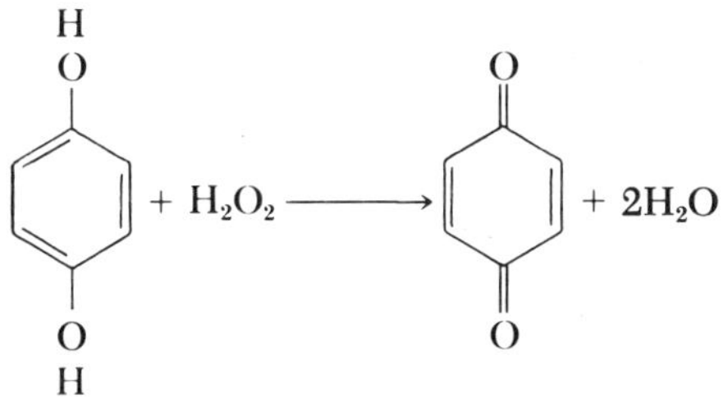

Peroxidase activity has been demonstrated in tissue sections primarily by the benzidine method (Adler and Adler, 1904). The disadvantage to the benzidine method is that the oxidized reaction product of benzidine fades within a few weeks. Numerous modifications of the method have been made but none are totally satisfactory. Straus (1964) modified the method to obtain a more stable blue reaction product and the results seemed reasonably satisfactory.

Graham and Karnovsky (1966) developed a method using 3,3'-diaminobenzidine (DAB). Although the method was primarily intended for ultrastructural demonstration of peroxidase, it is readily adapted for light microscopy. As a result of enzyme action, DAB is reduced and becomes an osmophilic polymer. Because the sections are treated with osmium tetroxide, the polymerized DAB is visualized as a black pigment, which is very permanent. Very precise enzyme localization can be realized with this method.

After studying some of the factors affecting horseradish peroxidase, Malmgren and Olsson (1977) presented an alternative method for demonstrating this enzyme, one that is more sensitive than the method of Graham and Karnovsky. Malmgren and Olsson used this method for detecting horseradish peroxidase during retrograde axonal transport studies, but it can also be used for any application in which horseradish peroxidase is used as a tracer.

Cytochrome Oxidase

Cytochrome oxidase is the terminal element of the electron transport system located in the mitochondria. It is the only element in the system that can readily reduce molecular oxygen.

Various types of cytochromes have been isolated and studied biochemically. Originally, only three types were recognized, and these were designated by the letters a, b, and c. Subsequently, other cytochromes have been isolated and characterized, and they were named with letters and subscripts in reference to the original three cytochromes depending on which of the three they most closely resemble. Cytochrome c is water-soluble; the other members can be solubilized with detergents from the mitochondrial walls for study.

Cytochromes are characterized and named by studying biochemically isolated members of the cytochrome chain. Cytochrome oxidase, on the other hand, is known primarily on the basis of functional studies in situ; it is able to reduce molecular oxygen readily. It probably represents a complex consisting of cytochromes a and a_3.

Cytochrome oxidase has been studied for almost 100 years, although for a long time it was known as indophenol oxidase or "Atmungsferment." The term *indophenol oxidase* was derived from the fact that indophenol pigment was formed in tissues from α-naphthol and N,N-dimethyl-p-phenylenediamine. The reaction was named the "nadi" reaction, a term derived from the first two letters of each compound (*na*phthol and *di*methyl-p-phenylenediamine). The identity of indophenol oxidase with cytochrome oxidase was not clearly established until about 1938. (See the discussion by Nachlas et al. [1958].) It is possible to demonstrate cytochrome oxidase with the nadi reaction, although the method is not entirely satisfactory. The problems include diffusion of the pigment, rapid fading, lipophilia, and lack of protein substantivity.

Modified nadi methods have been described by a number of workers, starting with Nachlas et al. (1958), who used 4-amino-1-N,N-dimethylnaphthylamine instead of dimethyl-p-phenylenediamine. The resulting purple pigment, indonaphthol, is finely granular, does not diffuse, and has a more stable color than the former pigment. However, the problem of lipophilia was not overcome and protein

substantivity was not improved. Burstone (1959, 1960, and 1961) substituted a series of reagents to improve the method. The color of the
reaction product was made more stable and protein substantivity was
increased. The method of Burstone (1961), using p-aminodiphenyl-
amine in place of α-naphthol and p-methoxy-p'-aminodiphe-
nylamine in place of p-phenylenediamine, is the most satisfactory
method of the nadi type.

The most recent method for the demonstration of cytochrome
oxidase (Seligman et al., 1968) is based on the oxidative polymerization of DAB:

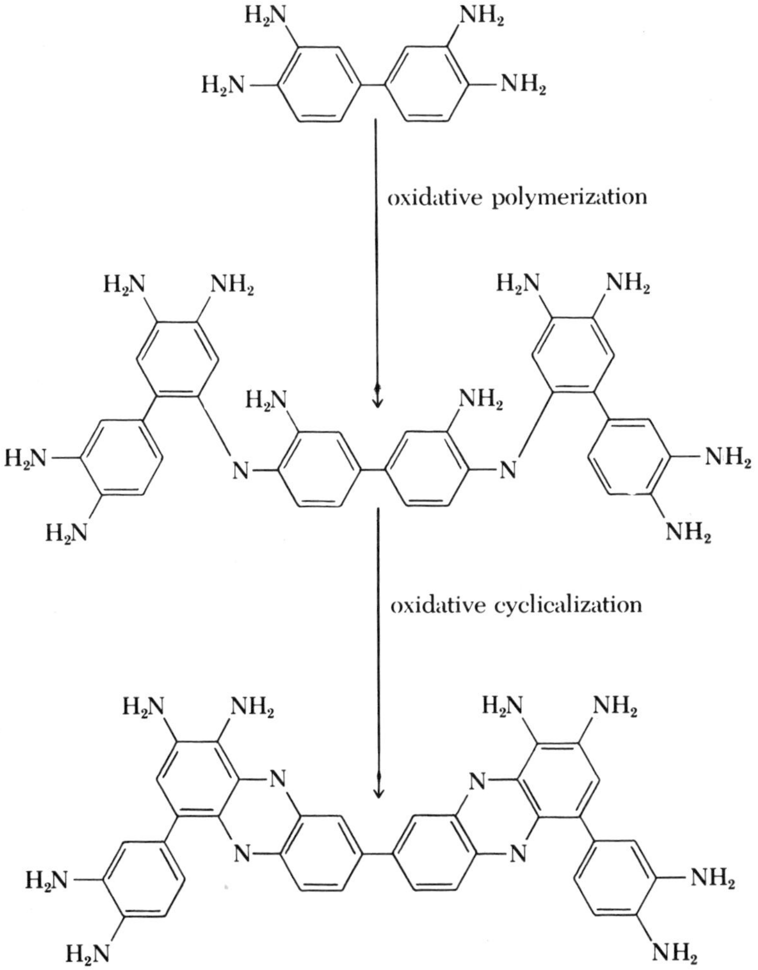

This method gives very precise localization of enzyme activity, and the brown reaction product is permanent. It is considered the method of choice.

Glucose 6-Phosphate Dehydrogenase

Glucose 6-phosphate dehydrogenase exists as several electrophoretically separable forms. Each species seems to have its own distinctive electrophoretic pattern of enzyme forms, one or more of which are sex-linked. In human liver, for instance, two forms were found by Ohno et al. (1966); one form is sex-linked and is specific for glucose 6-phosphate, while the form that is not sex-linked can oxidize galactose 6-phosphate as readily as glucose 6-phosphate. The deer mouse also has two forms, one sex-linked and another that is under the control of an autosomal gene (Shaw and Barto, 1965). In the rat liver, seven electrophoretically distinct forms—designated A, B, C, D, E, E', and F—were found by Hori and Matsui (1967 and 1968). The D form is greater in the female than in the male. Administering estradiol to males stimulates the D form and administering dihydroepiandrosterone to females suppresses it. In vitro, estradiol also inhibits the D form. Forms A and C are enzymes of the smooth endoplasmic reticulum, E' and F are mitochondrial enzymes, and B, D, and E are cytoplasmic enzymes. All forms are effectively inhibited by manganese sulfate.

Glucose 6-phosphate dehydrogenase catalyzes the formation of gluconolactone 6-phosphate. Thus the enzyme represents the gateway by which glucose may enter the pentose monophosphate shunt. Histochemically, either NAD or NADP may be used as the coenzyme, although the latter generally gives a stronger reaction. Undoubtedly, certain of the forms are NAD-linked and others NADP-linked.

Either the standard method for soluble dehydrogenases or the modified method of Rieder et al. (1978) may be used for demonstrating this enzyme.

6-Phosphogluconate Dehydrogenase

Another enzyme of the pentose monophosphate shunt, 6-phosphogluconate dehydrogenase, catalyzes the formation of ribulose 5-phosphate from gluconate 6-phosphate, with the liberation of CO_2. This enzyme is also dependent on NADP as a coenzyme. Like most enzymes of the pentose monophosphate shunt, it is soluble. It may be demonstrated by the standard method for soluble dehydrogenases or by the modified method of Rieder et al. (1978).

Malic Enzyme

Two forms of malic enzyme exist; one is found in the mitochondria of brain, heart, and adrenal cortex, while the other is a cytoplasmic enzyme found in the liver and adipose tissue. Both forms catalyze the reaction of malic acid to pyruvic acid and CO_2. Both forms are also dependent on the coenzyme NADP and are therefore connected with lipogenesis. Either the standard method for soluble dehydrogenases or the modified method of Rieder et al. (1978) should be used to demonstrate malic enzyme.

α-Glycerophosphate Dehydrogenase (Soluble)

The cytoplasmic enzyme α-glycerophosphate dehydrogenase catalyzes the following reaction:

α-glycerophosphate + NAD ⇌ dihydroxyacetone phosphate + NADH

The equilibrium lies far to the left, greatly favoring the formation of α-glycerophosphate. The physiological significance of the enzyme is not fully understood, but part of its function may be the formation of glycerol for triglyceride synthesis. It may also be involved in an NAD-NADH "shuttle" between the cytoplasm and mitochondria.

α-Glycerophosphate dehydrogenase is a sulfhydryl enzyme that is very effectively inhibited by p-chloromercuribenzoate and N-ethyl maleimide; its activity is increased by ethylenediaminotetraacetate (EDTA). It can be demonstrated with the standard method for soluble dehydrogenases.

α-Glycerophosphate Dehydrogenase (Mitochondrial)

α-Glycerophosphate dehydrogenase is a bound mitochondrial enzyme that catalyzes the interconversion of L-glycerol-3-phosphate to dihydroxyacetone phosphate. It is a flavoprotein, and nucleotide coenzymes do not participate in the electron transfer; in this respect it is similar to succinate dehydrogenase. The enzyme is readily inhibited by dicumarol.

The standard method for insoluble dehydrogenases can be used to demonstrate α-glycerophosphate dehydrogenase; the enzyme reluctantly reduces tetrazolium compounds. Wattenberg and Leong (1960) have shown that the reduction of tetrazolium compounds by both α-glycerophosphate dehydrogenase and succinate dehydrogenase is enhanced by the addition of menadione. In rat brain, the effect of menadione on α-glycerophosphate dehydrogenase is much greater than on succinate dehydrogenase (Hess and Pearse, 1961a).

Glyceraldehyde Phosphate Dehydrogenase

Glyceraldehyde phosphate dehydrogenase catalyzes an important step in the glycolysis pathway. Glyceraldehyde 3-phosphate is reversibly catalyzed to 1,3-diphosphoglycerate:

$$\underset{\substack{|\\ \text{HCOH}\\ |\\ \text{H}_2\text{COPO}_3\text{H}_2}}{\text{CHO}} + \text{NAD} + \text{H}_3\text{PO}_4 \rightleftharpoons \underset{\substack{|\\ \text{C=O}\\ |\\ \text{HCOH}\\ |\\ \text{H}_2\text{COPO}_3\text{H}_2}}{\text{OPO}_3\text{H}_2} + \text{NADH}$$

Because the aldehyde becomes phosphorylated, inorganic phosphate is an absolute requirement.

The enzyme is NAD-linked but the NAD is rather tightly bound to the enzyme, very unlike the situation with other NAD-linked dehydrogenases. It is a sulfhydryl enzyme and is therefore inhibited by heavy metal ions, by the usual sulfhydryl inhibitors, and particularly by iodoacetate.

Glyceraldehyde phosphate dehydrogenase was demonstrated by Himmelhoch and Karnovsky (1961), who were obliged to prepare their own substrate because it was not available commercially. It is now commercially available but it is rather costly and somewhat unstable. Although the preparation of the substrate is fairly inconvenient, some investigators may still prefer this option.

Himmelhoch and Karnovsky (1961) reported that a final substrate concentration of 0.15mM was adequate but 0.36mM was optimal; a concentration of 0.45mM caused considerable inhibition. The enzyme is activated by the addition of EDTA in the incubation medium.

Lactate Dehydrogenase

The interconversion of lactic acid and pyruvic acid is catalyzed by lactate dehydrogenase, a reaction that is linked to anaerobic metabolism. Under anaerobic conditions, this reaction temporarily takes the place of the citric acid cycle of mitochondria. Both pyruvic acid and NADH are normally removed from the metabolic pool by the citric acid cycle under aerobic conditions. Under anaerobic conditions, pyruvic acid is converted to lactic acid with the formation of NAD from NADH. Lactic acid is not further metabolized, but it must be reconverted to pyruvic acid in another tissue in which oxygen is available.

The subunit nature of certain enzymes is best appreciated in lactate dehydrogenase. Each molecule is made up of four subunits held together by forces other than covalent bonds, since they can be dissociated with a high concentration of guanidine. There are two types of electrophoretically distinct subunits, designated M and H. Hence a single lactate dehydrogenase molecule may consist of four M subunits (M_4) or four H subunits (H_4), or it may be a hybrid of M_3H_1, M_2H_2, or M_1H_3. The M_4 and H_4 forms of lactate dehydrogenase predominate in skeletal muscle and heart tissue, respectively, which is the reason for the designations M and H.

The M_4 and H_4 lactate dehydrogenase have differing enzymatic properties, which reflect their intended function. The H_4 form is inhibited by very low concentrations of pyruvate, making it useful to tissues that must convert lactic acid to pyruvic acid. The M_4 form is inhibited only by high concentrations of pyruvic acid, making it useful to tissues that have short bursts of anaerobic metabolism.

Diffusion is difficult to control in histochemical methods for lactate dehydrogenase because the enzyme is particularly soluble. This problem has led some workers to favor semipermeable membrane systems or the use of gelatinous incubation media. Lactate dehydrogenase may also be demonstrated with the standard method for soluble dehydrogenases, which is certainly easier to perform than the more specialized techniques.

Because of the difference in behavior of the H_4 and M_4 forms of lactate dehydrogenase, it is possible to demonstrate the two forms differentially. The groundwork for this technique was developed by Brody and Engel (1964) and it was refined by McMillan (1967) and Jacobsen (1969).

Isocitrate Dehydrogenases

There are two isocitric dehydrogenases that have distinctly different physicochemical properties; one is a bound mitochondrial enzyme and is NAD-dependent, while the other is a soluble cytoplasmic enzyme and is NADP-dependent. Both enzymes catalyze the conversion of isocitric acid to α-ketoglutarate and CO_2. Both are activated by manganese ions, although zinc ions are nearly as effective with the NADP enzyme and magnesium ions are nearly as effective with the NAD enzyme. Isocitric dehydrogenase activity may be demonstrated with the standard method for soluble dehydrogenase by the method of Rieder et al. (1978). If there is particular interest in the soluble, NAD-dependent enzyme, magnesium should be replaced with manganese or even zinc.

Succinate Dehydrogenase

Succinate dehydrogenase is one of the enzymes of the citric acid cycle, and it catalyzes the conversion of succinate to fumarate. The enzyme is tightly bound to the inner mitochondrial membrane. FAD serves as the coenzyme but unlike NAD and NADP, FAD is tightly bound to the enzyme molecule. In histochemical procedures, it is unnecessary to add FAD to the incubation medium.

It was once supposed that the tetrazolium compounds would accept electrons directly from the enzyme-$FADH_2$ complex. Experimental evidence, however, indicates that electrons are transferred only by an intermediate electron carrier (Horowitz et al., 1967). If no electron carriers are added, the transfer is probably accomplished by coenzyme Q already in the tissue, but the presence of coenzyme Q cannot be depended upon. Phenazine methosulfate, menadione, or coenzyme Q should be added to the incubation medium to transfer electrons to the tetrazolium compound.

Malate Dehydrogenase

Malate dehydrogenase catalyzes the interconversion of malic acid and oxaloacetic acid; it is an NAD-linked, citric acid cycle enzyme. It should not be confused with malic enzyme, an NADP-linked cytoplasmic enzyme that oxidizes malic acid to pyruvic acid.

Malate dehydrogenase should be considered a soluble enzyme even though it is found in the mitochondria. It can be demonstrated with the standard method for soluble dehydrogenases.

β-Hydroxybutyrate Dehydrogenase

β-Hydroxybutyrate dehydrogenase catalyzes the conversion of β-hydroxybutyric acid to acetoacetic acid. The reaction is reversible, and under normal circumstances the formation of β-hydroxybutyric acid may be more significant than the forward reaction. Acetoacetone is a normal by-product of fatty acid catabolism. β-Hydroxybutyric acid is an end product and can be eliminated only by excretion through the kidney or by reconversion to acetoacetic acid.

Both acetoacetic acid and β-hydroxybutyric acid are known as ketone bodies. Excessive formation of ketone bodies (or ketosis) is seen in severe diabetes or starvation, when the body must depend on catabolism of fat for its source of energy.

The standard method for insoluble dehydrogenases may be used to demonstrate β-hydroxybutyrate dehydrogenase.

Glutamate Dehydrogenase

The interconversion of glutamic acid and α-ketoglutarate, an extremely important reaction in intermediate metabolism, is catalyzed by glutamate dehydrogenase. Not only is glutamic acid one of the 20 amino acids of proteins, but it is the common denominator in the synthesis and degradation of several other amino acids. This reaction is also the major point for the interconversion of amino acid, nitrogen, and ammonia.

The enzyme is almost universally distributed throughout the plant and animal kingdoms. Either NAD or NADP may serve as the coenzyme. The enzyme utilizes NAD during biosynthetic functions and NADP during catabolic functions. Another glutamate dehydrogenase found in the liver utilizes only NAD. It has an equilibrium far to the right, favoring formation of α-ketoglutarate. Glutamate dehydrogenases are bound mitochondrial enzymes. The enzymes are activated by magnesium ions ($5mM$) and by adenosine diphosphate (ADP) ($2mM$) (Jarvie and Ottaway, 1975). The standard method for insoluble dehydrogenases may be used to demonstrate glutamate dehydrogenase activity, but it is advisable to add either magnesium sulfate or ADP for enhanced activity.

UDP-Glucose Dehydrogenase

UDP-glucose dehydrogenase catalyzes an important step in the synthesis of glucuronic acid, which is an important "building block" of chondroitin sulfate and hyaluronic acid. The reaction involves the conversion of uridine diphosphoglucose to uridine diphospho-glucuronic acid:

UDP-glucose + 2NAD → UDP-glucuronic acid + 2NADH

The enzyme catalyzes two consecutive oxidation events; for each mole of substrate, two moles of NAD are required.

Methods for determining this enzyme were published by Balogh and Cohen (1961) and by Hess and Pearse (1961b). The former method used a pH of 8.4, making it vulnerable to "nothing dehydro-genase" interference; however, in the latter method, a pH of 7.4 was used. Hess and Pearse (1961b) used MTT as the tetrazolium, and they found that menadione was required for the transfer of electrons. Stiller and Gorski (1969) studied the factors influencing UDP-glucose dehy-drogenase. They used an incubation medium of pH 7.0 to 7.2 to avoid the "nothing dehydrogenase" phenomenon.

Alcohol Dehydrogenase

Alcohols are reversibly oxidized to aldehydes by alcohol dehydroge-nase. Mammalian alcohol dehydrogenase is an NAD-linked metal-loenzyme containing 2 gram-atoms of zinc per mole. The zinc plays an integral part in the removal of hydrogens from the substrate. The enzyme can be inhibited with EDTA, which chelates the zinc.

The enzyme has a broad substrate specificity, oxidizing many pri-mary and secondary but not tertiary alcohols (Winer, 1958). Emulsions of long-chain alcohols may even serve as substrates (Schöpp and Rothe, 1975). For ethanol oxidation (the forward reaction) the enzyme has a surprisingly high pH optimum—9.3 for human liver alcohol de-hydrogenase (Blair and Vallee, 1966) and 9.7 for the enzyme from rat brain (Tabakoff and Erwin, 1970). However, for aldehyde reductase activity (the reverse reaction), the pH optimum of human liver alcohol dehydrogenase is pH 6.75. The physiological significance of these widely divergent pH optima is not known.

Alcohol dehydrogenase can be demonstrated histochemically with the standard method for soluble dehydrogenases. Even though the pH optimum for ethanol oxidation is high, sufficient activity remains at pH 7.4 to demonstrate the enzyme readily.

3-β-Hydroxysteroid Dehydrogenase

Several chemical reactions involving steroid compounds are dehydrogenase reactions. Enzymes catalyzing such reactions should be demonstrable using nitro BT and the appropriate substrate. The enzyme of greatest histochemical interest is 3-β-hydroxysteroid dehydrogenase, an NAD-linked enzyme found in the adrenal cortex, ovary (corpus luteum), testis (cells of Leydig), and placenta. Wattenberg (1958) and Levy et al. (1959) published histochemical methods for demonstrating this enzyme. Either dehydroepiandrosterone or 5-pregnanolone can serve as the substrate, but the former is generally preferred. It is necessary to dissolve the substrate in an organic solvent because steroids are poorly soluble in water. Acetone, dimethylformamide, and propylene glycol have been used; however, propylene glycol should be avoided, because it may serve as the substrate for "secondary" alcohol dehydrogenase. Wattenberg (1958) used an incubation medium at pH 8, but most workers since have preferred a lower pH to avoid the risk of getting a "nothing dehydrogenase" reaction. The method presented on p. 307 is a slight modification of the method of Levy et al. (1959).

METHOD FOR DOPA-OXIDASE
(Modified from Becker et al., 1935; Rappaport, 1955)

TISSUES:
Either use fresh-frozen cryostat sections (may be fixed lightly with cold formalin) or fix in cold 10 percent formalin overnight and use paraffin sections.

PREPARATION OF INCUBATION MEDIUM:
Dissolve 100 mg DL-dihydroxyphenylalanine (DOPA) in 100 ml 0.1M phosphate buffer, pH 7.4.

PROCEDURE:
(1) Deparaffinize sections and bring to water.
(2) Place sections in incubation medium for 12 hours or more.
(3) Change to fresh incubation medium and continue to incubate for up to 12 more hours.

(4) Wash in water.
(5) Counterstain nuclei if desired.
(6) Wash in water.
(7) Dehydrate in graded alcohols, clear in xylene, and mount in synthetic medium.

RESULTS:
Dark brown or black granules indicate the sites of DOPA-oxidase.

TETRAZOLIUM METHOD FOR MONOAMINE OXIDASE
(Glenner et al., 1957)

TISSUES:
Use fresh-frozen cryostat sections.

PREPARATION OF INCUBATION MEDIUM:

Tryptamine HCl	25 mg
Sodium sulfate	4 mg
Nitro blue tetrazolium	5 mg
Phosphate buffer, $0.1M$, pH 7.6	5 ml
Distilled water	15 ml

PROCEDURE:
(1) Incubate for 30 to 45 minutes at 37°C.
(2) Rinse in distilled water.
(3) Dehydrate through graded acetone-xylene series and mount with synthetic mountant.

RESULTS:
Blue deposits of formazan indicate sites of monoamine oxidase activity.

DAB METHOD FOR PEROXIDASE
(Graham and Karnovsky, 1966)

TISSUES:
Fix tissues with formol-calcium for 24 hours. (Graham and Karnovsky used a paraformaldehyde-glutaraldehyde fixative to preserve the ultrastructure.) Use 8-μ cryostat sections.

PREPARATION OF INCUBATION MEDIUM:
Prepare a saturated solution of 3,3′-diaminobenzidine by shaking 2 to 3 mg of the compound with 10 ml 0.05M tris buffer, pH 7.6. Filter the solution and add 0.01 percent hydrogen peroxide (0.1 ml of a 1 percent solution).

PROCEDURE:
(1) Incubate the sections for 20 minutes.
(2) Wash in three changes of distilled water.
(3) Postfix in 1.3 percent osmium tetroxide in S-collidine buffer, pH 7.2, containing 5 percent sucrose for 90 minutes. (Other buffers can also be used.)
(4) Dehydrate through graded alcohols, clear xylene, and mount with a synthetic medium.

RESULTS:
Black deposits of the osmicated polymer indicate the sites of enzyme activity.

ALTERNATIVE DAB METHOD FOR PEROXIDASE
(Malmgren and Olsson, 1977)

TISSUES:
Perfuse experimental animals through the vascular system with 1.5 percent glutaraldehyde in 0.1M phosphate buffer, pH 7.4. Fix the tissues for an additional 4 hours in the same fixative. Wash in sucrose-phosphate buffer overnight. Use frozen sections.

PREPARATION OF INCUBATION MEDIUM:
Dissolve 20 mg 3,3′-diaminobenzidine·4HCl in 10 ml 0.1M sodium cacodylate buffer, pH 5.1, and add 0.1 ml of a 1 percent solution of hydrogen peroxide.

PROCEDURE:
(1) Incubate sections in the dark at room temperature for 30 minutes.
(2) Wash in cacodylate buffer.
(3) Allow sections to air-dry.

(4) Dehydrate for 2 minutes in absolute ethanol.

(5) Clear in xylene and mount with a synthetic medium.

RESULTS:

A brown reaction product indicates the sites of peroxidase activity. A Kodak Wratten No. 46 filter enhances the contrast.

DAB METHOD FOR CYTOCHROME OXIDASE
(Seligman et al., 1968)

TISSUES:

Use cryostat sections either of fresh-frozen tissues or of tissues fixed with cold formol-calcium.

INCUBATION MEDIUM:

Phosphate buffer, 0.05M, pH 7.4	9.0 ml
3,3'-Diaminobenzidine·4HCl	5.0 mg
Catalase (20 μg/ml)	1.0 ml
Cytochrome c (Sigma, type II)	10.0 mg
Sucrose	750 mg

PROCEDURE:

(1) Incubate for up to 1 hour at room temperature.

(2) Wash in water.

(3) Dehydrate through graded series of alcohols, clear in xylene, and mount with synthetic resin.

CONTROLS:

Prepare incubation medium as described previously but add 6.5 mg (0.01M) potassium cyanide.

RESULTS:

The brown deposits of polymerized DAB indicate sites of cytochrome oxidase activity. Control sections should be negative.

STANDARD METHOD FOR DEMONSTRATING BOUND DEHYDROGENASES

ENZYMES DEMONSTRATED BY THIS METHOD:

Enzyme	Abbreviation
Succinate dehydrogenase	SD
α-Glycerophosphate dehydrogenase (mitochondrial)	α-GD(m)
Glutamate dehydrogenase	GLD
β-Hydroxybutyrate dehydrogenase	HBD
NADH diaphorase	NADII-D
NADPH diaphorase	NADPII-D

TISSUES:

Use fresh-frozen cryostat sections.

PREPARATION OF STOCK SUBSTRATE SOLUTIONS:

Enzyme	Substrate	Mol. Wt.	Final Concen- tration	Amount in 8 ml H$_2$O	Neutral- ize with	Add H$_2$O to Make
SD	DiNa succinate·6H$_2$O	270	2.5M	6.75 gm	1N HCl	10 ml
α-GD(m)	DiNa-α-glycero- phosphate·6H$_2$O	306	1.0M	3.05 gm	1N HCl	10 ml
GLD	Na-L-Glutamate	187	1.0M	1.87 gm	1N HCl	10 ml
HBD	Na-β-Hydroxybutyrate	126	1.0M	1.26 gm	1N HCl	10 ml

PREPARATION OF BASIC MEDIUM:

Tris buffer, 0.2M, pH 7.4	2.5 ml
Nitro blue tetrazolium (4 mg/ml)	2.5 ml
MgCl$_2$, 0.05M	1.0 ml
Distilled water	3.0 ml

PREPARATION OF INCUBATION MEDIUM:

Enzyme	Basic Medium	Stock Substrate Solution	Distilled Water	Coenzyme	
SD	9 ml	1 ml	—	—	
α-GD(m)	9 ml	1 ml	—	—	
GLD	9 ml	1 ml	—	NAD,	20 mg
HBD	9 ml	1 ml	—	NAD,	20 mg
NADH-D	9 ml	—	1 ml	NADH,	20 mg
NADPH-D	9 ml	—	1 ml	NADPH,	20 mg

For the enzymes SD and α-GD(m), the basic medium should be saturated with menadione. Add 50 mg menadione to 10 ml of the basic medium and shake intermittently for an hour. Filter before use.

PROCEDURE:
(1) Cover cryostat sections with the incubation medium of choice and incubate at room temperature for up to 1 hour.
(2) Remove the medium and fix with formol-saline for 15 minutes.
(3) Wash in running tap water for 3 minutes.
(4) Counterstain with 2 percent methyl green (chloroform-extracted) for 2 to 3 minutes.
(5) Wash in distilled water for 2 minutes.
(6) Dehydrate with graded alcohols, clear with xylene, and mount with a synthetic resin.

RESULTS:
Deposits of the blue formazan indicate sites of enzyme activity. Nuclei are stained green.

STANDARD METHOD FOR DEMONSTRATING SOLUBLE DEHYDROGENASES

ENZYMES DEMONSTRATED BY THIS METHOD:

Enzyme	Abbreviation
Malate dehydrogenase	MD
Malic enzyme	ME
Isocitrate dehydrogenase	ICD
Lactate dehydrogenase	LD
Glucose 6-phosphate dehydrogenase	GPD
6-Phosphogluconate dehydrogenase	6-PD
Alcohol dehydrogenase	AD
α-Glycerophosphate dehydrogenase (soluble)	α-GPD(s)

TISSUES:
Use fresh-frozen cryostat sections.

PREPARATION OF STOCK SUBSTRATE SOLUTIONS:

Enzyme	Substrate	Mol. Wt.	Final Concen- tration	Amount in 8 ml H₂O	Neutral- ize with	Add H₂O to Make
MD	MonoNa-L-malate	156	1M	1.56 gm	40% NaOH	10 ml
ME	MonoNa-L-malate	156	1M	1.56 gm	40% NaOH	10 ml
ICD	TriNa-DL-isocitrate	258	1M	2.58 gm	1N HCl	10 ml
LD	Na-DL-lactate	112	1M	1.25 ml	—	10 ml
GPD	Na-Glucose-6-phos-phate	282	1M	2.82 gm	1N HCl	10 ml
6-PD	TriNa-6-phospho-gluconate	342	1M	3.42 gm	1N HCl	10 ml
AD	Ethanol, absolute	46	1M	0.58 ml	—	10 ml
α-GPD(s)	DiNa-α-glycero-phosphate·6H₂O	306	1M	3.06 gm	1N HCl	10 ml

PREPARATION OF BASIC MEDIUM:

Tris buffer, 0.2M, pH 7.4	2.5 ml
Nitro BT, 4 mg/ml	2.5 ml
MgCl₂, 0.05M	1.0 ml
Distilled water	3.0 ml

PREPARATION OF INCUBATION MEDIA:

Enzyme	Basic Medium	Stock Substrate Solutions	Coenzyme	0.1M NaCN	Poly-vinyl Pyrrol-idone
MD	9 ml	1 ml	NAD, 20 mg	—	—
ME	9 ml	1 ml	NADP, 20 mg	—	0.75 gm
ICD	9 ml	1 ml	NAD or NADP, 20 mg	1 ml	0.75 gm
LD	9 ml	1 ml	NAD, 20 mg	—	0.75 gm
GPD	9 ml	1 ml	NADP, 20 mg	1 ml	0.75 gm
6-PD	9 ml	1 ml	NADP, 20 mg	1 ml	0.75 gm
AD	9 ml	1 ml	NAD, 20 mg	—	—
α-GPD	9 ml	1 ml	NAD, 20 mg	—	0.75 gm

For the enzyme α-GPD, use a basic medium in which the tris buffer is replaced with 0.06M phosphate buffer, pH 7.4.

PROCEDURE:

(1) Cover cryostat sections with the incubation medium of choice and incubate at room temperature for up to an hour.

(2) Remove the medium and fix with formol-saline for 15 minutes.

(3) Wash in running tap water for 3 minutes.

(4) Counterstain with 2 percent methyl green (chloroform-extracted) for 2 to 3 minutes.

(5) Wash in distilled water for 2 minutes.

(6) Dehydrate with graded alcohols, clear with xylene, and mount with a synthetic resin.

RESULTS:

Deposits of the blue formazan indicate sites of dehydrogenase activity.

METHOD FOR SOLUBLE DEHYDROGENASES
(Rieder et al., 1978)

ENZYMES DEMONSTRATED BY THIS METHOD:

Enzyme	Abbreviation
Glucose 6-phosphate dehydrogenase	G6PDH
6-Phosphogluconate dehydrogenase	6PGDH
Malic enzyme	ME
Isocitrate dehydrogenase	ICDH

TISSUES:

Use fresh-frozen cryostat sections cut at 5 to 20 μ. Dry sections for 5 minutes at 37°C. Defat sections by treating them sequentially with 100 percent acetone (5 minutes at 0°C), 100 percent chloroform (10 minutes at −20°C), and 100 percent acetone (1 minute at 0°C). Air-dry for 3 minutes at 37°C.

PREPARATION OF STOCK SUBSTRATE SOLUTIONS:

For G6PDH—152 mg disodium glucose 6-phosphate in 5.0 ml distilled water (0.1M)

For 6PGDH—189 mg trisodium 6-phosphogluconate in 5.0 ml distilled water (0.1M)

For ME—780 mg sodium L-malate in 5.0 ml distilled water (0.1M). Adjust to pH 7.4.

For ICDH—78 mg trisodium-DL-isocitrate in 5.0 ml distilled water (0.05M).

PREPARATION OF PVA SOLUTION:
Suspend 20 gm PVA (05/140) in 50 ml 0.2*M* tris-HCl buffer, pH 7.4. Dissolve by heating to 95°C and stirring. Readjust the pH to 7.4.

PREPARATION OF INCUBATION MEDIUM:
Combine the following items in order:

Nitro blue tetrazolium (45 mg in 1.1 ml distilled water)	1.1 ml
Tris-HCl buffer, 0.2*M*, pH 7.4	1.3 ml
$MgCl_2$, 0.05*M* (50 mg $MgCl_2 \cdot 6H_2O$ in 5.0 ml distilled water)	1.0 ml
Stock substrate solution	1.0 ml
NADP, 0.04*M* (63 mg NADP, disodium salt in 2.0 ml distilled water)	0.2 ml
PVA solution	5.0 ml

The items of the medium to this point may be combined and stored at −70°C. Just before use, the pH of the medium should be adjusted to 7.4, and the following items should be added:

NaN_3, 0.5*M* (65 mg in 2.0 ml distilled water)	0.2 ml
PMS, 0.0163*M* (10 mg in 2.0 ml distilled water)	0.2 ml

PROCEDURE:
(1) Incubate sections in the dark for 3 to 15 minutes at 37°C.
(2) Wash in 0.9 percent NaCl solution (37°C) for 1 minute.
(3) Postfix in formol-calcium containing 7.5 percent PVP for 20 minutes at 0°C.
(4) Rinse in distilled water for 2 to 3 minutes.
(5) Mount in glycerin jelly.

CONTROLS:
If control procedures are desired, either the coenzyme or the substrate should be left out of the medium.

RESULTS:
Formazan deposits indicate sites of dehydrogenase activity. Control sections should be negative.

METHOD FOR UDP-GLUCOSE DEHYDROGENASE
(Stiller and Gorski, 1969)

TISSUES:
Use fresh-frozen cryostat sections.

PREPARATION OF INCUBATION MEDIUM:

UDP-glucose, $5mM$	1.0 ml
Tris-HCl buffer, $0.2M$, pH 7.2	2.5 ml
KCN ($0.1M$)	1.0 ml
Nitro BT, 1 mg/ml	2.5 ml
NAD, $1.25mM$ (final concentration)	8.5 mg

Check the pH and adjust to 7.0 to 7.2 with tris stock. Add distilled water to make 10 ml.

PROCEDURE:
(1) Incubate for 60 minutes at 37°C.
(2) Wash briefly in distilled water.
(3) Postfix in formol-calcium for 1 hour.
(4) Wash briefly in distilled water.
(5) Mount in glycerin jelly.

CONTROLS:
Incubate sections in a medium as above, but omit substrate.

RESULTS:
Formazan deposits indicate sites of enzyme activity. Control sections should be negative.

METHOD FOR ISOENZYMES OF LACTATE DEHYDROGENASE
(Jacobsen, 1969)

TISSUES:
Use fresh-frozen cryostat sections.

PREPARATION OF STOCK SOLUTION A:
To 60 ml of 0.05M tris buffer, pH 7.0, add 24 gm polyvinyl alcohol and
dissolve by warming to 90°C. Cool and adjust the pH to 7.2.

PREPARATION OF STOCK SOLUTION B:

Tris buffer, 0.05M, pH 7.0	10.0 ml
Sodium DL-lactate	0.5 ml (0.4M)
Nitro blue tetrazolium	80.0 mg (9.8mM)
Sodium cyanide	4.9 mg (10mM)
Phenazine methasulfate	2.0 mg (0.66mM)

Stock solution B is used to demonstrate total lactate dehydrogenase
activity. In order to demonstrate predominately the M-form, add urea
to a concentration of 6.5M (the final concentration in the incubation
will be 3.25M). In order to demonstrate predominately the H-form,
reduce the concentration of lactate to 20mM (final concentration
10mM).

PREPARATION OF INCUBATION MEDIUM:
Combine equal volumes of stock solutions A and B on a 40°C water
bath. Add NAD at the rate of 1 mg/ml.

PROCEDURE:
(1) Treat sections with acetone for 10 minutes at 0° to 4°C.
(2) Treat with chloroform for 10 minutes at −15°C.
(3) Treat briefly with acetone at 0° to 4°C and air-dry the sections.
(4) Incubate sections for 5 minutes (for the M-form) or for 24 minutes
    (for the H-form).
(5) Wash sections with lukewarm water.
(6) Fix for 15 minutes in formol-calcium.
(7) Remove monoformazans with brief wash in acetone.
(8) Wash in distilled water.
(9) Mount with glycerin jelly.

CONTROLS:
For α-hydroxyacid oxidase, incubate sections in a medium lacking
NAD. For total lactate dehydrogenase activity, incubate sections in a

medium containing $0.01M$ p-chloromercuribenzoate (PCMB). For "nothing dehydrogenase," incubate sections in a medium without substrate.

RESULTS:
Deposits of the blue formazan indicate the sites of lactate dehydrogenase activity. Any activity obtained without NAD is attributed to α-hydroxyacid oxidase and must be subtracted from the test section to estimate the true amount of lactate dehydrogenase activity. The addition of PCMB should completely inhibit enzyme activity. Controls incubated in a medium without substrate should be negative (that is, it should not show "nothing dehydrogenase").

METHOD FOR GLYCERALDEHYDE 3-PHOSPHATE DEHYDROGENASE
(Himmelhoch and Karnovsky, 1961)

TISSUES:
Use fresh-frozen sections.

PREPARATION OF INCUBATION MEDIUM:

Substrate solution, 15 to $36mM$	1.9 ml
NAD, 5.0 mg/ml	1.6 ml
Nitro BT, 5.0 mg/ml	1.3 ml
Phosphate buffer, $0.2M$, pH 7.2	3.0 ml
Disodium EDTA, 6.0 mg/ml	0.6 ml
Distilled water	1.6 ml

This solution (without the NAD) can be stored at $-20°C$ for at least 1 month without deterioration.

PROCEDURE:
(1) Fix frozen sections in acetone for 20 minutes at 4°C.
(2) Rinse quickly in 0.85 percent saline.
(3) Incubate for 3 to 20 minutes at 37°C.
(4) Rinse briefly in 0.85 percent saline.
(5) Fix for 20 minutes in formol-calcium at 4°C.

(6) Rinse in 0.85 percent saline.

(7) Mount in glycerin jelly.

RESULTS:

Sites of formazan deposits indicate sites of enzyme activity.

METHOD FOR 3-β-HYDROXYSTEROID DEHYDROGENASE
(Modified from Levy et al., 1959)

TISSUES:

Use fresh-frozen cryostat sections. Mount the sections on glass slides or on coverslips.

INCUBATION MEDIUM:

Dehydroepiandrosterone, $10mM$ in acetone	0.5 ml
Nicotinamide, 1.6 mg/ml	1.0 ml
Nitro BT, 1 mg/ml	1.5 ml
NAD, 3 mg/ml	1.2 ml
Phosphate buffer, $0.1M$, pH 7.1 to 7.4	5.8 ml

PROCEDURE:

(1) Incubate for 5 minutes to 1 hour at 37°C.

(2) Rinse in distilled water.

(3) Fix for 30 minutes in a mixture containing 50 percent ethanol and 10 percent formalin.

(4) Wash in distilled water for 5 minutes.

(5) Mount with glycerin jelly.

RESULTS:

Deposits of the formazan dye indicate sites of enzyme activity.

REFERENCES

Adler, O., and Adler, R. (1904).   Über das Verhalten gewisser organischer Verbindungen gegenüber Blut mit besonderer Berücksichtigung des Nachweises von Blut. *Hoppe Seylers Z. Physiol. Chem.* 41:59–67.

Altman, F. P. (1976a).   Tetrazolium salts: A consumer's guide. *Histochem. J.* 8:471–485.

Altman, F. P. (1976b).   The quantification of formazans in tissue sections by microdensitometry. II. The use of BPST, a new tetrazolium salt. *Histochem. J.* 8:501–506.

Altman, F. P. (1977).   The use of industrial chemicals for research purposes. *Histochem. J.* 9:247–248.

Altman, F. P., and Chayen, J. (1965).   Retention of nitrogenous material in unfixed sections during incubation for histochemical demonstration of enzymes. *Nature* 207:1205–1206.

Altman, F. P., and Chayen, J. (1966).   The significance of a functioning hydrogen-transport system for the retention of "soluble" dehydrogenases in unfixed sections. *J. R. Microsc. Soc.* 85:175–180.

Atkinson, E., Melvin, S., and Fox, S. W. (1950).   Some properties of 2,3,5-triphenyltetrazolium chloride and several iodo derivatives. *Science* 111:385–387.

Balogh, K., and Cohen, R. B. (1961).   Histochemical localization of uridine diphosphoglucose dehydrogenase in cartilage. *Nature* 192:1199–1200.

Becker, S. W., Praver, L. L., and Thatcher, H. (1935).   An improved (paraffin section) method for the dopa reaction. *Arch. Dermatol. Syph.* 31:190–195.

Blair, A. H., and Vallee, B. L. (1966).   Some catalytic properties of human liver alcohol dehydrogenase. *Biochemistry* 5:2026–2034.

Block, B. (1917).   Das Problem der Pigmentbildung in der Haut. *Arch. Dermatol.* 124:129–207.

Brody, I. A., and Engel, W. K. (1964).   Isozyme histochemistry: The display of selective lactate dehydrogenase isozymes in sections of skeletal muscle. *J. Histochem. Cytochem.* 12:687–695.

Burstone, M. S. (1959).   New histochemical techniques for the demonstration of tissue oxidase (cytochrome oxidase). *J. Histochem. Cytochem.* 7:112–122.

Burstone, M. S. (1960).   Histochemical demonstration of cytochrome oxidase with new amine reagents. *J. Histochem. Cytochem.* 8:63–70.

Burstone, M. S. (1961).   Modifications of histochemical techniques for the demonstration of cytochrome oxidase. *J. Histochem. Cytochem.* 9:59–65.

Fahimi, H. D. (1969).   Cytochemical localization of peroxidatic activity of catalase in rat hepatic microbodies (peroxisomes). *J. Cell Biol.* 43:275–288.

Fahimi, H. D., and Amarasingham, C. R. (1964).   Cytochemical localization of lactic dehydrogenase in white skeletal muscle. *J. Cell Biol.* 22:29–48.

Farber, E., Sternberg, W. H., and Dunlap, C. E. (1956).   Histochemical localization of specific oxidative enzymes. I. Tetrazolium stains for diphosphopyridine nucleotide diaphorase and triphosphorpyridine nucleotide diaphorase. *J. Histochem. Cytochem.* 4:254–265.

Glenner, G. G., Burtner, H. J., and Brown, G. W. (1957).   The histochemical demonstration of monoamine oxidase activity by tetrazolium salts. *J. Histochem. Cytochem.* 5:591–600.

Graham, R. C., and Karnovsky, M. J. (1966).   The early stages of absorption of injected horse radish peroxidase in the proximal tubules of the mouse kidney: Ultrastructural cytochemistry by a new technique. *J. Histochem. Cytochem.* 14:291–302.

Hess, R., and Pearse, A. G. E. (1961a).   Histochemical and homogenization studies of mitochondrial α-glycerophosphate dehydrogenase in the nervous system. *Nature* 191:718–719.

Hess, R., and Pearse, A. G. E. (1961b). Histochemical demonstration of uridine diphosphate glucose dehydrogenase. *Experientia* 17:317–318.

Himmelhoch, S. R., and Karnovsky, M. J. (1961). The histochemical demonstration of glyceraldehyde-3-phosphate dehydrogenase activity. *J. Biophys. Biochem. Cytol.* 9:573–581.

Hori, S. H., and Matsui, S.-I. (1967). Effects of hormones on hepatic glucose-6-phosphate dehydrogenase of rat. *J. Histochem. Cytochem.* 15:530–534.

Hori, S. H., and Matsui, S.-I. (1968). Intracellular distribution of electrophoretically distinct forms of hepatic glucose-6-phosphate dehydrogenase. *J. Histochem. Cytochem.* 16:62–63.

Horwitz, C. A., Benitez, L., and Bray, M. (1967). The effect of coenzyme Q on the histochemical succinic tetrazolium reductase reaction: A histochemical study. *J. Histochem. Cytochem.* 15:216–224.

Jacobsen, N. O. (1969). The histochemical localization of lactic dehydrogenase isoenzymes in the rat nephron by means of an improved polyvinyl alcohol method. *Histochemistry* 20:250–265.

Jarvie, D. R., and Ottaway, J. H. (1975). Localization in cardiac muscle of some enzymes related to glutamate metabolism. *Histochem. J.* 7:165–178.

Kalina, M., and Gahan, P. B. (1965). A quantitative study of the validity of the histochemical demonstration for pyridine nucleotide-linked dehydrogenases. *Histochemistry* 5:430–436.

Kalina, M., and Gahan, P. B. (1968). A gelatin film method for improved histochemical localization of dehydrogenases in plant cells. *Stain Technol.* 43:51–57.

Koelle, G. B., and Valk, A. de T. (1954). Physiological implications of the histochemical localization of monoamine oxidase. *J. Physiol.* 126:434–447.

Levy, H., Deane, H. W., and Rubin, B. L. (1959). Visualization of steroid-3β-ol-dehydrogenase activity in tissues of intact and hypophysectomized rats. *Endocrinology* 65:932–943.

Manns, E. (1972). The histochemical localization of glucose dehydrogenase activity in sheep liver. *Histochem. J.* 4:25–33.

Malmgren, L., and Olsson, Y. (1977). A sensitive histochemical method for light- and electron-microscopic demonstration of horseradish peroxidase. *J. Histochem. Cytochem.* 25:1280–1283.

McMillan, P. J. (1967). Differential demonstration of muscle and heart type lactic dehydrogenase of rat muscle and kidney. *J. Histochem. Cytochem.* 15:21–31.

Nachlas, M. M., Tsou, K.-C., de Souza, E., Cheng, C.-S., and Seligman, A. M. (1957). Cytochemical demonstration of succinic dehydrogenase by the use of a new p-nitrophenyl substituted ditetrazole. *J. Histochem. Cytochem.* 5:420–436.

Nachlas, M. M., Crawford, D. T., Goldstein, T. P., and Seligman, A. M. (1958). The histochemical demonstration of cytochrome oxidase with a new reagent for the Nadi reaction. *J. Histochem. Cytochem.* 6:445–456.

Ohno, S., Payne, H. W., Morrison, M., and Beutler, E. (1966). Hexose-6-phosphate dehydrogenase found in human liver. *Science* 153:1015–1016.

Oster, K. A., and Schlossman, N. C. (1942). Histochemical demonstration of amine oxidase in the kidney. *J. Cell Comp. Physiol.* 20:373–378.

Padykula, H. A. (1952).   The localization of succinic dehydrogenase in tissue sections of the rat. *Am. J. Anat.* 91:107–145.

Pearse, A. G. E. (1957).   Intracellular localisation of dehydrogenase systems using monotetrazolium salts and metal chelation of their formazans. *J. Histochem. Cytochem.* 5:515–527.

Rappaport, B. Z. (1955).   A semiquantitative dopa reaction by use of frozen-dried skin. *Arch. Pathol.* 60:444–450.

Rieder, H., Teutsch, H. F., and Sasse, D. (1978).   NADP-dependent dehydrogenases in rat liver parenchyma. I. Methodological studies on the qualitative histochemistry of G6PDH, 6PGDH, malic enzyme and ICDH. *Histochemistry* 56:283–298.

Rutenburg, A. M., Gofstein, R., and Seligman, A. M. (1950).   Preparation of a new tetrazolium salt which yields a blue pigment on reduction and its use in the demonstration of enzymes in normal and neoplastic tissues. *Cancer Res.* 10:113–121.

Schöpp, W., and Rothe, U. (1975).   Kinetische Untersuchungen zum Umsatz längerkettiger Alkohole durch Hefe-Alkoholdehydrogenase im Bereich des Übergangs von der echten Lösung zur Emulsion bzw. Suspension. *Acta Biol. Med. Ger.* 34:197–201.

Seligman, A. M., Karnovsky, M. J., Wasserkrug, H. L., and Hanker, J. S. (1968).   Nondroplet ultrastructural demonstration of cytochrome oxidase activity with a polymerizing osmiophilic reagent, diaminobenzidine (DAB). *J. Cell Biol.* 38:1–14.

Shaw, C. R., and Barto, E. (1965).   Autosomally determined polymorphism of glucose-6-phosphate dehydrogenase in *Peromyscus*. *Science* 148:1099–1100.

Shelton, E., and Schneider, W. C. (1952).   On the usefulness of tetrazolium salts as histochemical indicators of dehydrogenase activity. *Anat. Rec.* 112:61–81.

Stiller, D., and Gorski, J. (1969).   Untersuchungen zur Histotopochemie der Uridindiphosphatglucose-Dehydrogenase. *Acta Histochem.* 32:356–375.

Straus, W. (1964).   Factors affecting the cytochemical reaction of peroxidase with benzidine and the stability of the blue reaction product. *J. Histochem. Cytochem.* 12:462–469.

Tabakoff, B., and Erwin, V. G. (1970).   Purification and characterization of a reduced nicotinamide adenine dinucleotide phosphate-linked aldehyde reductase from brain. *J. Biol. Chem.* 245:3263–3268.

Wattenberg, L. W. (1958).   Microscopic histochemical demonstration of steroid-3β-ol dehydrogenase in tissue sections. *J. Histochem. Cytochem.* 6:225–232.

Wattenberg, L. W., and Leong, J. L. (1960).   Effects of coenzyme Q_{10} and menadione on succinic dehydrogenase activity as measured by tetrazolium salt reduction. *J. Histochem. Cytochem.* 8:296–303.

Winer, A. D. (1958).   A note on the substrate specificity of horse liver alcohol dehydrogenase. *Acta Chem. Scand.* 12:1695–1696.

16

MISCELLANEOUS ENZYMES

Certain enzymes that will be considered in this chapter do not logically fit into any other chapter. Many other enzymes that can be demonstrated histochemically could also have been included here. However, it is not in keeping with the purpose of this book to include all the histochemically demonstrable enzymes; it was necessary to narrow the number down to a relatively few. The selection, as elsewhere, was based on the relevance and importance of the enzyme and on the reliability and simplicity of the histochemical method. Even so, the selection will all too often seem arbitrary.

CARBONIC ANHYDRASE

Carbonic anhydrase is a zinc metalloenzyme that catalyzes the hydration of carbon dioxide to form carbonic acid:

$$CO_2 + H_2O \rightarrow H_2CO_3$$

Its existence in red blood cells and its role in CO_2 transport is well known. The enzyme is also present in the parietal cells of the stomach, where it is undoubtedly involved in the formation of HCl. It is also present in salivary glands, pancreas, and kidney, in the bladder of some animals, and in frog skin. In addition to catalyzing the hydration of CO_2, the enzyme also catalyzes the hydration of aldehydes, and it also acts as a weak esterase.

Kurata (1953) described a method for the histochemical demonstration of carbonic anhydrase that used sodium carbonate as the substrate. Liberated carbonate ions were trapped and precipitated with manganese ions. The manganese carbonate precipitate was then visualized with periodate. The method was considered unsatisfactory by some workers because there were troublesome precipitates and because of a nonspecific reaction that could not be eliminated with known inhibitors of the enzyme. Also, the nature of the histochemical reaction was not understood. Häusler (1958) improved the method by using a cobalt salt instead of manganese, and the cobalt precipitate was visualized with ammonium sulfide. Hansson (1967) further modified the method and used the specific inhibitor, acetazolamide, to prepare controls. Rosen (1970) again modified the procedure by simply allowing the incubation medium to stand exposed to air for 20 minutes to allow a carbonate (and pH) gradient to develop. However, despite all the modifications and improvements, criticism of the method persisted. Muther (1972) pointed out that the reaction catalyzed by the enzyme proceeds nonenzymatically as well, although it is very slow; he thought that the histochemical reaction might not represent enzymatic activity at all. The validity of the method was eloquently defended by Rosen and Musser (1972) against the criticism of Muther (1972). The method of Rosen (1970) can therefore be recommended for demonstrating carbonic anhydrase.

PHOSPHOGLUCOMUTASE
Phosphoglucomutase catalyzes the interconversion of glucose 1-phosphate and glucose 6-phosphate. Phosphoglucomutase has an important function in glycogen metabolism, because glucose 1-phosphate is converted to uridine diphosphoglucose (UDP-glucose), which is the substrate for glycogen synthesis. Glucose 1-phosphate is also the immediate product of glycogen degradation.

Meijer (1967) published an indirect method for demonstrating phosphoglucomutase. The substrate, glucose 1-phosphate, is enzymatically converted to glucose 6-phosphate, which in turn serves as the substrate for exogenous glucose 6-phosphate dehydrogenase. The final reaction product is a formazan by the usual dehydrogenase method using nitro blue tetrazolium (nitro BT). A refined version of the method, published by Yano (1968), used certain activators and an inhibitor.

GLUCOSE PHOSPHATE ISOMERASE
Glucose phosphate isomerase catalyzes the conversion of fructose 6-phosphate to glucose 6-phosphate. This enzyme is present in many tissues at high concentrations. Glucose phosphate isomerase is of great importance because it regulates the amount of glucose 6-phosphate that may enter the glycolysis pathway. The alternative pathway for glucose 6-phosphate is the phosphogluconate pathway. It is interesting, therefore, that two intermediates of this alternative pathway—namely, erythrose 4-phosphate and 6-phosphogluconate—are potent inhibitors of glucose phosphate isomerase.

Meijer and Bloem (1969) described an indirect method for demonstrating glucose phosphate isomerase. Fructose 6-phosphate, which is used as the substrate, is enzymatically converted to glucose 6-phosphate. Exogenous glucose 6-phosphate dehydrogenase is added to the incubation medium and the remainder of the procedure is similar to the one for phosphoglucomutase, as described previously.

GLYCOGEN PHOSPHORYLASE

The normal function of glycogen phosphorylase is to degrade glycogen to form glucose 1-phosphate. The degradation reaction is reminiscent of a hydrolysis reaction except that phosphoric acid is introduced rather than water. The reaction is reversible, although, under physiological conditions, glycogen degradation is highly favored. The two most important depots of glycogen are the liver and skeletal muscles, and these are therefore the two most important tissues of phosphorylase.

Even though under physiological conditions the degradation of glycogen is favored, the conditions can be altered in such a way that synthesis of glycogen is favored. Histochemical methods for demonstrating this enzyme capitalize on the enzyme's ability to synthesize glycogen. Tissue sections are incubated under conditions that favor the synthesis of glycogen by phosphorylase. The two approaches that can be used are described in the following sections.

Demonstration of Glycogen

The newly formed glycogen can be demonstrated by a number of existing methods—the periodic acid–Schiff (PAS) reaction, Best's carmine, and iodine have been used for this purpose. The iodine method, developed by Takeuchi and Kuriaki (1955), is the most satisfactory, because it distinguishes between newly formed glycogen (blue-black) and preexisting glycogen (mahogany color). The other two methods do not distinguish between the two types of glycogen, and the sections must therefore be preincubated with amylase to deplete the section of preexisting glycogen. However, the iodine method has the disadvantage that it does not result in permanently stained sections and must be further processed to render them permanent.

The method of Takeuchi and Kuriaki (1955) has been modified by the addition of polyvinyl pyrrolidone (PVP) in the incubation medium as well as by increasing the substrate concentration tenfold (Eränkö and Palkama, 1961). This is a reliable method and has been widely used.

Metal Precipitation Methods

The other approach involves the formation of precipitates from the phosphate ions that are liberated during the synthesis of glycogen. Hori (1964) applied the lead precipitation principle of the Gomori phosphatase technique and visualized the precipitate with ammonium sulfide. It was shown, however, by Lindberg and Palkama (1972) and Lindberg (1973a) that lead ions seriously inhibit phosphorylase activity. Lindberg (1973b) therefore developed a method that uses ferric ions to precipitate the liberated phosphate ions. At $5mM$, ferric ions also strongly inhibit the enzyme. However, only $3mM$ of ferric ions is required to precipitate the phosphate satisfactorily, and at that concentration the inhibition is not serious. The precipitate was again visualized with ammonium sulfide.

Lindberg's method (1973b) for demonstrating phosphorylase has merit. However, the method of Eränkö and Palkama (1961) is still the method of choice for this purpose.

TRANSAMINASES

Transaminases are a group of enzymes concerned with transferring amino groups from amino acids to α-ketoglutarate—or vice versa, since the reactions are reversible. This is a very important type of reaction in the synthesis of amino acids. Pyridoxal phosphate always serves as the coenzyme and is directly involved in transferring the amino group. Glutamate oxaloacetate transaminase, which has been studied more than other transaminases, catalyzes the following reaction:

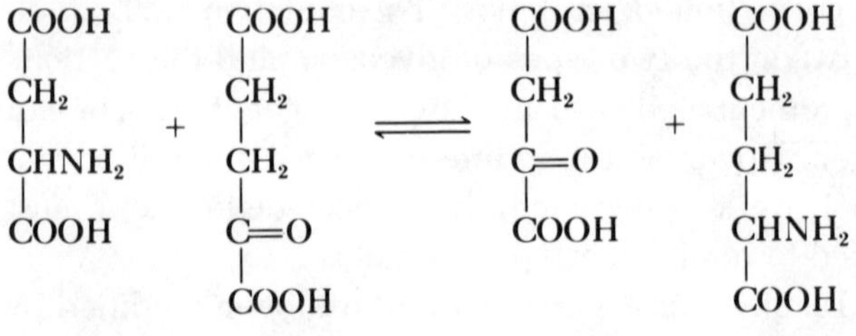

$$
\begin{array}{cccc}
\text{COOH} & \text{COOH} & \text{COOH} & \text{COOH} \\
| & | & | & | \\
\text{CH}_2 & \text{CH}_2 & \text{CH}_2 & \text{CH}_2 \\
| \quad + \quad | & \rightleftharpoons & | \quad + \quad | \\
\text{CHNH}_2 & \text{CH}_2 & \text{C=O} & \text{CH}_2 \\
| & | & | & | \\
\text{COOH} & \text{C=O} & \text{COOH} & \text{CHNH}_2 \\
 & | & & | \\
 & \text{COOH} & & \text{COOH} \\
\text{L-aspartate} & \alpha\text{-ketoglutarate} & \text{oxaloacetate} & \text{L-glutamate}
\end{array}
$$

The enzyme requires two substrates, either the two compounds on the left of the equation or the two on the right.

A histochemical reaction could be designed if one of the four compounds involved in the reaction could be precipitated and visualized. It turns out that oxaloacetate can be readily precipitated with lead, and Lee and Torack (1968) used this fact as a basis for a histochemical method for this enzyme. Aspartate and α-ketoglutarate are supplied as substrates for the enzyme, and the oxaloacetate is precipitated with Pb^{2+} as it is formed.

The concentration of Pb^{2+} and the pH of the incubation medium are of critical importance. At a slightly acid pH, Pb^{2+} may promote nonenzymatic decarboxylation of oxaloacetate. If the concentration of Pb^{2+} is too high, it may be difficult to hold it in solution. These two factors must therefore be carefully controlled.

An indirect method that has been described by Dikow and Lolova (1974) can be used to demonstrate a number of different transaminases. An appropriate amino acid and α-ketoglutarate are provided as substrates. As glutamate is formed by the action of the transaminase, it serves as the substrate for exogenous glutamate dehydrogenase. The remainder of the reaction is similar to other indirect methods using exogenous dehydrogenases. Many different transaminases can be demonstrated merely by adding the appropriate amino acid to the incubation medium. By using the appropriate amino acids as substrates, Dikow and Lolova (1974) were able to demonstrate the transaminases of aspartate, alanine, glycine, isoleucine, leucine, phenylalanine, serine, tryptophan, tyrosine, and valine.

UDP-GALACTOSE-4-EPIMERASE

UDP-galactose-4-epimerase catalyzes the interconversion of UDP-glucose and UDP-galactose, and it is therefore sometimes called UDP-glucose-4-epimerase. The reaction is essential for the body's utilization of galactose such as that derived from ingesting lactose (milk sugar). Nicotinamide-adenine dinucleotide (NAD) is an absolute requirement for this enzyme. The reaction evidently consists of two separate events, with a keto-sugar and NADH as intermediates.

Diculescu et al. (1968) have demonstrated this enzyme with a method similar to their earlier method for UDP-glucuronate epimerase (see the next section). They assumed that sometime during the reaction an electron was available that could reduce nitro BT to a formazan. Since the reaction is reversible, either UDP-glucose or UDP-galactose could be used as the substrate. However, a much higher concentration of UDP-glucose than of UDP-galactose is required for a positive reaction. This is undoubtedly because the reaction equilibrium lies in favor of UDP-glucose formation. Furthermore, if UDP-glucose were used, the dehydrogenase might oxidize the substrate and contribute to formazan formation.

Diculescu et al. (1968) found intense reactions in the fibroblasts of skin and umbilical cord, in the periportal areas of the liver, and in the renal collecting tubules. Other types of connective tissue showed less intense reactions.

UDP-GLUCURONATE-5-EPIMERASE

UDP-glucuronate-5-epimerase catalyzes the epimerization of UDP-glucuronate to form UDP-iduronate, which is a constituent of dermatan sulfate. Diculescu and Onicescu (1966) described a method for demonstrating this enzyme. By supplying NAD and nitro BT in the incubation medium, a formazan was deposited at the enzyme sites. Reactions were found in connective tissue cells and in the basal laminae.

METHOD FOR CARBONIC ANHYDRASE
(Hansson, 1967; Rosen, 1970)

TISSUES:
Use either fresh-frozen sections or frozen sections of aldehyde-fixed tissues. Pick up sections on a Millipore filter (25 μ thick with 0.45-μ pores).

PREPARATION OF INCUBATION MEDIUM:
Prepare the following solution:

$CoSO_4$, $1.75mM$
H_2SO_4, $53.0mM$
$NaHCO_3$, $15.7mM$
KH_2PO_4, $11.7mM$

Allow the medium to stand for 20 minutes before incubation.

PROCEDURE:
(1) Float the Millipore filter containing the sections on top of the incubation medium. Incubate for 6 to 12 minutes.
(2) Wash the sections in a $0.67mM$ solution of KH_2PO_4, pH 5.0.
(3) Treat with 0.6 percent ammonium sulfide for 1 minute.
(4) Rinse in saline.
(5) Mount with glycerin jelly, or dehydrate, clear, and mount with Permount.

CONTROLS:
Prepare an incubation medium as described previously but containing in addition $10^{-5}M$ acetazolamide.

RESULTS:
Areas of carbonic anhydrase activity are yellowish brown or black. All activity should have been prevented in sections incubated in the control medium.

METHOD FOR PHOSPHOGLUCOMUTASE
(Meijer, 1967; Yano, 1968)

TISSUES:
Use fresh-frozen sections. Fix sections for 30 minutes in acetone at 0° to 4°C.

INCUBATION MEDIUM:

Disodium glucose 1-phosphate	80 mg
Nicotinamide-adenine dinucleotide phosphate (NADP)	2.5 mg
Adenosine triphosphate (ATP)	5.0 mg

$MgCl_2 \cdot 6H_2O$	12.5 mg
L-Histidine	18.0 mg
Nitro BT, 1 mg/ml	5.0 ml
Imidazole buffer, $40mM$, pH 7.4	10.0 ml
Gelatin, 3 percent solution	5.0 ml
Glucose 6-phosphate dehydrogenase, 1 mg/ml	0.04 ml
Distilled water to make 25 ml	

PROCEDURE:
(1) Incubate sections for 1 hour at 37°C.
(2) Immerse in 10 percent formalin for 10 minutes.
(3) Wash in water.
(4) Mount with polyvinyl pyrrolidone mounting medium.

CONTROLS:
Prepare two control media as indicated above except that in the first medium the substrate (glucose 1-phosphate) is omitted, and in the second medium $1mM$ beryllium sulfate (an inhibitor of phosphoglucomutase) is added.

RESULTS:
Formazan deposits present in the test but absent in both controls can be regarded as sites of phosphoglucomutase activity.

METHOD FOR GLUCOSE PHOSPHATE ISOMERASE
(Meijer and Bloem, 1969)

TISSUES:
Use fresh-frozen sections. Fix sections in acetone for 30 minutes at −25°C.

PREPARATION OF INCUBATION MEDIUM:

Disodium D-fructose-6-phosphate	10 mg
NADP, 5 mg/ml	0.6 ml
Nitro BT, 5 mg/ml	0.5 ml
Imidazole buffer, $0.15M$, pH 7.0	2.0 ml

Gelatin, 3 percent solution	3.0 ml
Glucose 6-phosphate dehydrogenase, 1 mg/ml	0.02 ml

Adjust the pH to 7.0. Prepare immediately before use.

PROCEDURE:
(1) Incubate sections for 1 hour at 37°C.
(2) Fix in 10 percent neutral buffered formalin for 30 minutes.
(3) Mount in glycerin jelly.

CONTROL:
Prepare an incubation medium in which the substrate (fructose 6-phosphate) is omitted.

RESULTS:
Formazan deposits present in the test sections but absent in the control sections indicate sites of enzyme activity.

IODINE METHOD FOR GLYCOGEN PHOSPHORYLASE
(Eränkö and Palkama, 1961)

TISSUES:
Mount fresh-frozen cryostat sections on glass slides and allow them to dry.

PREPARATION OF INCUBATION MEDIUM:

Acetate buffer, 0.1M, pH 5.9	10 ml
Glucose 1-phosphate	100 mg
Adenosine 5-phosphate	10 mg
Glycogen	2 mg
Sodium fluoride	180 mg
Polyvinyl pyrrolidone	900 mg
Insulin, 40 IU/ml	1 drop

Make up fresh each time and filter immediately before use.

PREPARATION OF STAINING SOLUTION:

Iodine crystals	1 gm
Potassium iodide	2 gm
Sucrose	32.7 gm
Distilled water	300 ml

PROCEDURE:

(1) Incubate sections for 15 minutes to 3 hours at 37°C.
(2) Allow sections to dry.
(3) Immerse in 40 percent ethanol for 2 minutes.
(4) Allow to dry.
(5) Immerse in 0.32M sucrose.
(6) Stain for 5 minutes.
(7) Mount with iodine glycerin (1:5 mixture).

CONTROLS:

Sections are treated similarly except that glucose 1-phosphate is omitted from the incubation medium.

RESULTS:

Blue or blue-violet granules indicate sites of phosphorylase activity. The controls should be completely negative.

FERRIC ION PRECIPITATION METHOD FOR GLYCOGEN PHOSPHORYLASE
(Lindberg, 1973b)

TISSUES:

Use cryostat sections of fresh-frozen tissues, 20 to 30 μ thick.

PREINCUBATION MEDIUM:

Acetate buffer, 0.1M, pH 5.9	10 ml
Absolute ethanol	2 ml
Polyvinyl pyrrolidone	900 mg
Sodium fluoride	180 mg

INCUBATION MEDIUM:

Acetate buffer, 0.1M, pH 5.9	10 ml
Absolute ethanol	2 ml
Polyvinyl pyrrolidone	900 mg
Dithiothreitol	22.2 mg
$FeCl_2 \cdot 4H_2O$	7.14 mg
Glucose 1-phosphate	100 mg

PROCEDURE:

(1) Place sections in preincubation medium for 30 minutes at 37°C.
(2) Wash in 3 changes (2 minutes each) of preincubation medium in which sodium fluoride was omitted.
(3) Place in incubation medium for 2 hours at 37°C.
(4) Wash in 3 changes (3 minutes each) of distilled water.
(5) Place in 1 percent ammonium sulfide (freshly prepared) for 2 minutes.
(6) Wash in distilled water.
(7) Mount in glycerin jelly.

CONTROLS:

Prepare another preincubation medium as previously described, and add 20mM dichlorophenoxyacetic acid, an inhibitor of phosphorylase. To prepare control sections, substitute this medium in step 1 of the incubation procedure.

RESULTS:

A yellowish brown to black precipitate indicates the sites of phosphorylase activity. Control sections should be completely negative.

METHOD FOR TRANSAMINASES
(After Dikow and Lolova, 1974)

TISSUES:

Use fresh-frozen sections.

PREPARATION OF INCUBATION MEDIUM:

Amino acid of choice, 1.0M	1.0 ml
α-Ketoglutarate, 0.5M	0.2 ml
NAD	30 mg
Nitro BT	10 mg
Polyvinyl pyrrolidone	100 mg
Glutamate dehydrogenase, 1 mg/ml	0.5 ml
Diaphorase, 1 mg/ml	0.1 ml
Phosphate buffer, 0.1M, pH 7.6	7.0 ml

Adjust the pH to 7.6 if necessary and add distilled water to make 10 ml.

PROCEDURE:

(1) Incubate sections for 20 to 60 minutes at 37°C.
(2) Postfix in 10 percent neutral buffered formalin overnight.
(3) Wash in water.
(4) Dehydrate through graded alcohols, clear in xylene, and mount.

CONTROLS:

Prepare an incubation medium in which the amino acid substrate or the α-ketoglutarate or both are replaced with water, and substitute in the incubation procedure.

RESULTS:

Deposits of the blue formazan dye indicate sites of transferase activity. Control sections may show diffuse rose color but should have no deposits of blue formazan.

METHOD FOR UDP-GLUCURONATE EPIMERASE
(Diculescu and Onicescu, 1966)

TISSUES:

Use fresh-frozen cryostat sections. Mount on clean glass slides and dry briefly.

PREPARATION OF INCUBATION MEDIUM:
The following reagents are dissolved in $0.1M$ tris buffer, pH 7.1 to 7.2, at the concentrations indicated:

UDP-glucuronic acid	$1mM$
NAD	$5mM$
$MgCl_2$	$1mM$
Ethylenediaminotetraacetate (EDTA)	$5mM$
Nitro BT	$10mM$

PROCEDURE:
(1) Incubate sections for 45 minutes at 37°C.
(2) Rinse briefly.
(3) Fix in 10 percent formalin for 1 to 2 hours.
(4) Mount with Apathy's syrup.

CONTROLS:
Prepare two incubation media, one in which the substrate UDP-glucuronic acid is omitted and another in which p-chloromercuroben-zoate, $20mM$ (final concentration), is added.

RESULTS:
Deposits of the blue formazan dye represent sites of enzyme activity. Both controls should be negative.

METHOD FOR UDP-GALACTOSE EPIMERASE
(Diculescu et al., 1968)

TISSUES:
Use fresh-frozen cryostat sections. Mount on clean glass slides and dry briefly.

PREPARATION OF INCUBATION MEDIUM:
The following components are dissolved in $0.1M$ tris buffer, pH 7.4, at the concentrations indicated.

UDP-galactose	$2mM$
NAD	$5mM$
NaCl	$40mM$
Nitro BT	$10mM$

PROCEDURE:

(1) Incubate sections for 45 minutes at 37°C.

(2) Rinse quickly.

(3) Fix in 10 percent formalin for 1 to 2 hours.

(4) Mount with Apathy's syrup.

RESULTS:

Deposits of the blue formazan dye represent sites of enzyme activity.

REFERENCES

Diculescu, I., and Onicescu, D. (1966).   The histochemistry of epimerases. I. Uridine diphosphate glucuronic acid epimerases. *Acta Histochem.* 25:242–250.

Diculescu, I., Onicescu, D., Szegli, G., and Dumitrescu, A. (1968).   Histochemistry of the epimerases. III. Uridine diphosphogalactose-4-epimerase. *Histochemistry* 14:143–148.

Dikow, A., and Lolova, I. (1974).   Histochemischer Nachweis der Aminotransferasen. I. Methode zur histochemischen Nachweis einiger Aminotransferasen in Rattenorganen. *Acta Histochem.* 51:102–108.

Eränkö, O., and Palkama, A. (1961).   Improved localization of phosphorylase by the use of polyvinyl pyrrolidone and high substrate concentration. *J. Histochem. Cytochem.* 9:585.

Hansson, H. P. J. (1967).   Histochemical demonstration of carbonic anhydrase activity. *Histochemistry* 11:112–128.

Häusler, G. (1958).   Zur Technik und Spezifität des histochemischen Carboanhydrasenachweises im Modellversuch und in Gewebsschnitten von Rattennieren. *Histochemistry* 1:29–47.

Hori, S. H. (1964).   Cytological phosphorylase locations in rat liver and muscle as shown by a lead precipitation method. *Stain Technol.* 39:275–278.

Kurata, Y. (1953).   Histochemical demonstration of carbonic anhydrase activity. *Stain Technol.* 28:231–233.

Lee, S. H., and Torack, R. M. (1968).   A biochemical and histochemical study of glutamic oxalacetic transaminase activity of rat hepatic mitochondria fixed in situ and in vitro. *J. Cell Biol.* 39:725–732.

Lindberg, L. A. (1973a).   Lead and some other metals in the histochemical demonstration of rat liver glycogen phosphorylase activity. *Histochemistry* 36:347–353.

Lindberg, L. A. (1973b).   Histochemical demonstration of rat liver glycogen phosphorylase activity with iron (Fe^{++}). *Histochemistry* 36:355–365.

Lindberg, L. A., and Palkama, A. (1972).   The effect of some factors on the histochemical demonstration of liver glycogen phosphorylase activity. *J. Histochem. Cytochem.* 20:331–335.

Meijer, A. E. F. H. (1967).   Histochemical method for the demonstration of the activity of phosphoglucomutase. *Histochemistry* 8:248–251.

Meijer, A. E. F. H., and Bloem, J. H. (1969).   Histochemical method for the demonstration of the activity of glucosephosphate isomerase. *Acta Histochem.* 32:110–116.

Muther, T. F. (1972).   A critical evaluation of the histochemical methods for carbonic anhydrase. *J. Histochem. Cytochem.* 20:319–330.

Rosen, S. (1970).   Localization of carbonic anhydrase activity in transporting urinary epithelia. *J. Histochem. Cytochem.* 18:668–670.

Rosen, S., and Musser, G. L. (1972).   Observations on the specificity of newer histochemical methods for the demonstration of carbonic anhydrase activity. *J. Histochem. Cytochem.* 20:951–954.

Takeuchi, T., and Kuriaki, H. (1955).   Histochemical detection of phosphorylase in animal tissues. *J. Histochem. Cytochem.* 3:153–160.

Yano, Y. (1968).   Improved method for histochemical demonstration of phosphoglucomutase. *Acta Histochem. Cytochem.* 1:186–196.

17

QUANTITATIVE HISTOCHEMISTRY

Histochemical research is generally concerned with an analysis of elements on the tissue, cellular, or subcellular level. Because of the minute quantities of the elements of interest, quantitation of histochemical reactions is difficult, and conventional methods of chemical analysis are often of little aid.

The earliest application of quantitative histochemical methods was directed toward the measurement of DNA in single cells by Caspersson and co-workers during the late 1930s and the 1940s. In fact, more papers have been published and continue to be published about this subject than about any other in quantitative histochemistry. However, quantitative histochemical methods can be useful tools in investigating many other histochemical problems, and in recent years quantitative histochemical methods have been applied to many studies that do not involve nucleic acids.

VISUAL METHODS

The most popular, or at least the most frequently used, system of histochemical quantitation involves grading the strengths of the reactions somewhat subjectively during visual examination. A system of plusses is used to represent the strengths of the reactions: 4+ describes a very intense reaction, 1+ describes a weak reaction, and 0 (or a minus sign) is assigned to a negative reaction. This must surely be the crudest and the most primitive method of histochemical quantitation. The merit of the system is that no special instrumentation is required, which accounts for the frequency with which it is used. Anyone can use the system with a little practice. However, the relativeness and the subjectivity inherent in this system limit the confidence that can be placed in the results.

Another approach to quantitative histochemistry has been a type of comparimetry. Such a procedure was used by Benditt and Arase (1958) to quantitate enzyme histochemical reactions. First, a series of standards was prepared by incubating sections for varying periods of time. The experimental sections were then blindly matched with the standard series to get an estimate of the relative reaction intensity. A similar system was used by Jeffree (1970) for estimating phosphatases in osteoclasts. The same principle was used by Goldstein (1963) to study the thermodynamics of azure A staining, although the standards series was obtained differently. A 1-percent staining solution of azure A was prepared, and it was then serially diluted, 1 : 1 each time, until 14 different concentrations of stain were obtained. The concentrations of azure A thus ranged from 1/2 percent to 1/16,384 percent. Serial sections were then stained to equilibrium, one section in each of the concentrations, to obtain the series of standards. The experimental sections were then blindly matched with the standards as before.

Since the human eye has to serve as the "photometer" in such procedures, the usable range of reaction intensities is somewhat limited. Reactions that are too intense or too weak could not be estimated with any reliability.

Hopsu and Glenner (1965) found the method of Benditt and Arase (1958) totally unsatisfactory and developed a different type of visual system. Tissue sections were frequently removed and observed under the microscope during the course of incubation. The period of time required to obtain a barely perceptible reaction was noted and recorded as the experimental parameter. The subjectivity inherent in this system still limits its precision and is probably the reason why the method is rarely used.

INSTRUMENTAL METHODS

Instruments for the measurement of histochemical stain intensities range from very crude ones to ones that are extremely sophisticated. An instrument that provides measurements at one wavelength or with a fixed portion of the spectrum is called a microdensitometer, whereas one that is capable of continuously varying the wavelength is called a microspectrophotometer. Any such instrument requires a few essential elements: a good quality microscope, a means for limiting the viewing field, a stable source of illumination, and a mechanism for measuring the light intensity. The crudest system might consist of a simple microscope with an ordinary illuminator, using the field diaphragm for limiting the viewing field, and a cadmium sulfide photoresistor light meter connected to a milliammeter for measuring the light intensity. This sort of system is comparatively easy to build and has been very useful in quantitating autoradiograms (Haas et al., 1975). However, such systems are not frequently used for estimating staining intensities because of their relative insensitivity.

The system already described could be improved tremendously by introducing either a color filter or an interference filter into the illumination beam in order to select only the portion of the spectrum that corresponds to the absorption peak of the stain. The sensitivity could be further increased by the following steps: (1) The cadmium sulfide cell can be replaced with a photomultiplier tube with good sensitivity throughout the visible range. (2) A stabilized power supply, such as a voltage regulator, can be used to stabilize the intensity of the illumination. (3) A small-diameter aperture can be inserted into the illumination beam to limit the viewing field. These improvements would probably yield a reasonably accurate microdensitometer.

It would be possible to replace the illumination system completely with a continuously variable monochromer, which would confer the advantage of a very narrow band of the spectrum selectable from any portion of the visible spectrum. This advantage, however, is obtainable only at considerable expense. Usually the high cost of commercial microspectrophotometers makes it difficult to justify purchasing them. They are extremely useful, however, in making accurate measurements of stained tissue sections and particularly the spectra of tissue components stained with metachromatic dyes.

Instruments of this kind have been used to good advantage for estimating intensities of many different histochemical reactions. By using a specially built miniature incubation chamber, the intensities of enzyme histochemical reactions have been monitored during incubation (Täljedal, 1970; Bacsy and Rappay, 1971; Rost et al., 1973; Jarrett and Please, 1970) to study the kinetics of the enzymes. Alternatively, the reaction can be done with conventional incubation, and measurements would then be obtained after the sections are mounted. With the latter method, Cabrini et al. (1969) quantitated the reaction product of succinic dehydrogenase.

Distributional Error

When a beam of light is passed through a uniformly stained section, some light is absorbed but a portion of it is transmitted. It is the transmitted light, or *transmittance,* that is measured by the photomultiplier tube. However, it is the amount of absorbed light, or *absorption,* that is linearly proportional to the concentration of absorbing material. Absorption is a function of the logarithm of transmittance, and it can easily be calculated. However, normally the conversion is made electronically by the instrument and the readout reflects absorption directly.

The problem of distributional error arises when the photomultiplier tube measures the light flux transmitted through an inhomogeneously stained field. For example, assume that half of the measured field has one stain intensity and the other half a different stain intensity. The absorption that will be recorded is lower than would be found if both fields were measured separately and averaged, even though each density is well within the measurable range. The photomultiplier tube will read transmittance from the two respective areas, T_1 and T_2. When measured together the absorption will be:

$$O.D. = \log \frac{100}{\frac{1}{2}(T_1 + T_2)}$$

When measured separately and averaged, the absorption will be:

$$O.D. = \frac{1}{2}\left(\log \frac{100}{T_1} + \log \frac{100}{T_2}\right)$$

These two equations are not the same. If the difference between T_1 and T_2 is small, the error will be small. But if the difference is great, the error will be quite large. For instance, if the transmittances are 90 percent and 1 percent, the error will be 66.6 percent (Altman, 1975).

A method has been developed independently by Patau (1952) and Ornstein (1952) for estimating the "distributional error." The method is known as the "two wavelengths" method, since measurements of the same area of tissue are made with two wavelengths. It is preferable to select the two wavelengths, γ_1 and γ_2, in such a way that their extinction coefficients, E_1 and E_2, have a proportionality of $1:2$. Other proportions could be selected but this proportion simplifies the calculation of the error. γ_1 could be selected at an absorption maximum of the dye and γ_2 could be selected at some other point of the absorption spectrum such that E_2 is only one-half as great as E_1. If it becomes necessary to use the two wavelengths method, the paper by Mendelsohn (1958) should be consulted.

Another method for overcoming the problem of distributional error is to scan the area of interest with a spot of light near the limit of light resolution—that is, 0.25 to 0.50 microns. Successive horizontal scan lines can be used until the whole area of interest has been covered. In effect, an area only the size of the spot is measured at any one instant. Information from the whole area is electronically integrated to give a single reading. Many of the commercial instruments now available have provision for scanning.

Spectrophotometry of Large Uniform Sections

Methods have been devised for obtaining the spectra of histochemical materials with the object of eliminating the need for a microspectrophotometer. Rosenberg (1971) wanted to study the spectra of metachromatically stained cartilage; he used bovine nasal cartilage, which gave him sections of fairly uniform tissue at least a square centimeter in size. These sections could be placed into an ordinary spectrophotometer by simply modifying the cuvette holder to accommodate the slide.

In a series of papers, Tas (1975), Tas and Greenen (1975), and Tas and Roozemond (1973) studied some characteristics of metachromatic dyes with a somewhat similar system. However, instead of using histological sections, they embedded polyanions into a thin polyacrylamide film on a histological slide to simulate a histological section. These slides were then stained and analyzed with a spectrophotometer as is done in the method of Rosenberg.

ANALYSIS OF WHOLE SECTIONS

It is sometimes useful to analyze the total reaction product of a whole section. Enzyme reaction products are often readily soluble in certain solvents so that they can be eluded from the tissue and measured with a densitometer or spectrophotometer. Hopsu and McMillan (1964) used this method to quantitate an azo-dye reaction product of an esterase. Formazan from the tetrazolium method for dehydrogenases can be quantified in this way (McCabe et al., 1965; Altman, 1969a, 1969b, and 1972; Butcher and Altman, 1973).

If the reaction product contains a metal, the reaction product of the whole section can be quantified with an atomic absorption spectrophotometer (Rosenquist and Rosenquist, 1974). After the desired histochemical reaction has been carried out, the sections are removed from the slide, weighed, and dissolved in nitric acid. The milieu is then analyzed for the metal of interest. The method works well with alcian blue (copper), Hale's colloidal iron method (iron), and the Gomori reaction for phosphatases (lead). The procedure is sensitive and can be performed with great accuracy. Since this method involves the complete destruction of the section, it is highly desirable to photograph it before analysis.

PHOTOGRAPHIC MICRODENSITOMETRY

Photographic microdensitometry has provided a way of quantifying histochemical reactions without the aid of a microdensitometer or a microspectrophotometer. The process essentially involves registering the intensity of the histochemical reaction on a photographic negative, printing an enlarged positive, and determining the amount of silver or pigment in the print. This opens the possibility of magnifying the area of interest to a size on the photographic print that is large enough to facilitate analysis. The silver (from black and white prints) or the dye (from color prints) is either measured densitometrically in the print or it is eluded and measured with standard analytical chemical methods.

It is of course essential that linearity exist between the density of the tissue section and the density of the photographic film. If the logarithm of the exposure is plotted against the density of the film, a sigmoid curve is obtained (Fig. 17-1); this is called the characteristic curve, or the Hurter-Driffield curve, and it varies from one type of film to another. Typically, the middle portion of the curve is straight, and that portion is useful for the purpose of photographic microdensitometry. It is generally possible to establish good linearity between the density of the tissue sections and the straight middle portion of the characteristic curve. Mitchell (1977) made an extensive study of the suitability of various photographic films for photographic microdensitometry.

Caspersson (1936) seems to have been the first to use this type of system. He obtained black and white photographic (ultraviolet) images and measured the silver gravimetrically, and he studied the ul-

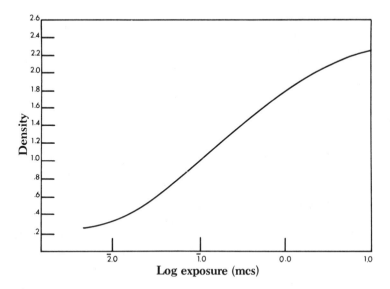

Figure 17-1. The characteristic, or Hurter-Driffield, curve for Kodak Panatomic X film (mcs = meter candle seconds).

traviolet absorption of nuclei with this method. Niemi (1958) used a similar method to determine hemoglobin content of red blood cells. He also discussed many of the theoretical and practical problems of photographic densitometry with black and white negatives and prints. Since the measurement of silver is generally difficult and inaccurate, Ornstein (1952) described the use of color positives from which the dye can be eluded and measured with a colorimeter. Den Tonkelaar and van Duijn (1964a) described in detail the practical application of such a method and used it (1964b) to determine the DNA content of nuclei. Negative images of the nuclei were obtained, and color prints were made. The dye was eluded and the concentration measured colorimetrically.

Troyer and Rosenquist (1975) developed another method in which the silver was eluded from the negatives and measured with an atomic absorption spectrophotometer, which eliminates the need for printing positives; Panatomic X 35 mm film was used for this purpose. All the experimental readings (including controls and blanks) can be made on a single 20- or 36-exposure roll of film. In this way, small variations that might occur during developing of separate films will not affect the results.

The usefulness of this method has been amply demonstrated in a study of the kinetics of alkaline phosphatase of the epiphyseal cartilage (Troyer et al., 1976) and in evaluation of the staining intensities of cartilage of arthritic rats (Nusbickel and Troyer, 1976).

REFERENCES

Altman, F. P. (1969a).   The quantitative elution of nitro-blue formazan from tissue sections. *Histochemistry* 17:319–326.

Altman, F. P. (1969b).   The use of eight different tetrazolium salts for a quantitative study of pentose shunt dehydrogenation. *Histochemistry* 19:363–374.

Altman, F. P. (1972).   Quantitative dehydrogenase histochemistry with special reference to pentose-shunt dehydrogenases. *Progr. Histochem. Cytochem.*, Gustav Fischer Verlag, Stuttgart; 4:225–273.

Altman, F. P. (1975).   Quantitation in histochemistry: A review of some commercially available microdensitometers. *Histochem. J.* 7:375–395.

Bacsy, E., and Rappay, G. (1971).   Microspectrophotometric data on the kinetics of acid phosphatase activity in cells removed from the peritoneal cavity of the rat. *Histochem. J.* 3:193–199.

Benditt, E. P., and Arase, M. (1958).   Enzyme kinetics in a histochemical system. *J. Histochem. Cytochem.* 6:431–434.

Butcher, R. G., and Altman, F. P. (1973).   Studies on the reduction of tetrazolium salts. II. The measurement of the half reduced and fully reduced formazans of neotetrazolium chloride in tissue sections. *Histochemistry* 37:351–363.

Cabrini, R. L., Viñuales, E. J., and Itoiz, M. E. (1969).   A microspectrophotometric method for histochemical quantitation of succinic dehydrogenase. *Acta Histochem.* 34:287–291.

Caspersson, T. (1936).   Über den chemischen Aufbau der Strukturen des Zellkernes. *Skand. Arch. Physiol.* 73 (Suppl. 8):1–151.

den Tonkelaar, E. M., and van Duijn, P. (1964a).   Photographic colorimetry as a quantitative cytochemical method. I. Principles and practice of the method. *Histochemistry* 4:1–9.

den Tonkelaar, E. M., and van Duijn, P. (1964b).   Photographic colorimetry as a quantitative cytochemical method. II. Determination of relative amounts of DNA in cell nuclei. *Histochemistry* 4:10–15.

Goldstein, D. J. (1963).   An approach to the thermodynamics of histological dyeing, illustrated by experiments with Azure A. *Q. J. Microsc. Sci.* 104:413–439.

Haas, R. A., Robertson, D. M., and Meyers, N. (1975).   Microscope densitometer system for point measurement of autoradiograms. *Stain Technol.* 50:139–141.

Hopsu, V. K., and Glenner, G. G. (1965).   A histochemical enzyme kinetic system applied to the trypsin-like amidase and esterase activity in human mast cells. *J. Cell Biol.* 17:503–520.

Hopsu, V. K., and McMillan, P. J. (1964).   Quantitative characterization of a histochemical enzyme system. *J. Histochem. Cytochem.* 12:315–324.

Jarrett, A., and Please, N. W. (1970).   A quantitative histochemical technique for the estimation of azo dye coupling reactions. *Histochem. J.* 2:297–313.

Jeffree, G. M. (1970).   The histochemical differentiation of various phosphatases in a population of osteoclasts by a simultaneous coupling method using different diazonium salts, with observations on the presence of inhibitors in stable diazonium salts. *Histochem. J.* 2:231–242.

McCabe, M., Maple, A. J., and Jones, G. R. N. (1965).   Variations in dehydrogenase and neotetrazolium reductase (diaphorase) activities in various buffers studied quantitatively in frozen sections. *J. Histochem. Cytochem.* 13:541–546.

Mendelsohn, M. L. (1958).   The two-wavelength method of microspectrophotometry. II. A set of tables to facilitate the calculations. *J. Biophys. Biochem. Cytol.* 4:415–424.

Mitchell, J. P. (1977).   Choosing materials for photographic colorimetry of sectioned tissue. *Histochem. J.* 9:767–779.

Niemi, M. (1958).   Cytophotometry by silver analysis of photographs. *Acta Anat.* (Suppl. 34):1–92.

Nusbickel, F. R., and Troyer, H. (1976).   Histochemical investigation of adjuvant-induced arthritis. *Arth. Rheum.* 19:1339–1346.

Ornstein, L. (1952).   The distributional error in microspectrophotometry. *Lab. Invest.* 1:250–265.

Patau, K. (1952).   Absorption microphotometry of irregular shaped objects. *Chromosoma* 5:341–362.

Rosenberg, L. (1971).   Chemical basis for the histological use of Safranin O in the study of articular cartilage. *J. Bone Joint Surg.* [Am.] 53:69–82.

Rosenquist, T. H., and Rosenquist, J. W. (1974).   A procedure for the use of atomic absorption spectrophotometry in quantitative histochemistry. *J. Histochem. Cytochem.* 22:104–109.

Rost, F. W. D., Bollmann, R., and Moss, D. W. (1973).   Characterization of alkaline phosphatase in tissue sections by microspectrofluorometry. *Histochem. J.* 5:567–575.

Täljedal, I.-B. (1970).   Direct fluorophotometric recording of enzyme kinetics in cryostat sections: Glucose-6-phosphate dehydrogenase in the endocrine mouse pancreas. *Histochemistry* 21:307–313.

Tas, J. (1975).   Histochemical conditions influencing metachromatic staining: A comparative study by means of a model system of polyacrylamide films. *Histochem. J.* 7:1–19.

Tas, J., and Greenen, L. H. M. (1975).   Microspectrophotometric detection of heparin in mast cells and basophilic granulocytes stained metachromatically with Toluidine Blue O. *Histochem. J.* 7:231–248.

Tas, J., and Roozemond, R. C. (1973).   Direct recording of metachromatic spectra in a model system of polyacrylamide films. *Histochem. J.* 5:425–436.

Troyer, H., and Rosenquist, T. H. (1975).   Atomic absorption spectrophotometry applied to photographic densitometry. *J. Histochem. Cytochem.* 23:941–944.

Troyer, H., Fisher, O. T., and Rosenquist, T. H. (1976).   Bone alkaline phosphatase kinetics studied by a new method. *Histochemistry* 50:251–259.

Apathy's Syrup

Mix 50 gm gum arabic with 50 gm sucrose dry, and then dissolve the mixture in 100 ml distilled water by warming to 55°C while shaking frequently. Add 15 mg merthiolate or 100 mg thymol as a preservative.

Canada Balsam

Dissolve the Canada balsam resin in xylene until it is a thick, syrupy mixture.

Farrant's Medium

Arsenic trioxide	1 gm
Gum arabic	50 gm
Glycerol	50 ml
Distilled water	50 ml

Glycerin Jelly Mounting Medium

Soak 40 gm gelatin in 210 ml distilled water for 2 hours. Add 250 ml glycerol and 5 gm phenol, and heat gently for 10 to 15 minutes while stirring. Store in a refrigerator and warm when needed.

Holt's Gum Sucrose Solution

Mix 30 gm sucrose with 1 gm gum arabic in the dry state (this is to prevent the gum arabic from caking). Add water to make 100 ml.

Karo Syrup

Mix equal volumes of corn syrup and water.

Mayer's Hematoxylin

Dissolve 0.1 gm hematoxylin and 5 gm aluminum ammonium sulfate in 100 ml distilled water. Add 20 mg sodium iodate ($NaIO_3$). As a preservative, 5 gm chloral hydrate should be added. Can be used immediately after filtering. Shelf life is 2 to 3 months.

Methyl Green Counterstain

Dissolve 1 gm methyl green in 100 ml of $0.1M$ veronal acetate buffer, pH 4.0. Some workers prefer to use methyl green that has been chloroform-extracted, as described on page 157.

Acetazolamide   (5-acetamido-1,3,4-thiadiazole-2-sulfonamide)

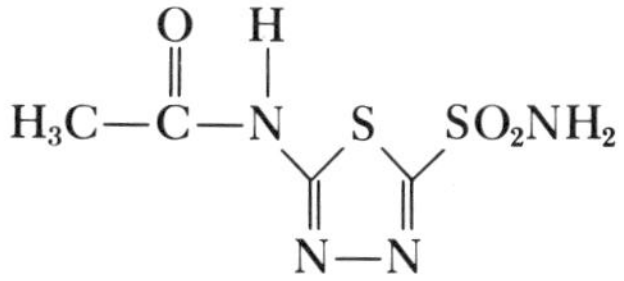

$$H_3C-\underset{\underset{\displaystyle O}{\|}}{C}-\underset{\underset{\displaystyle H}{|}}{N}\diagdown S\diagup SO_2NH_2$$
$$N-N$$

Mol. wt. 222.25

A specific inhibitor of carbonic anhydrase. As a drug, it is a diuretic and is used in the treatment of glaucoma.

Acetic acid, glacial

CH_3COOH

Mol. wt. 60.05   Sp. gr. 1.053

An irritant; causes burns of the skin and has a pungent odor. Used as a solvent for many organic compounds. Completely miscible in water and many organic solvents. To make an approximately 1.0M solution, dissolve 58 ml in water and bring to 1 liter with water.

Acetoacetic acid

One of the ketone bodies and a normal by-product of fatty acid metabolism. It is the reaction product of β-hydroxybuterate dehydrogenase.

Acetone

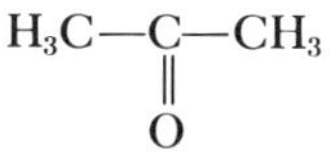

$$H_3C-\underset{\underset{\displaystyle O}{\|}}{C}-CH_3$$

Mol. wt. 58.08

A volatile, inflammable organic solvent. It is the simplest aliphatic ketone.

Acetylcholine chloride

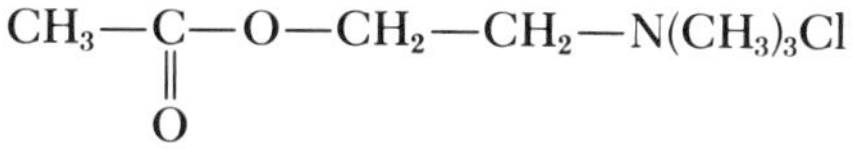

$$CH_3-\underset{\underset{\displaystyle O}{\|}}{C}-O-CH_2-CH_2-N(CH_3)_3Cl$$

Mol. wt. 181.66

Plays a role in the synaptic transmission of nerve impulses and therefore has numerous physiological effects. Used as a substrate for cholinesterase.

Acetyl-β-methylcholine chloride

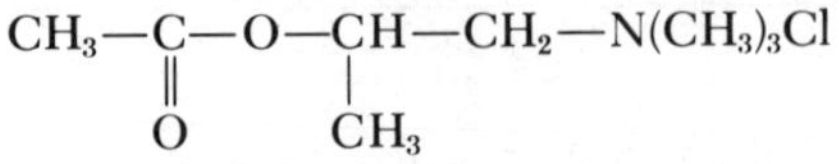

$$CH_3-\underset{\underset{O}{\|}}{C}-O-\underset{\underset{CH_3}{|}}{CH}-CH_2-N(CH_3)_3Cl$$

Mol. wt. 195.69
Can be used as a substrate for acetylcholinesterase.

Acetylthiocholine iodide

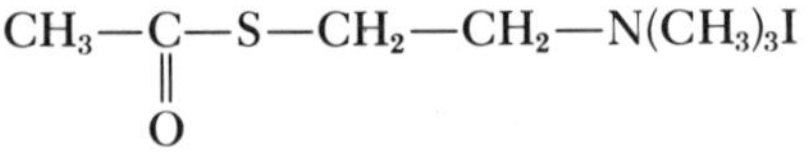

$$CH_3-\underset{\underset{O}{\|}}{C}-S-CH_2-CH_2-N(CH_3)_3I$$

Mol. wt. 289.18
Used as a substrate for cholinesterases.

Acrolein

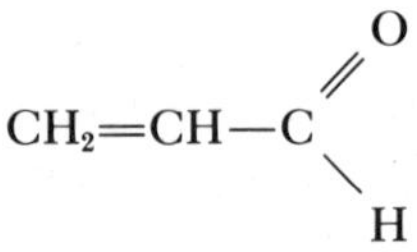

$$CH_2{=}CH-C\overset{\displaystyle\nearrow O}{\underset{\displaystyle\searrow H}{}}$$

Mol. wt. 56.06
An unsaturated aldehyde with a disagreeable odor. Its fumes are poisonous and tear-evoking. It is used as a fixative.

Adenosine diphosphate, trisodium

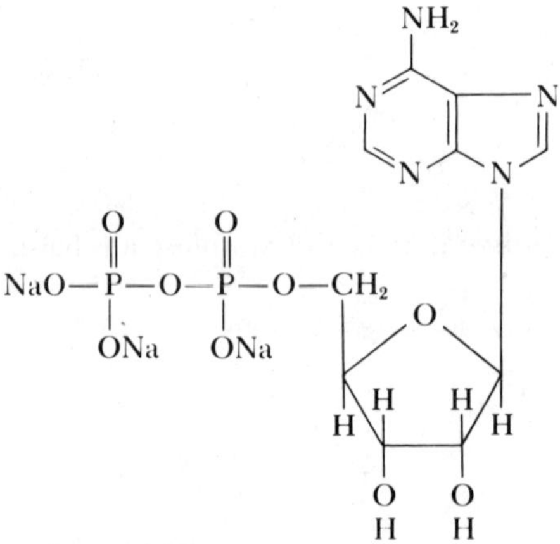

Mol. wt. 493.15
Used as a substrate for ADPase.

Adenosine-5'-monophosphate, disodium

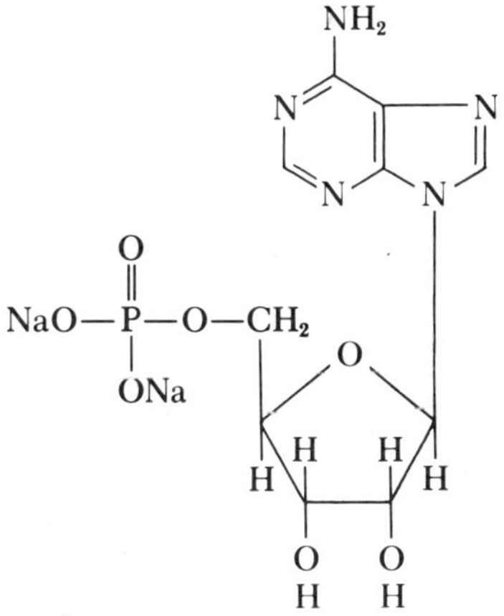

Mol. wt. 391.19
Used as a substrate for 5'-nucleotidase.

Adenosine triphosphate, disodium

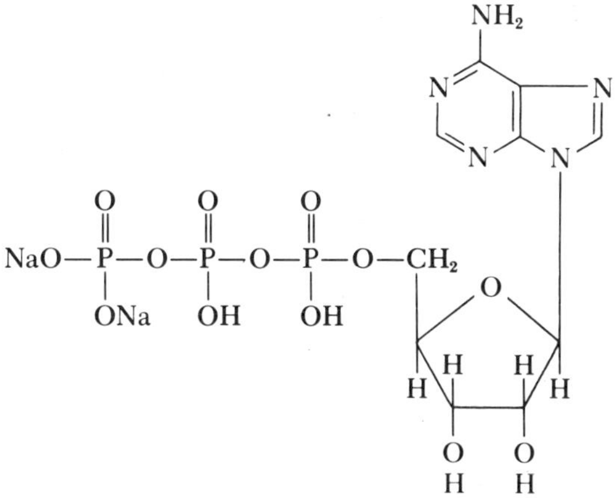

Mol. wt. 551.15
A normal constituent of most biological tissues. It is said to be a high-energy compound because upon hydrolysis of each of the phosphate bonds, 7500 calories are released. Used as a substrate for adenosine triphosphate.

Alanyl-β-naphthylamide

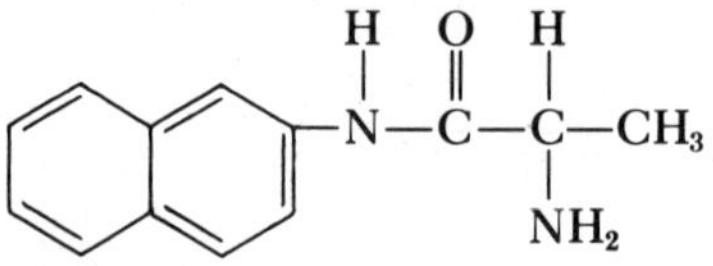

Mol. wt. 214.27

Used as a substrate for aminopeptidase.

Alcian blue 8GX

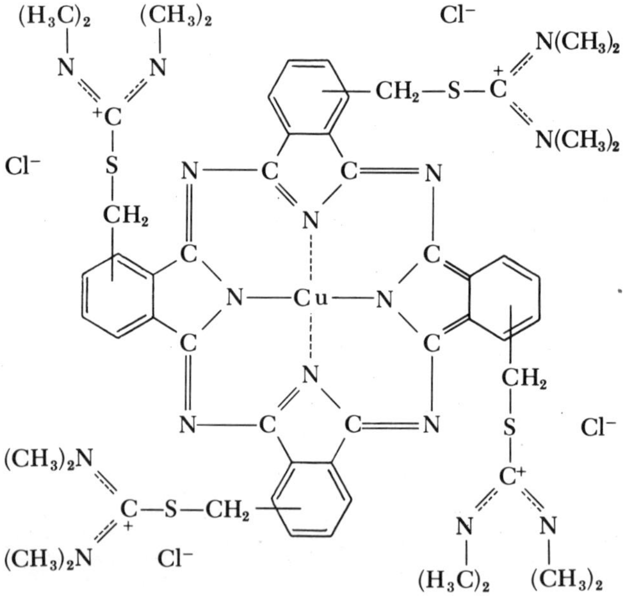

Mol. wt. 1298.87   C.I. No. 74240

A copper-containing phthalocyanin dye containing four isothio-uronium groups. Used as a stain for acidic carbohydrates.

Alcohols

Any of a group of aliphatic compounds containing a hydroxyl group. *Alcohol* is frequently used to mean ethanol (ethyl alcohol). An alcohol is called *primary* if the hydroxylated carbon is bound to one other carbon; *secondary* if it is bound to two other carbons; and *tertiary* if it is bound to three other carbons. The primary aliphatic alcohols form a series with ascending specific gravities and boiling points. The lower members of the series are colorless volatile liquids, the middle ones have an oily characteristic, and the higher ones are paraffinlike without taste or color.

Alizarin red S

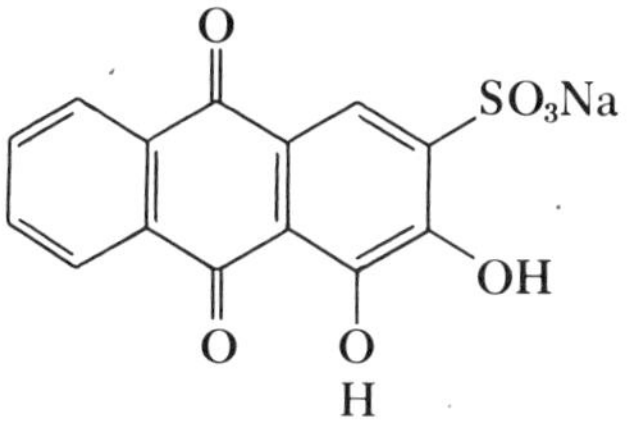

O
SO₃Na
OH
O   O
H

Mol. wt. 342.26   C.I. No. 58005

Used as a stain for calcium. Also used as an acid-base indicator.

4-Amino-1-N,N-dimethylnaphthylamine

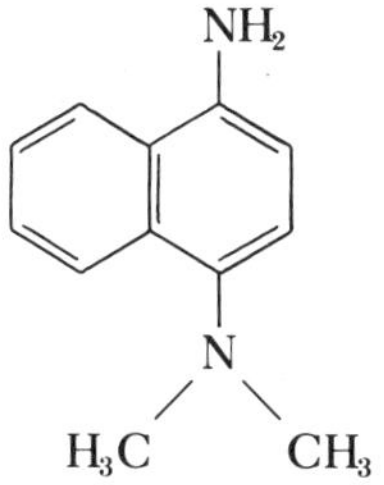

NH₂

N

H₃C   CH₃

Mol. wt. 186.26

Used as a substitute for dimethyl-p-phenylenediamine in the Nadi method for demonstrating cytochrome oxidase.

2-Amino-2-methyl-1-propanol (isobutanolamine)

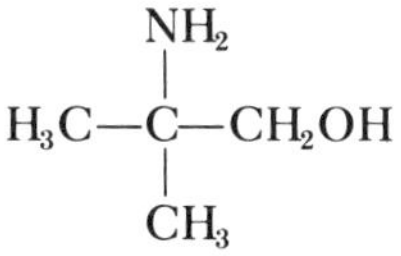

NH₂

H₃C—C—CH₂OH

CH₃

Mol. wt. 89.14

Used as a buffer in a method for demonstrating calcium-activated adenosine triphosphatase.

3-Amino-1,2,4-triazole

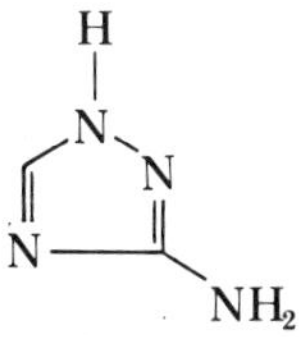

H

N N
N
N
NH₂

Mol. wt. 84.08

Used as an inhibitor for certain enzymes.

Ammonium hydroxide

NH_4OH

Mol. wt. 35.05

Sp. gr. 0.89–0.90

Forms a strong base in water (called ammonia); 68.0 ml dissolved in 1 liter of water makes approximately a 1.0M solution.

Ammonium sulfamate

$NH_4O_3SNH_2$

Mol. wt. 114.13

Used as one of the reagents in the diazotization-coupling method for demonstrating tyrosine.

Ammonium sulfide

$(NH_4)_2S$

Mol. wt. 68.15

Available in 20 to 24% solution. Used to convert metal salts to black metal sulfides, generally as the last step in the lead precipitation methods for demonstrating phosphatases.

Antiperoxidase

An antibody against peroxidase.

Argentophilic

This term refers to the affinity of certain cells for silver salts. Such cells can be demonstrated by simply reducing the silver salts, taken up by the cells, to black metallic silver.

Autolysis

Postmortem cellular degeneration and disintegration caused by lysosomal or other enzymes secreted by the cells themselves.

Azure A

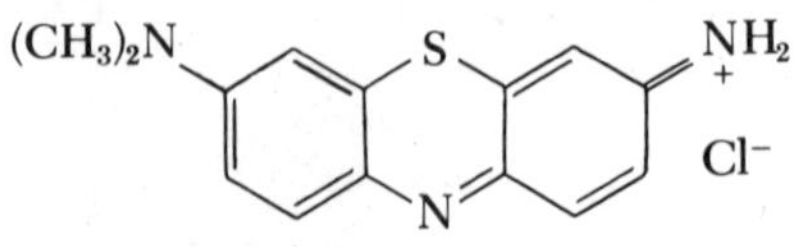

Mol. wt. 291.79

C.I. No. 52005

A basic metachromatic dye of the thiazine group.

Basic fuchsin

C.I. No. 42500

A mixture of triarylmethane dyes, consisting primarily of pararosanilin, rosanilin, and new fuchsin. Used in the preparation of Schiff's reagent.

Benzene

C_6H_6

Mol. wt. 78.11

A colorless, flammable liquid used as an apolar solvent. Miscible in ethanol but not in water.

2-(2-Benzothiazolyl)-3-(4-phthalhydrazidyl)-5-styryl-tetrazolium chloride (BPST)

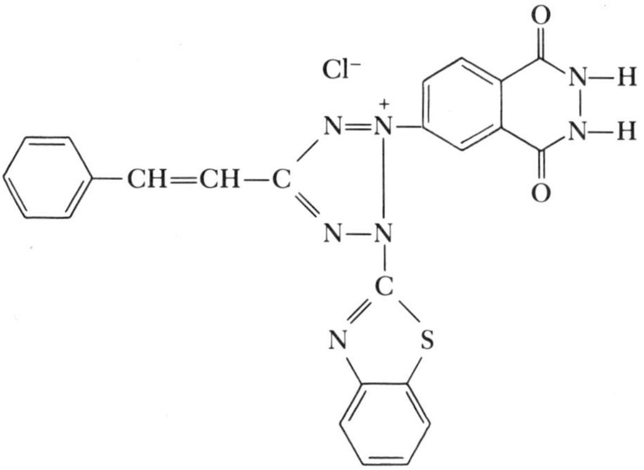

Mol. wt. 501.97

Recommended as an electron acceptor in histochemical reactions for dehydrogenase enzymes where the reaction will be quantified.

6-Benzoyl-2-naphthyl sulfate, potassium salt

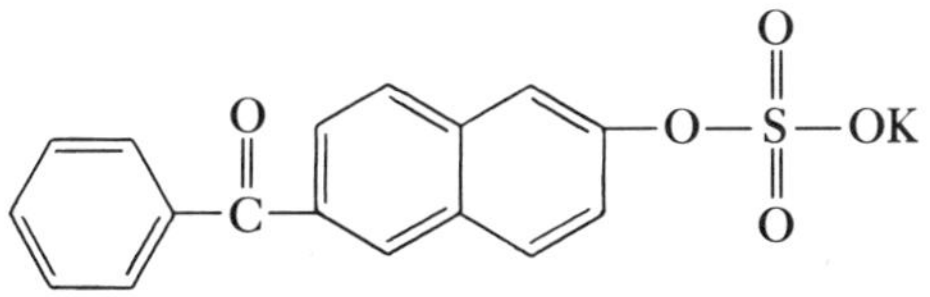

Mol. wt. 366.44

Used as a substrate for arylsulfatase.

Benzoylcholine chloride

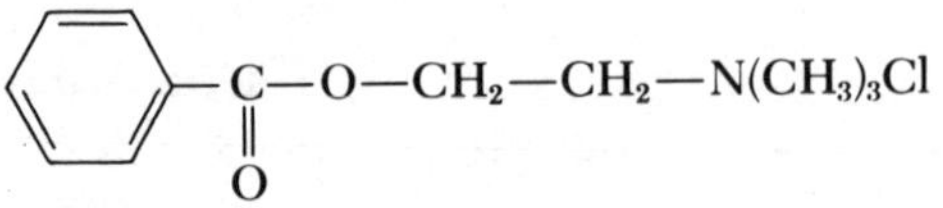

$$\text{C}_6\text{H}_5 - \underset{\underset{\text{O}}{\|}}{\text{C}} - \text{O} - \text{CH}_2 - \text{CH}_2 - \text{N(CH}_3)_3\text{Cl}$$

Mol. wt. 243.73
Selectively hydrolyzed by pseudocholinesterase.

Benzylamine

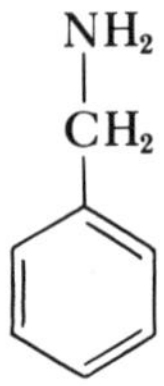

$$\underset{\text{CH}_2}{\overset{\text{NH}_2}{|}}$$

Mol. wt. 107.16
A substrate specific for monoamine oxidase, A form.

Beryllium sulfate, tetrahydrate
$BeSO_4 \cdot 4H_2O$
Mol. wt. 177.14
An inhibitor of phosphoglucomutase.

Betz cells
Giant neurons (up to 150μ) found in the motor cortex of the brain.

Bifunctional reagents
Certain reactive molecules provided with two functional groups (aldehydes, sulfhydryls), which can bind and cross-link two cellular sites.

Birefringence
When a tissue component splits light into two polarized forms mutually at right angles, this splitting, called birefringence, can be detected by the use of crossed polarizing filters.

1,5 bis(4-allyldimethylammoniumphenyl) pentan-3-one dibromide (284 C 51)

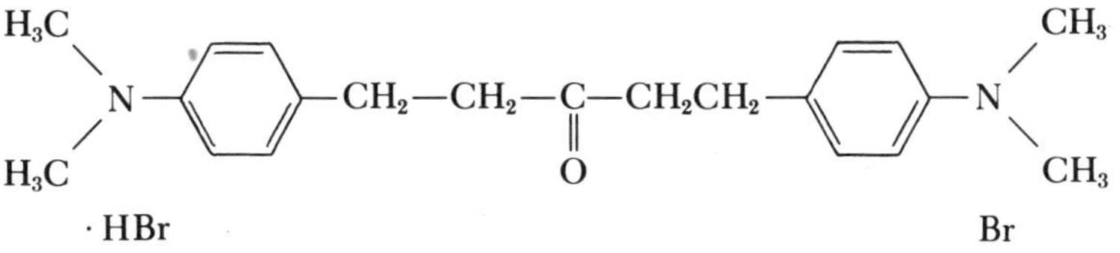

· HBr   Br

Mol. wt. 566.42

Used as an inhibitor of acetylcholinesterase.

Blue tetrazolium

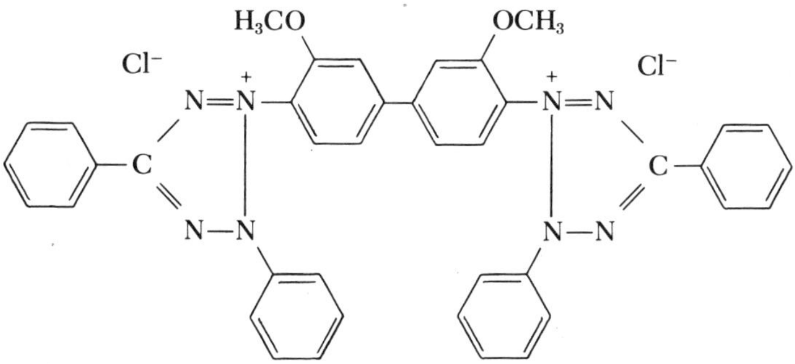

Mol. wt. 727.67

Formerly used as the electron acceptor in histochemical reactions for dehydrogenases.

Boric acid

H_3BO_3

Mol. wt. 61.84

Used in making borate buffers.

5-Bromo-4-chloroindoxyl acetate

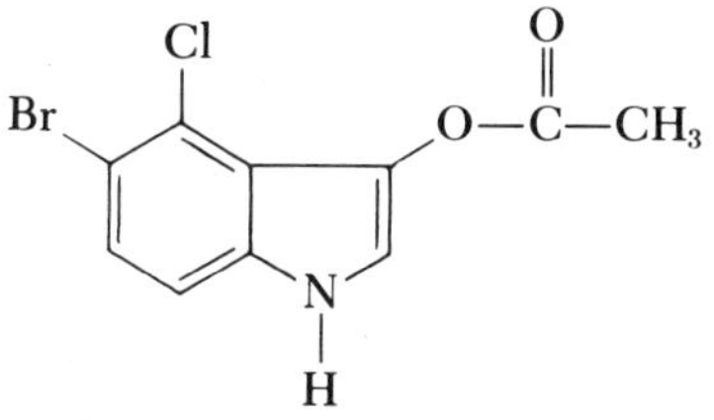

Mol. wt. 288.57

Used as a substrate for nonspecific esterase.

5-Bromo-4-chloroindoxyl-β-D-galactoside

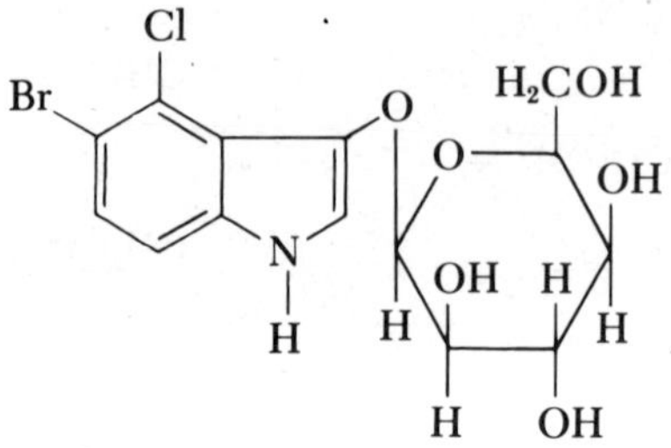

Mol. wt. 407.63
Used as a substrate for β-galactosidase.

Bromophenol blue

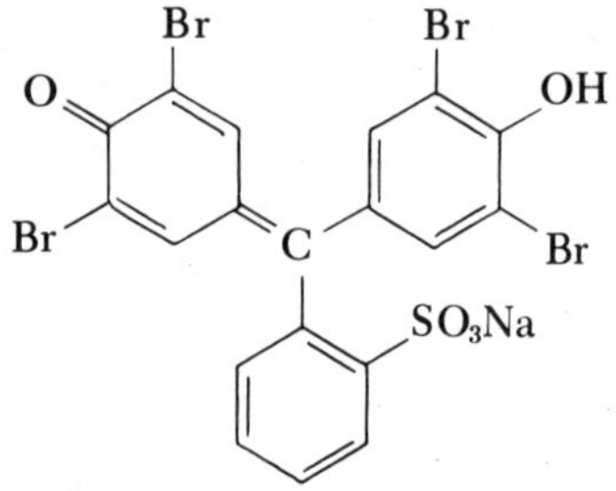

Mol. wt. 691.98
Used as a general protein stain.

***n*-Butyl alcohol** (*n*-butanol)

$H_3C—CH_2—CH_2—CH_2OH$

Mol. wt. 74.12
Used as a dehydrating agent in the methyl green–pyronin method for nucleic acids.

***tert*-Butyl alcohol** (*tert*-butanol)

$$
\begin{array}{c}
\quad\; CH_3 \\
\quad\; | \\
H_3C—C—OH \\
\quad\; | \\
\quad\; CH_3
\end{array}
$$

Mol. wt. 74.12
Used for dehydrating in the mercury-bromophenol blue method for demonstrating proteins.

Butyrylcholine chloride

$$H_3C—CH_2—CH_2—\overset{\overset{\displaystyle O}{\|}}{C}—O—CH_2—CH_2—N(CH_3)_3Cl$$

Mol. wt. 209.72

May be used as a substrate for demonstrating pseudocholinesterase selectively because this substrate is not hydrolyzed by acetyl-cholinesterase.

Butyrylthiocholine iodide

$$H_3C—CH_2—CH_2—\overset{\overset{\displaystyle O}{\|}}{C}—S—CH_2—CH_2—N(CH_3)_3I$$

Mol. wt. 317.23

Used as a substrate for cholinesterase.

Cacodylic acid

$(CH_3)_2AsOOH$

Mol. wt. 137.99

Used in making cacodylate buffers.

Calcium chloride, anhydrous or dihydrate

$CaCl_2$ or $CaCl_2 \cdot 2H_2O$

Mol. wt.: anhydrous, 110.99; dihydrate, 147.02

Used in preparing formol-calcium fixative and as the capture agent in the original Gomori method for demonstrating alkaline phosphatase. The anhydrous form may also be used as a drying agent because it is very hygroscopic.

Calcium phosphate

$CaHPO_4$

An insoluble compound formed as the primary precipitate in Gomori's original method for alkaline phosphatase.

Carbodiimides

A group of bifunctional compounds, several of which have been used as fixatives in immunohistochemical studies.

Carbonic acid

H_2CO_3

Formed when carbon dioxide is dissolved in water or when carbonate or bicarbonate salts are dissolved in water.

Carbon tetrachloride (tetrachloromethane)

CCl_4

Mol. wt. 153.83

Carcinogen. Upon decomposition it may form phosgene gas, which is highly poisonous. Used as a solvent.

Chloroform

$CHCl_3$

Mol. wt. 119.38

May form the highly poisonous phosgene gas upon decomposition. Used as a solvent.

_p_-Chloromercuribenzoic acid

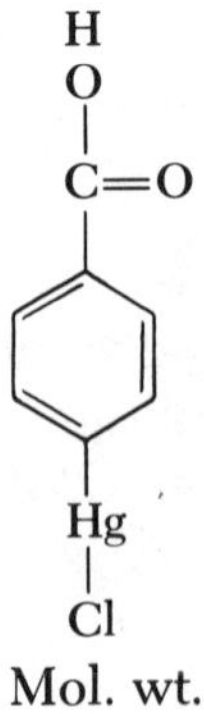

Mol. wt. 357.16

May be used to form mercaptan linkages with sulfhydryl groups.

Chrome potassium alum (chrom alum; chromium potassium sulfate)

$CrK(SO_4)_2 \cdot 12H_2O$

Mol. wt. 499.43

Used as one of the reagents in the performic acid-Schiff method for demonstrating unsaturated lipids.

Citric acid

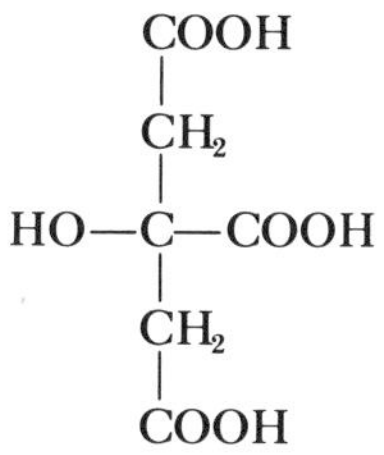

$$COOH$$
$$|$$
$$CH_2$$
$$|$$
$$HO-C-COOH$$
$$|$$
$$CH_2$$
$$|$$
$$COOH$$

Mol wt. 192.12

A normal component of the citric acid cycle. Used in preparing citrate buffer.

Clorgyline (N-methyl-N-propargyl-3-(2,4-dichlorophenoxy) propylamine)

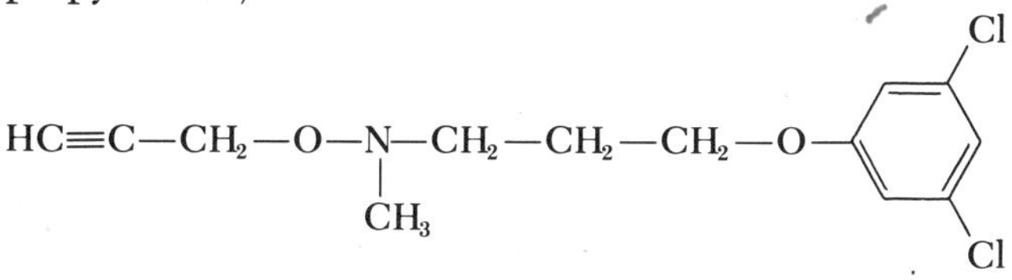

$$HC\equiv C-CH_2-O-N-CH_2-CH_2-CH_2-O-$$

with structure showing CH_3 and a dichlorophenyl ring with two Cl substituents.

Mol. wt. 288.18

A specific inhibitor of monoamine oxidase (A-form).

Cobalt nitrate (cobaltous nitrate)

$Co(NO_3)_2 \cdot 6H_2O$

Mol. wt. 291.05

A fire hazard when it comes in contact with organic substances. Used as a reagent in the original Gomori method for demonstrating alkaline phosphatase.

Cobalt phosphate

$Co_3(PO_4) \cdot 8H_2O$

An intermediate reaction product in the original Gomori method for alkaline phosphatase.

Cobalt sulfide

CoS

The final reaction product in the Gomori reaction for alkaline phosphatase.

Coenzyme Q

$CH_3C_6(O)_2(OCH_3)_2[CH_2CH{=}C(CH_3)CH_2]_nH$

Can be used as an electron carrier in the histochemical reaction for succinate dehydrogenase.

s-Collidine (syn-collidine; 2,4,6-trimethylpyridine)

Mol. wt. 121.18

An irritant. Used as a buffer.

Conjugation

The chemical process of joining two molecules through covalent bonds. Usually a third molecule is used to form the bond.

Copper acetate, monohydrate

$Cu(O_2CCH_3)_2 \cdot H_2O$

Mol. wt. 199.65

May be used for complexing and stabilizing the reaction product in the DHT method for histidine.

Copper glycinate

$Cu(NH_2CH_2COO)_2$

Mol. wt. 211.66

Blue crystals, which are soluble in most organic solvents but are only slightly soluble in water and alcohol. Used in a method for demonstrating cholinesterase.

Coupling

The process of one molecule reacting with another so that an insoluble end product is formed. Normally associated with enzyme histochemical reactions, in which it is the last part of the reaction. A highly insoluble, highly colored final reaction product is desired.

Cupric ferrocyanide (Hatchett's brown)

The final reaction product of certain histochemical procedures for hydrolases.

Dansylhydrazine

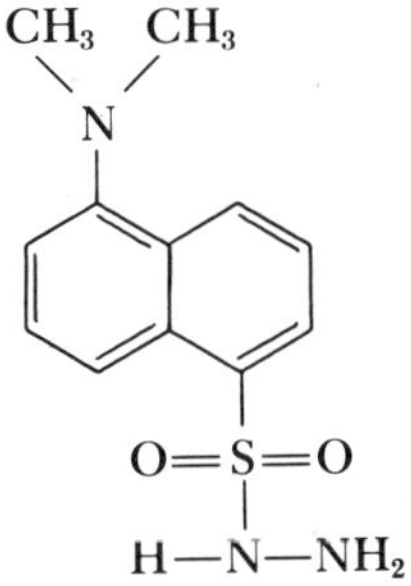

Mol. wt. 265.34

Used in a method for demonstrating sialic acid.

Dehydroepiandrosterone (5-pregnanolone)

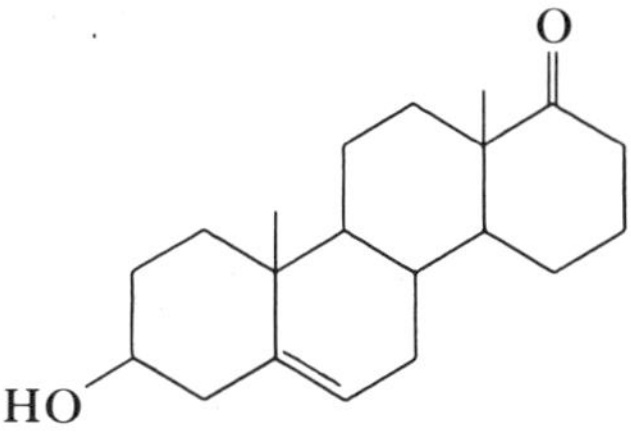

Mol. wt. 288.41

Used as a substrate for 3-β-hydroxysteroid dehydrogenase.

3,3'-Diaminobenzidine tetrahydrochloride· 2H$_2$O

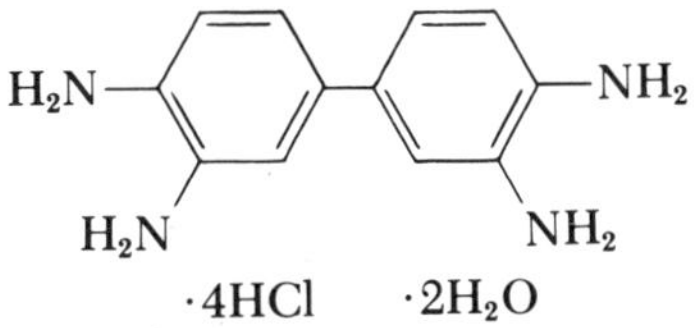

·4HCl   ·2H$_2$O

Mol. wt. 396.15

Carcinogen. Used as a substrate for peroxidase. It forms osmophilic polymers.

o-Dianisidine, tetrazotized (Fast blue B)

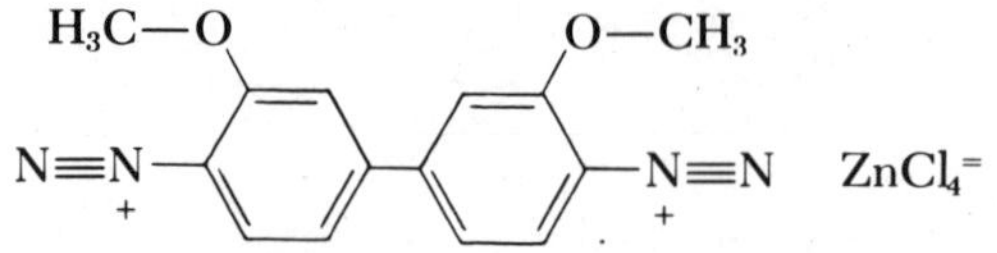

$$H_3C-O \qquad\qquad O-CH_3$$

$$N\equiv N-\text{(benzene ring)}-\text{(benzene ring)}-N\equiv N \quad ZnCl_4^=$$

Mol. wt. 268.28

Used as a chromogenic coupling agent. The compound is stable for only about 15 minutes in solution at room temperature.

p-Diazobenzenesulfonic acid

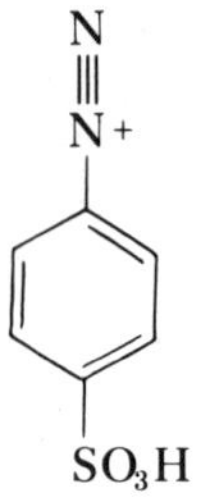

$$\begin{array}{c} N \\ \| \| \\ N+ \end{array}$$

(benzene ring)

SO_3H

Mol. wt. 185.17

Used in one method for demonstrating histidine. It is prepared when needed by diazotization of sulfanilic acid.

Diazo-1-H-tetrazole

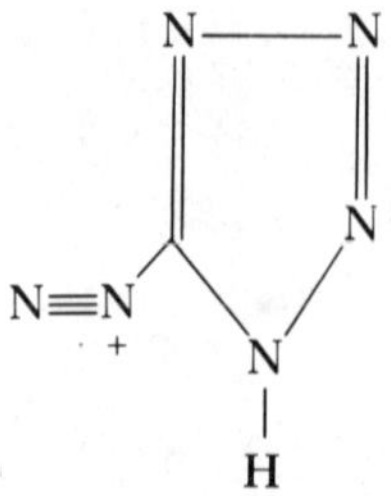

$$N\text{------}N$$

$$N\equiv N$$

$$N-H$$

Used as the principal reagent in the DHT reaction for histidine. It must be prepared by diazotization of amino-1-H-tetrazole. The diazo compound should not be allowed to precipitate because as a solid it is explosive.

Diazotization

The process of converting an amine group ($-NH_2$) to a diazo group ($-\overset{+}{N} N$) by treating with an acidic solution of sodium nitrite or with nitrous acid.

2,4-Dichloronaphthol

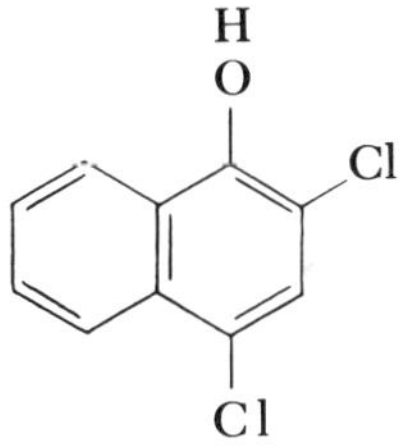

Mol. wt. 213.07
Used in a modified Sakaguchi method for demonstrating arginine.

Dichlorophenoxyacetic acid

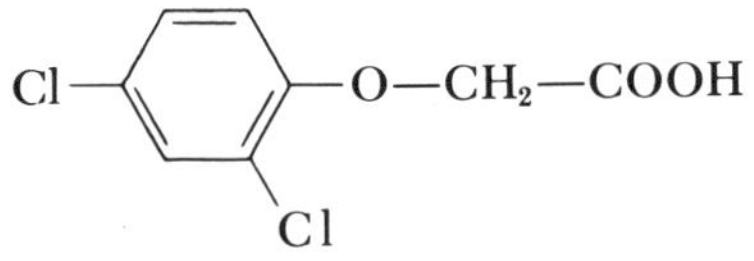

Mol. wt. 221.04
An inhibitor of phosphorylase.

Dicoumarol (3,3'-methylene bis-4-hydroxycoumarin)

$C_{19}H_{12}O_6$
Mol. wt. 336.30
A natural anticoagulant found in spoiled sweet clover. A potent inhibitor of mitochondrial α-glycerophosphate dehydrogenase.

Diethylacetamide

$$H_3C-\overset{\overset{\textstyle O}{\|}}{C}-N(C_2H_5)_2$$

Mol. wt. 115.18
Used as a solvent in the naphthol method for lipase.

Diethyl-*p*-nitrophenyl phosphate (E600)

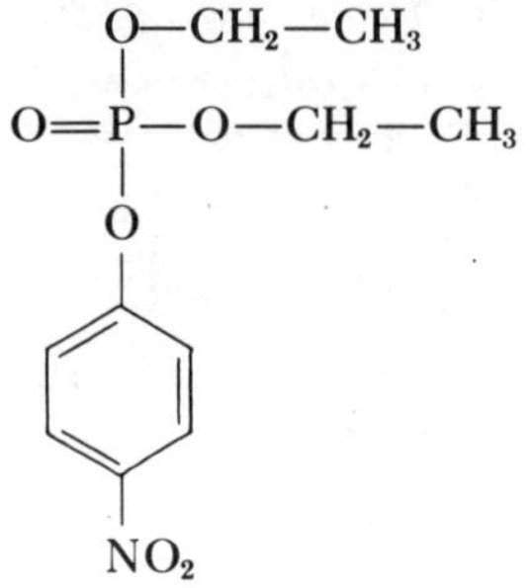

Mol. wt. 275.21
An inhibitor of B-esterases.

Differentiation

Destaining a tissue section. A tissue section is sometimes deliberately overstained and then carefully destained in such a way that certain tissue components selectively retain their stain and are therefore differentiated.

p,p-Difluoro-*m,m*-dinitrodiphenyl sulfone

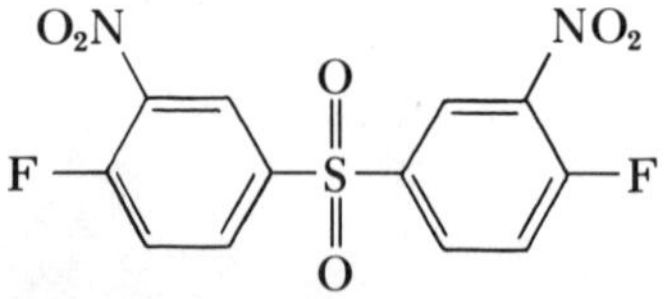

Mol. wt. 312.19
A bifunctional reagent used in conjugating proteins.

Digitonin

$C_{56}H_{92}O_{29}$
Mol. wt. 1229.35
Forms a birefringent crystalline complex with cholesterol.

2,2'-Dihydroxy-6,6'-dinaphthyl-disulfide

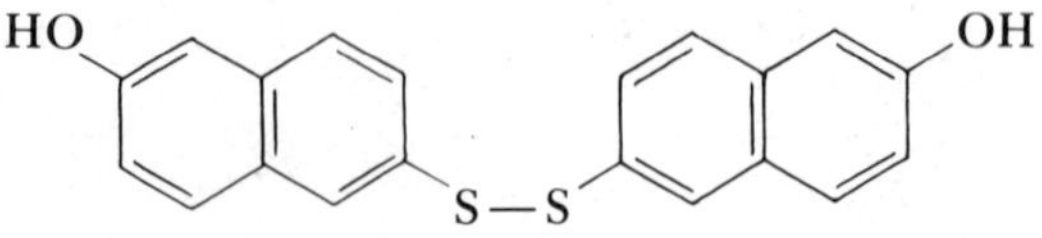

Mol. wt. 350.46
Used as the principal reagent in the DDD method for demonstrating sulfhydryl groups.

DL-Dihydroxyphenylalanine (DOPA)

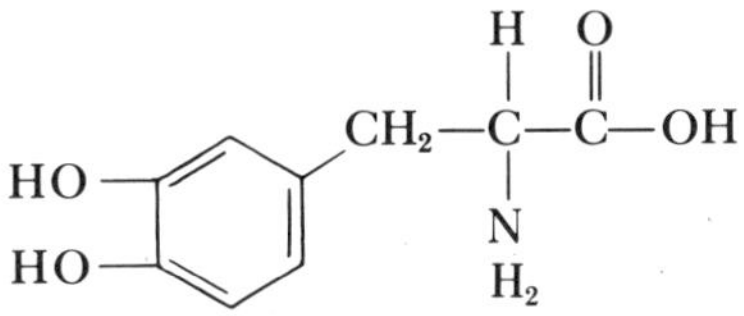

$$HO-C_6H_3(HO)-CH_2-\underset{\underset{H_2}{\overset{|}{N}}}{\overset{\overset{H}{|}}{C}}-\overset{\overset{O}{\|}}{C}-OH$$

Mol. wt. 197.19
Used as the substrate for DOPA oxidase.

Diimidoesters

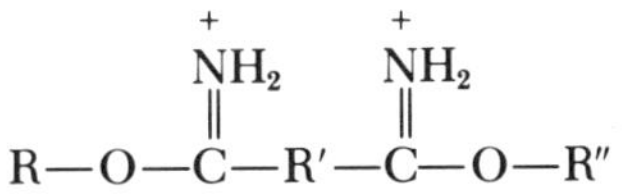

$$R-O-\overset{\overset{+}{\overset{NH_2}{\|}}}{C}-R'-\overset{\overset{+}{\overset{NH_2}{\|}}}{C}-O-R''$$

A group of bifunctional reagents used as fixatives in enzyme histochemistry and immunohistochemistry as well as in protein structural studies.

Diisopropyl fluorophosphate

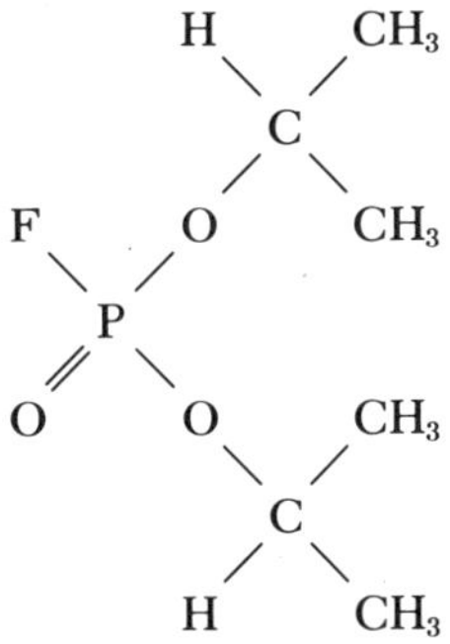

Mol. wt. 184.15
A potent poison. Used as an inhibitor of B-esterases and certain enzymes with serine-active centers.

Dimethoxypropane

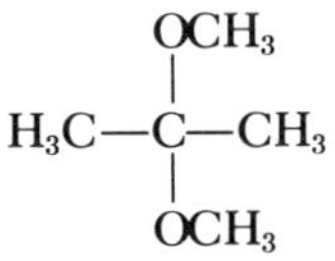

$$H_3C-\underset{\underset{OCH_3}{|}}{\overset{\overset{OCH_3}{|}}{C}}-CH_3$$

Mol. wt. 104.15
Serves as a dehydrating and clearing agent by chemically reacting with water.

Dimethylamine borane

$$H_3C$$
$$\diagdown$$
$$N\!-\!BH_3$$
$$\diagup \;\;|$$
$$H_3C\;\;H$$

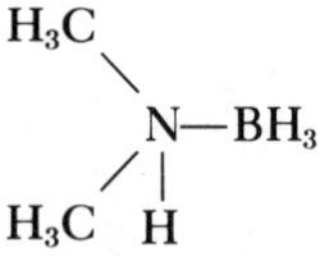

Mol. wt. 58.92

A reducing agent used to stabilize dansylhydrozone in a histochemical reaction for sialic acid.

Dimethylaminobenzaldehyde

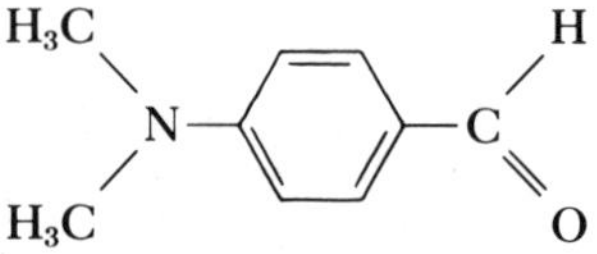

Mol. wt. 149.19

Used in a method for demonstrating tryptophan.

4-(p-Dimethylaminobenzene-azo)-phenylmercuric acetate

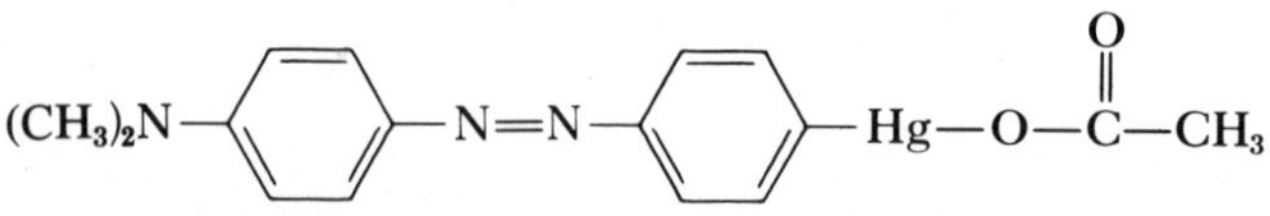

$$(CH_3)_2N\!-\!\langle\;\rangle\!-\!N\!=\!N\!-\!\langle\;\rangle\!-\!Hg\!-\!O\!-\!\overset{\displaystyle O}{\overset{\|}{C}}\!-\!CH_3$$

Mol. wt. 483.92

Used as a substitute for mercury orange in a method for demonstrating sulfhydryl groups.

Dimethyl formamide

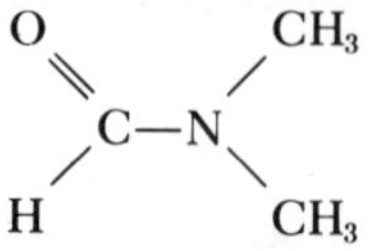

$$O\;\;\;\;\;\;\;\;CH_3$$
$$\diagdown\;\;\;\;\diagup$$
$$C\!-\!N$$
$$\diagup\;\;\;\;\diagdown$$
$$H\;\;\;\;\;\;\;\;CH_3$$

Mol. wt. 73.10

Used for dissolving many different organic compounds. It is also completely soluble in water.

N,N-Dimethyl-*m*-phenylenediamine·2HCl

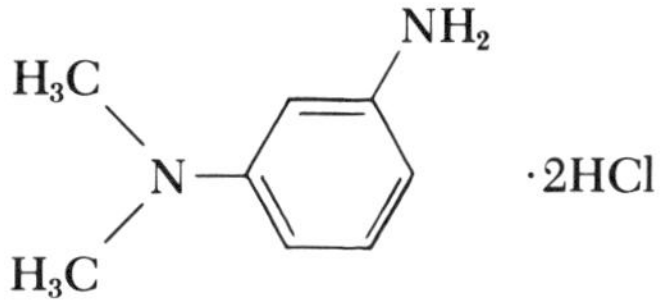

$$H_3C$$

NH$_2$

N— ·2HCl

$$H_3C$$

Mol. wt. 209.12

Used in the diamine methods for demonstrating carbohydrates.

N,N-Dimethyl-*p*-phenylenediamine·2HCl

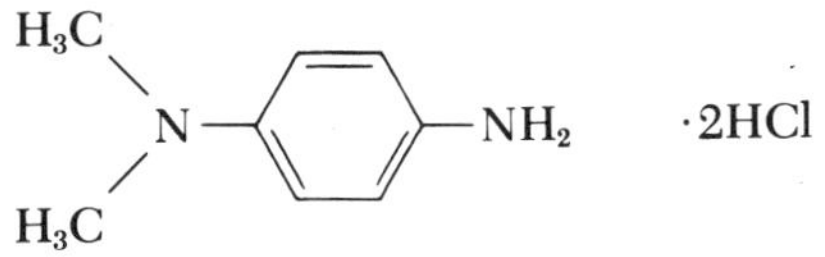

$$H_3C$$

N— —NH$_2$ ·2HCl

$$H_3C$$

Mol. wt. 209.12

Used in the diamine methods for demonstrating carbohydrates.

Dimethyl suberimidate·2HCl

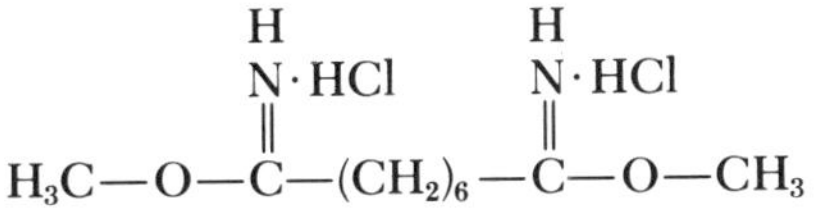

$$\begin{array}{ccc} & H & & H \\ & N\cdot HCl & & N\cdot HCl \\ & \| & & \| \\ H_3C-O-C-(CH_2)_6-C-O-CH_3 \end{array}$$

Mol. wt. 273.20

May be used as a fixative, especially for enzyme histochemistry and immunohistochemistry.

Dimethyl sulfoxide (DMSO)

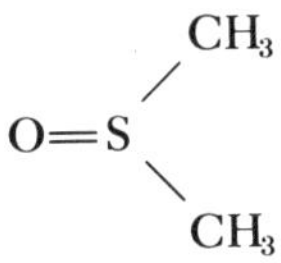

$$\begin{array}{c} CH_3 \\ / \\ O=S \\ \backslash \\ CH_3 \end{array}$$

Mol. wt. 78.13

Penetrates skin extremely rapidly. Used to dissolve, and promote penetration of certain organic compounds. One manufacturer adds it to paraffin to promote penetration of paraffin into tissues.

2,4-Dinitrofluorobenzene

Mol. wt. 186.10
Used as the main reagent in a method for demonstrating proteins.

Dioxane

Mol. wt. 88.11
Carcinogen. Used as a solvent.

Dithiothreitol (Cleland's reagent)

$$HS-CH_2-\overset{\overset{\displaystyle H}{\overset{\displaystyle |}{O}}}{\underset{\underset{\displaystyle H}{\underset{\displaystyle |}{}}}{C}}-\overset{\overset{\displaystyle H}{\overset{\displaystyle |}{O}}}{\underset{\underset{\displaystyle H}{\underset{\displaystyle |}{}}}{C}}-CH_2-SH$$

Mol. wt. 154.24
Used in the ferric ion precipitation method for demonstrating glyco-gen phosphorylase.

Dithizone (diphenyl thiocarbozone)

$$-\overset{\overset{\displaystyle H}{\overset{\displaystyle |}{}}}{N}-\overset{\overset{\displaystyle H}{\overset{\displaystyle |}{}}}{N}-\overset{\overset{\displaystyle S}{\overset{\displaystyle ||}{}}}{C}-N{=}N-$$

Mol. wt. 256.34
Used as the principal reagent in the method for demonstrating zinc.

Dopamine· HCl (3-hydroxytyramine hydrochloride)

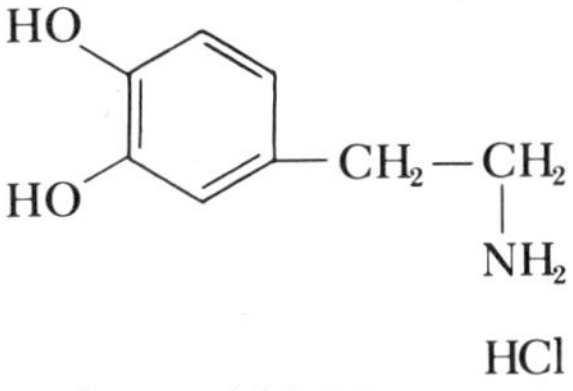

HO

HO —CH_2—CH_2
                |
              NH_2

HCl

Mol. wt. 189.64
Used as a substrate for monoamine oxidase.

Erythrose-4-phosphate

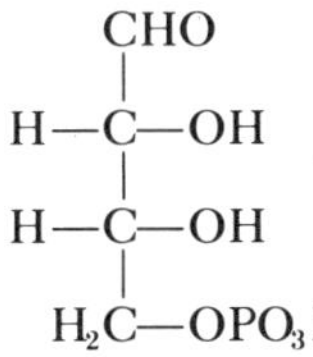

CHO
|
H—C—OH
|
H—C—OH
|
H_2C—OPO_3H_2

Mol. wt. 200.08
An inhibitor of glucose phosphate isomerase and an intermediate of the phosphogluconate pathway.

Eserine

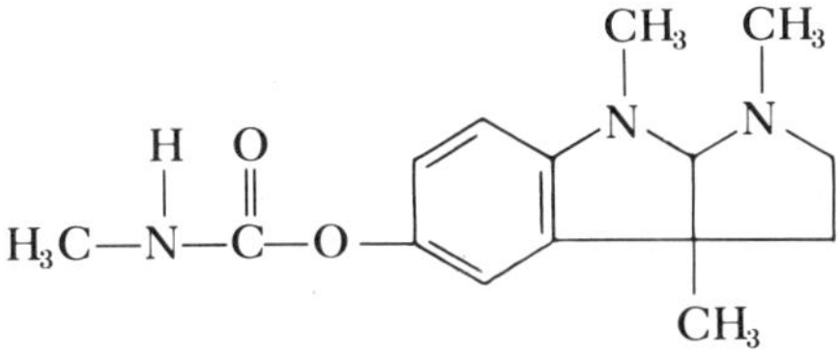

CH_3 CH_3

H   O
| ||
H_3C—N—C—O—

CH_3

Mol. wt. 648.6
Used to inhibit nonspecific esterase so that cholinesterases can be demonstrated selectively.

Ethylenediaminetetraacetic acid

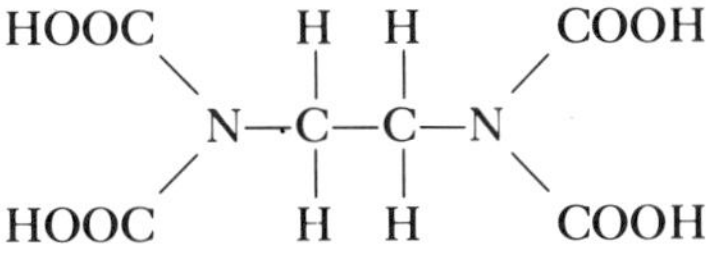

HOOC   H   H   COOH
     \ | | /
       N→C—C—N
     / | | \
HOOC   H   H   COOH

Mol. wt. 292.25
A chelating agent used in certain enzyme histochemical reactions as well as for decalcifying bones and teeth.

Ethylene glycol

$HOCH_2$—CH_2OH

Mol. wt. 62.07

Used as a solvent for fat stains.

N-Ethyl maleimide

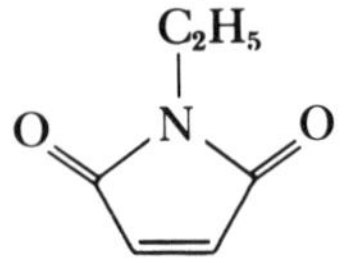

Mol. wt. 125.13

Reacts with sulfhydryl groups in tissues. Can be used to inactivate enzymes with sulfhydryl-active centers.

Fast black B

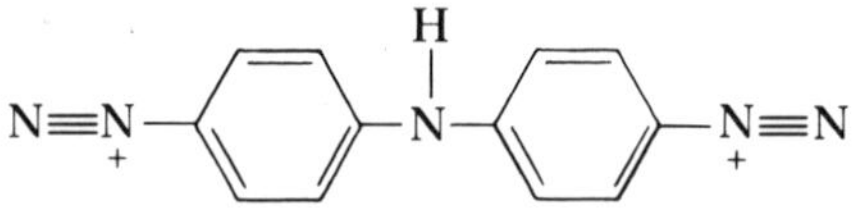

$ZnCl_4^-$

Mol. wt. of diazonium cation 223.24

Used as a coupling agent in histochemical reactions for hydrolases.

Fast blue BB

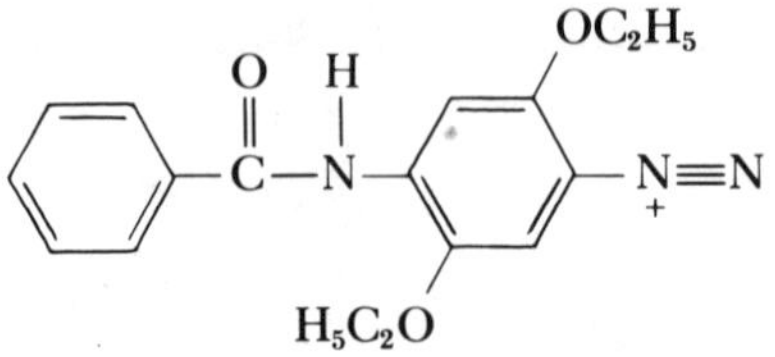

$(\frac{1}{2}ZnCl_4)^-$

Mol. wt. of diazonium cation 312.35

Used as a coupling agent in histochemical reactions for hydrolases.

Fast blue RR

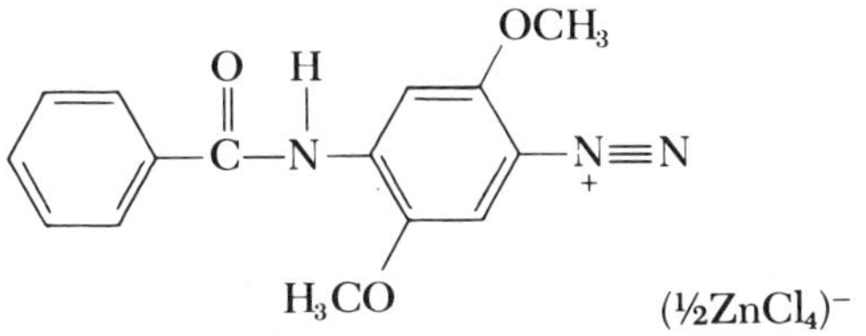

Mol. wt. of diazonium cation 284.30
Used as a coupling agent in histochemical reactions for hydrolases.

Fast blue VRT

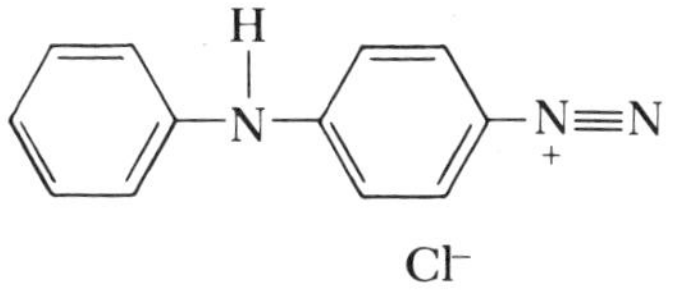

Mol. wt. of diazonium cation 196.23
Used as a coupling agent in histochemical reactions for hydrolases.

Fast Garnet GBC

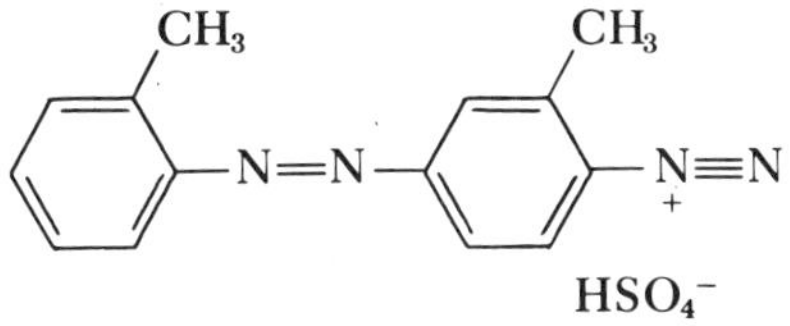

Mol. wt. of diazonium cation 237.29
Used as a coupling agent in histochemical reaction for hydrolases.

Fast green FCF

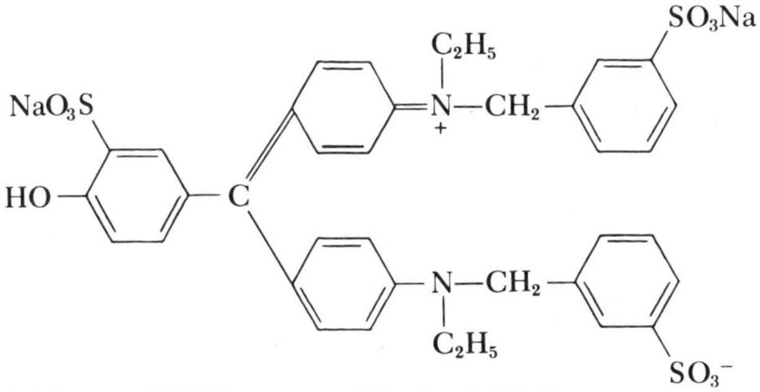

Mol. wt. 808.86   C.I. No. 42053
An acid dye of the triarylmethane series. Used as a cytoplasmic stain as well as in a staining procedure for histones.

Fast red LTR

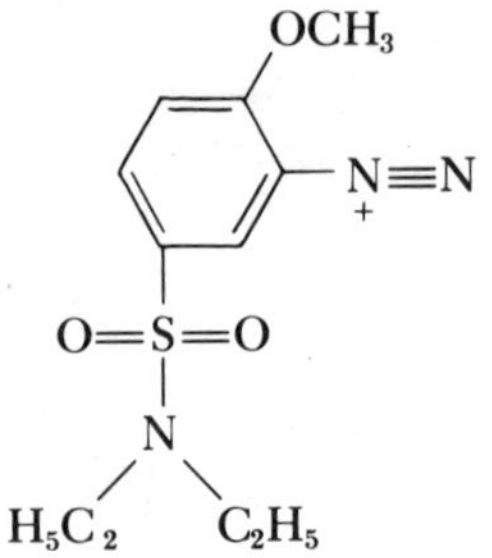

OCH$_3$

—N≡N$^+$

O=S=O

N

H$_5$C$_2$ C$_2$H$_5$

Mol. wt. of diazonium cation 270.33
Used as a coupling agent in histochemical reactions for hydrolases.

Fast red TR

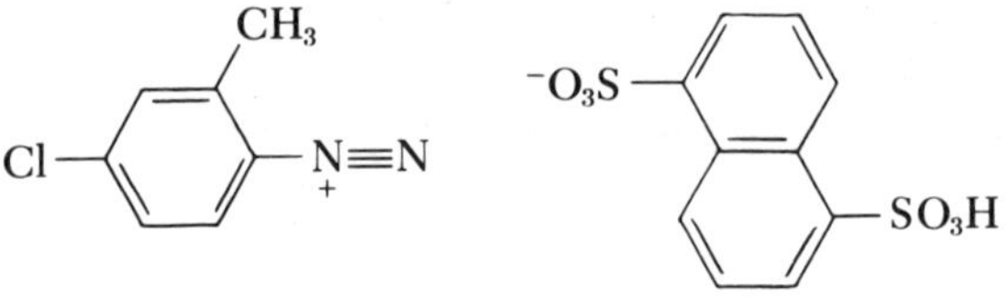

CH$_3$

Cl—   —N≡N$^+$

$^-$O$_3$S—   —SO$_3$H

Mol. wt. of diazonium cation 118.14
Used as a coupling agent in histochemical reactions for hydrolases.

Fast violet B

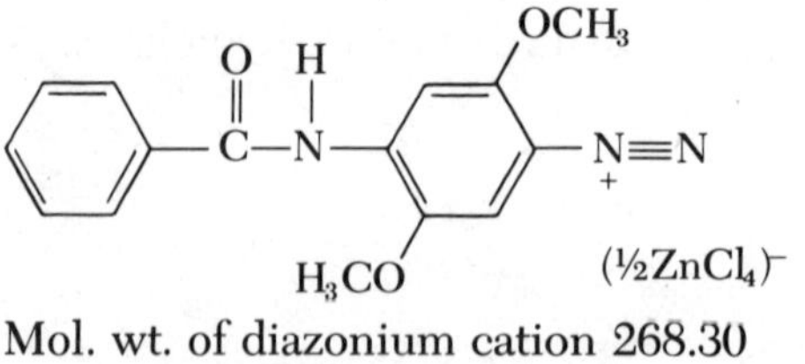

OCH$_3$

—C(=O)—N(H)—   —N≡N$^+$

H$_3$CO   (½ZnCl$_4$)$^-$

Mol. wt. of diazonium cation 268.30
Used as a coupling agent in histochemical reactions for hydrolases.

Ferric ammonium sulfate (ferric alum; iron alum)
FeNH$_4$(SO$_4$)$_2$·12H$_2$O
Mol. wt. 482.21
Used as one of the reagents in the modified Schultz method for demonstrating cholesterol.

Ferric ferricyanide

$Fe[Fe(CN)_6]$

An unstable compound used in the ferric ferricyanide method for demonstrating sulfhydryl groups of proteins.

Ferric ferrocyanide (prussian blue)

$Fe_4[Fe(CN)_6]_3$

A blue pigment formed as the final reaction product in (1) the ferric ferricyanide method for sulfhydryl groups, (2) the colloidal iron method for carbohydrates, (3) the Schmorl reaction for lipofuscins, and (4) Perls' method for tissue iron.

Fettrot

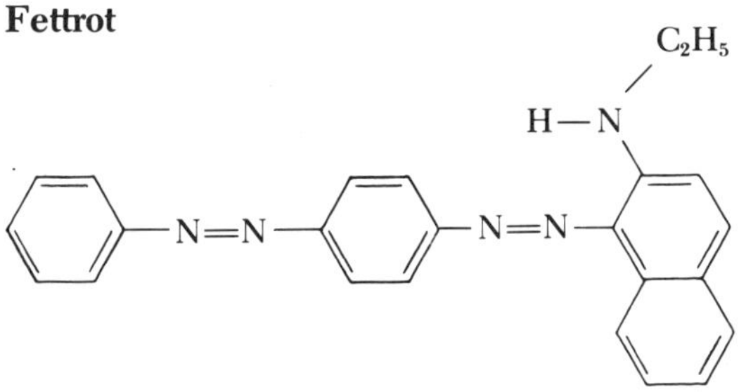

Mol. wt. 379.44
Used as a fat stain.

Flavin adenine dinucleotide (FAD)

The coenzyme for succinate dehydrogenase and mitochondrial α-glycerophosphate dehydrogenase.

Fluorescein isocyanate

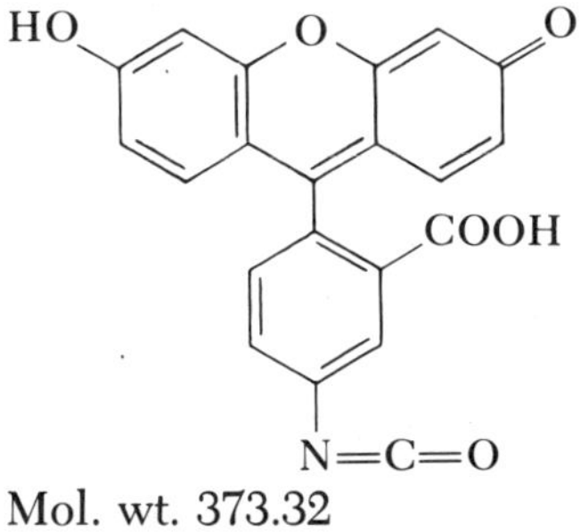

Mol. wt. 373.32
Formerly used as a fluorescent label conjugated to immunoglobulins.

Fluorescein isothiocyanate

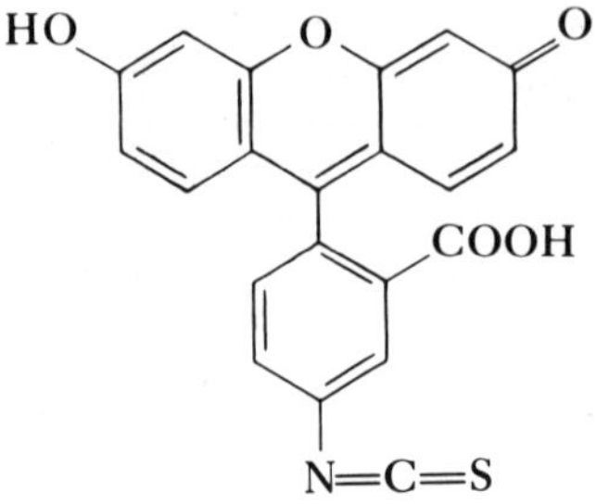

Mol. wt. 389.39
Used as a fluorescent label conjugated to immunoglobulins.

Fluorescence
Visible light emitted by a tissue component as a result of being irradiated with light of a shorter wave length, normally ultraviolet light.

Formaldehyde

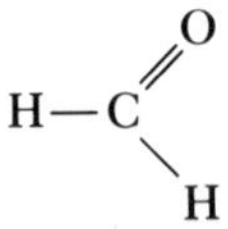

Mol. wt. 30.03
A gas that is highly soluble in water. It is commercially available as a 37 to 40% aqueous solution. Used as a fixative.

p-Formyl-benzoyl-N-OH-succinamide

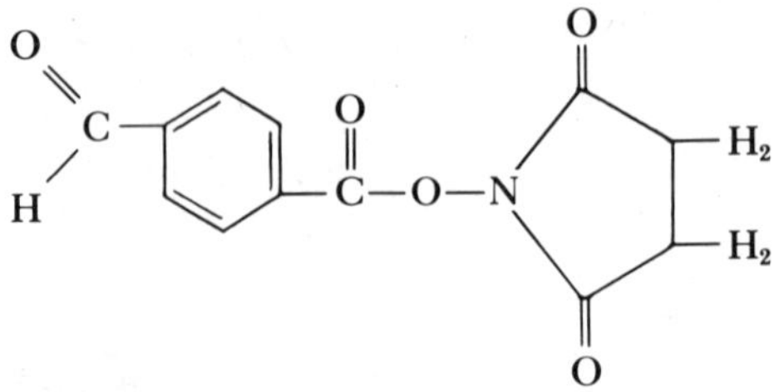

Mol. wt. 247.21
Used for conjugating microperoxidase with protein.

Fructose 6-phosphate, disodium salt

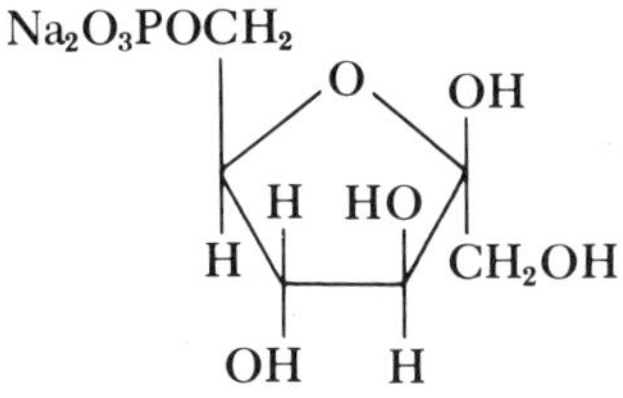

$Na_2O_3POCH_2$

Mol. wt. 304.10
Used as a substrate for glucose phosphate isomerase.

Fumaric acid

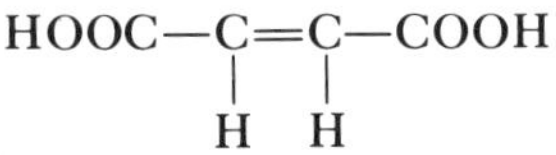

$$HOOC-\underset{\underset{H}{|}}{C}=\underset{\underset{H}{|}}{C}-COOH$$

Mol. wt. 116.08
The product of succinate dehydrogenase and an intermediate metabolite of the citric acid cycle.

Galactonolactone

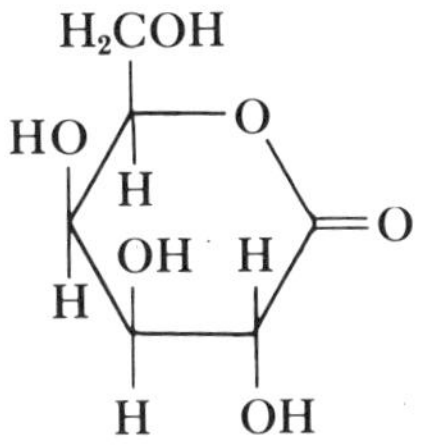

H_2COH

Mol. wt. 178.14
An inhibitor of β-galactosidase.

Glucuronic acid

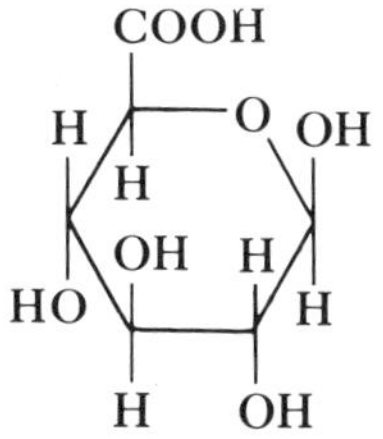

COOH

Mol. wt. 194.14
It is incorporated into glycosaminoglycans and exists in body fluids as conjugates of steroids and bilirubin.

Gluconolactone

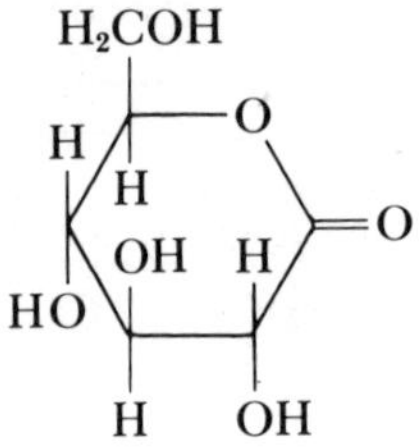

H$_2$COH

Mol. wt. 178.14

An inhibitor of the β-galactosidase found in the kidney and intestine but not of the β-galactosidase found in lysosomes.

Glucose l-phosphate, disodium salt

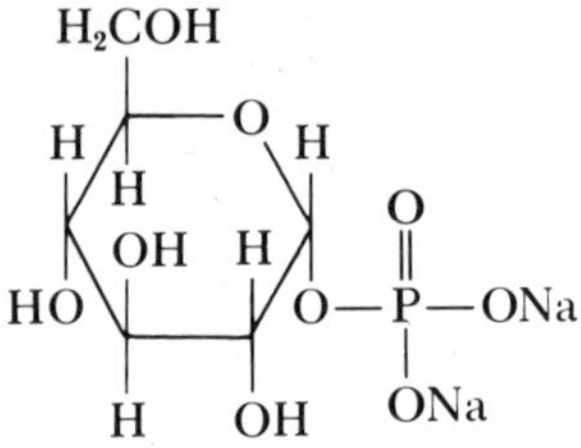

H$_2$COH

O—P—ONa
ONa

Mol. wt. 304.10

Used as a substrate for phosphoglucomutase and for glycogen phosphorylase.

Glucose 6-phosphate, monosodium salt

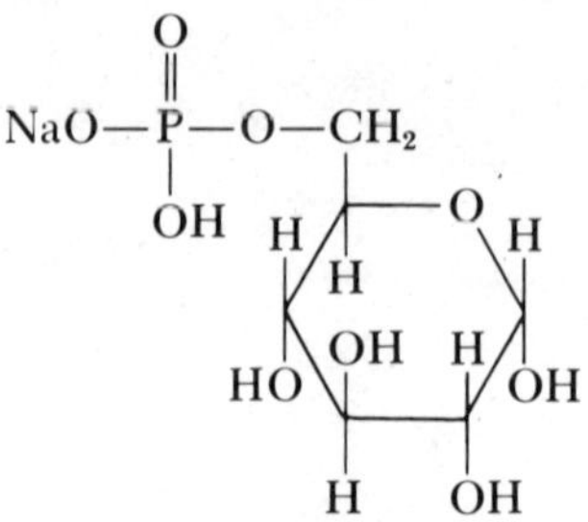

NaO—P—O—CH$_2$

Mol. wt. 282.12

Used as a substrate for glucose 6-phosphate dehydrogenase and for glucose 6-phosphatase.

L-Glutamic acid, monosodium salt

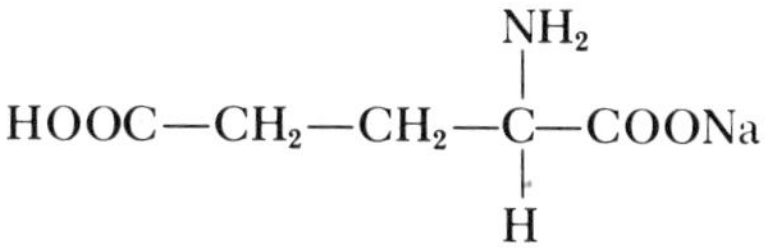

$$\text{HOOC}-\text{CH}_2-\text{CH}_2-\overset{\overset{\displaystyle \text{NH}_2}{|}}{\underset{\underset{\displaystyle \text{H}}{|}}{\text{C}}}-\text{COONa}$$

Mol. wt. 169.11

Used as a substrate for glutamate dehydrogenase.

Glutaraldehyde

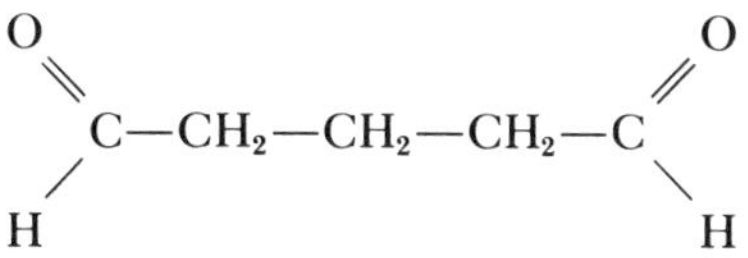

$$\underset{\text{H}}{\overset{\text{O}}{\diagdown}}\text{C}-\text{CH}_2-\text{CH}_2-\text{CH}_2-\text{C}\underset{\text{H}}{\overset{\text{O}}{\diagup}}$$

Mol. wt. 100.12

Used as a fixative, primarily in electron microscopic studies. It is also used for cross-linking proteins.

Glyceraldehyde-3-phosphate

$$\begin{array}{c} \text{CHO} \\ | \\ \text{H}-\text{C}-\text{OH} \\ | \\ \text{H}_2\text{C}-\text{O}-\overset{\overset{\displaystyle \text{O}}{\|}}{\underset{\underset{\displaystyle \text{OH}}{|}}{\text{P}}}-\text{OH} \end{array}$$

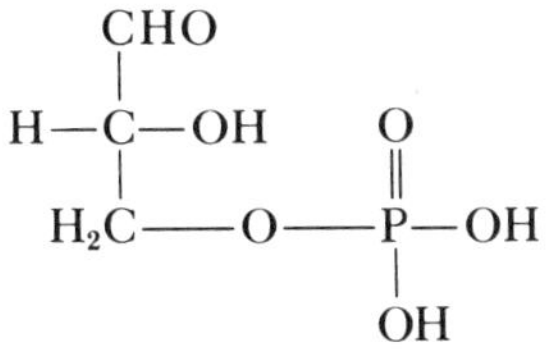

Mol. wt. 170.06

Used as the substrate for glyceraldehyde-3-phosphate dehydrogenase.

α-Glycerophosphate, disodium salt

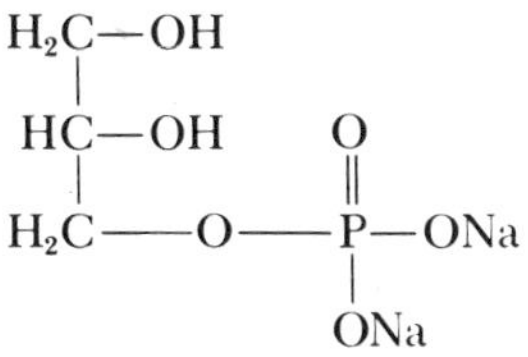

$$\begin{array}{c} \text{H}_2\text{C}-\text{OH} \\ | \\ \text{HC}-\text{OH} \\ | \\ \text{H}_2\text{C}-\text{O}-\overset{\overset{\displaystyle \text{O}}{\|}}{\underset{\underset{\displaystyle \text{ONa}}{|}}{\text{P}}}-\text{ONa} \end{array}$$

Mol. wt. 216.05 (anhydrous)

Used as a substrate for α-glycerophosphate dehydrogenase.

β-Glycerophosphate, disodium salt

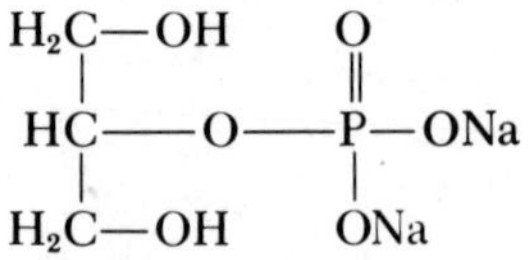

$$H_2C-OH \qquad O$$
$$HC{-}{-}O{-}{-}P-ONa$$
$$H_2C-OH \qquad ONa$$

Mol. wt. 216.05 (anhydrous)
Used as a substrate for acid phosphatase.

Glycine

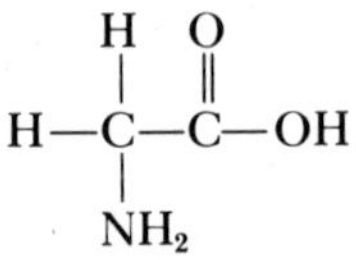

$$H{-}\underset{\underset{NH_2}{|}}{\overset{\overset{H}{|}}{C}}{-}\overset{\overset{O}{\|}}{C}{-}OH$$

Mol. wt. 75.06
Used in the preparation of glycine buffer.

Glycol methacrylate

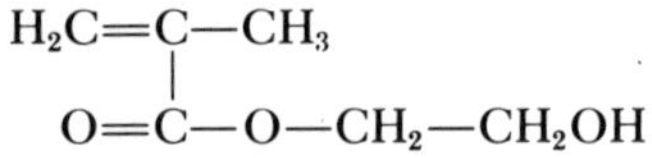

$$H_2C{=}C-CH_3$$
$$O{=}C-O-CH_2-CH_2OH$$

A water-soluble plastic that is used for embedding tissues for histochemistry and histological studies.

Gram's iodine

An aqueous solution of iodine crystals and potassium iodide. Used in the DHT method for histidine.

H-acid (8-amino-1-naphthol-3,6-disulfonic acid)

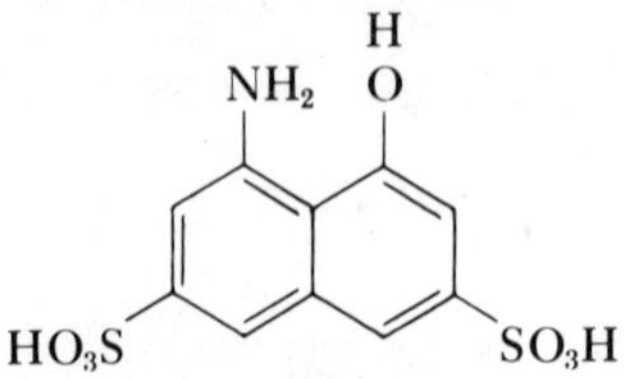

Mol. wt. 319.31
Used as a chromogenic coupling agent.

Hematoxylin

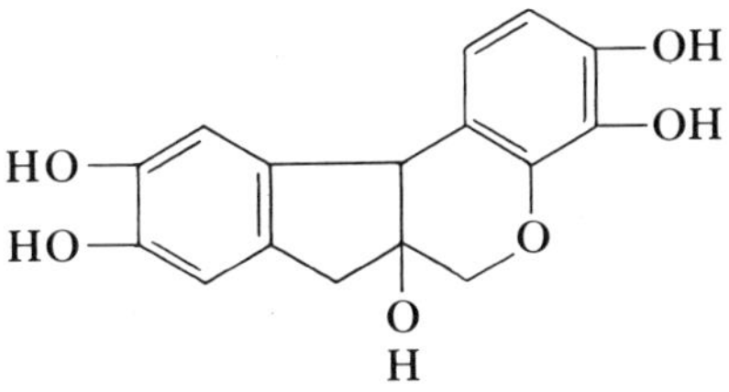

Mol. wt. 302.29

Used as a nuclear stain in routine histology as well as in methods for demonstrating phospholipids.

Hexazotization

Same chemical process as diazotization (converting amino groups to diazo groups) except that it is applied to triarylmethane dye with three amino groups.

Hydrogen peroxide

H_2O_2

Mol. wt. 34.02

Obtainable as a 30% or a 3% solution. It is used as the substrate for peroxidase but may also be used as a bleaching agent.

Hydroquinone

Mol. wt. 110.11

A reducing agent that is often used to reduce silver salts to metallic silver in photographic film or to reduce the silver salts produced in the von Kossa method for demonstrating calcium.

β-Hydroxybutyric acid

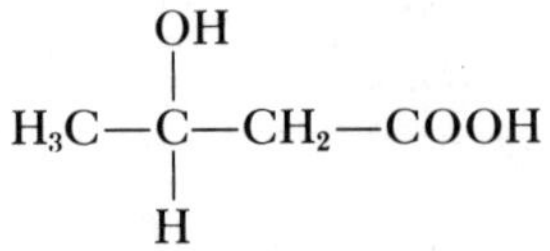

$$H_3C—\overset{\overset{\displaystyle OH}{|}}{\underset{\underset{\displaystyle H}{|}}{C}}—CH_2—COOH$$

Mol. wt. 104.11

Used as the substrate for hydroxybutyrate dehydrogenase.

3-Hydroxy-2-naphthoic acid hydrazide

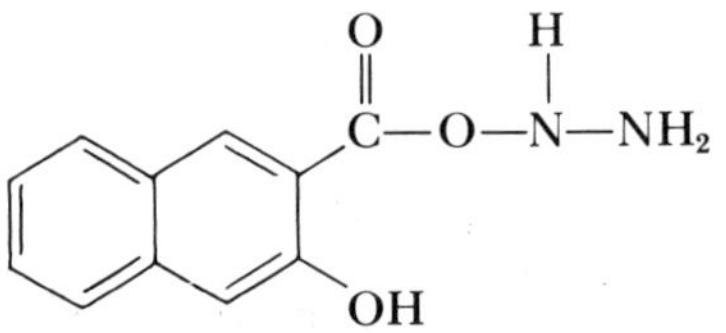

Mol. wt. 202.21

A reagent used in the original method for demonstrating monoamine oxidase.

8-Hydroxyquinoline glucuronide

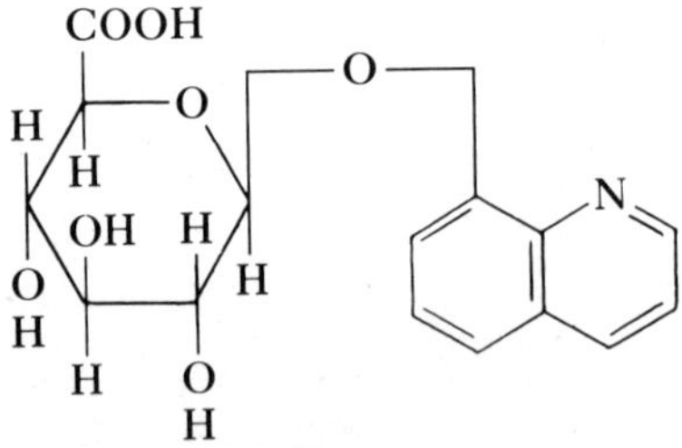

Mol. wt. 321.29

Used as a substrate for β-glucuronidase.

Indoxyl acetate

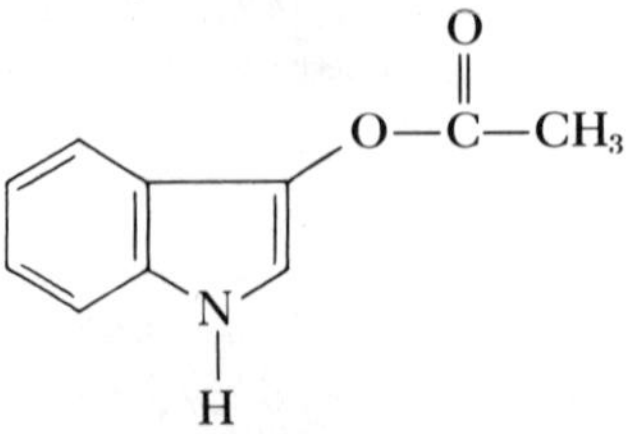

Mol. wt. 175.19

Used as a substrate for nonspecific esterase.

Indoxyl butyrate

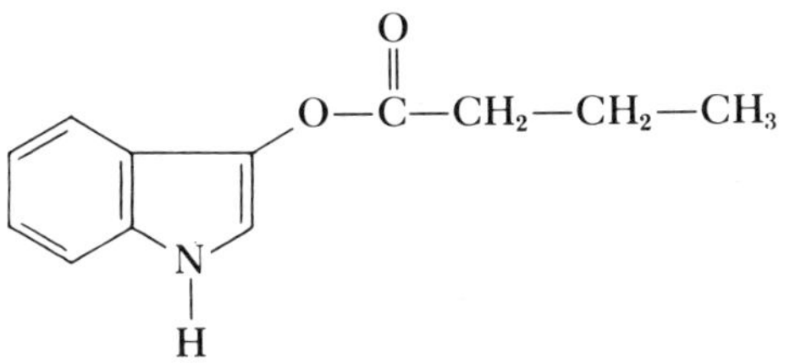

$$O$$
$$\parallel$$
$$O-C-CH_2-CH_2-CH_3$$

(indole ring with N—H)

Mol. wt. 203.24
Used as a substrate for esterases.

INT 2-(p-iodophenyl)-3-(p-nitrophenyl)-5-phenyl tetrazolium chloride

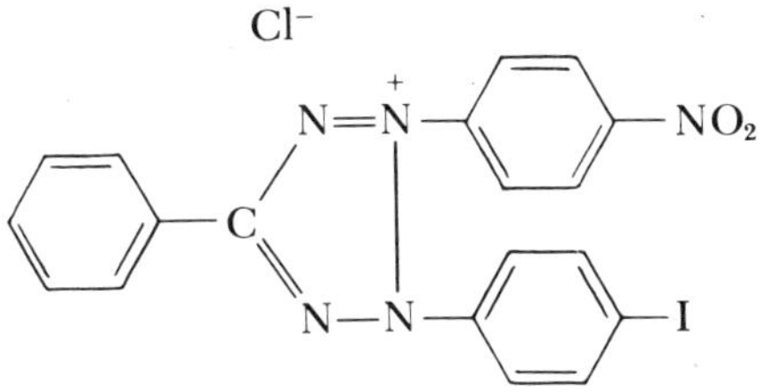

$$Cl^-$$

(tetrazolium ring structure: phenyl—C with N=N$^+$—phenyl—NO_2 and N—N—phenyl—I)

Mol. wt. 505.70
Formerly used as the electron acceptor in histochemical reactions for dehydrogenase enzymes.

Iodoacetate, sodium salt

$ICH_2-COONa$

Mol. wt. 207.94
Can react with sulfhydryl groups in tissues and may be used to inactivate enzymes with sulfhydryl-active centers.

Isocitric acid, trisodium salt

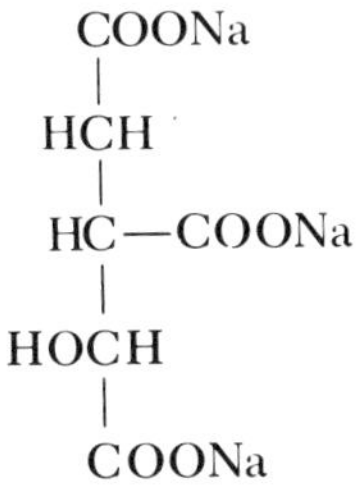

$$COONa$$
$$|$$
$$HCH$$
$$|$$
$$HC-COONa$$
$$|$$
$$HOCH$$
$$|$$
$$COONa$$

Mol. wt. 246.06
Used as the substrate for isocitric dehydrogenase.

Isopropanol (isopropyl alcohol)

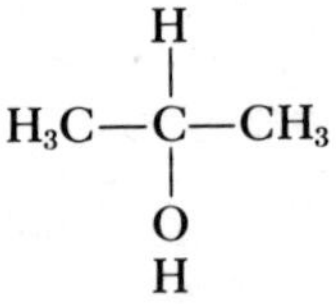

$$H_3C-\underset{\underset{H}{\overset{|}{O}}}{\overset{\overset{H}{|}}{C}}-CH_3$$

Mol. wt. 60.10
Used as a solvent for fat stains and other purposes.

α-Ketoglutaric acid

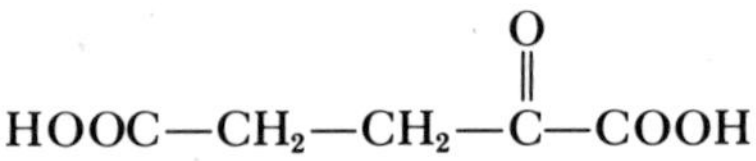

$$HOOC-CH_2-CH_2-\overset{\overset{O}{\parallel}}{C}-COOH$$

Mol. wt. 146.10
Used as the amino acceptor in the histochemical method for glutamate oxalate transaminase.

Kupffer cells
Phagocytic cells (macrophages) attached to the blood sinusoid walls of the liver.

Lactic acid

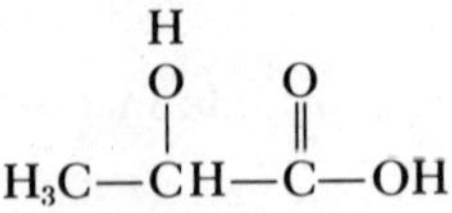

$$H_3C-\underset{\underset{H}{\overset{|}{O}}}{\overset{\overset{H}{|}}{C}}H-\overset{\overset{O}{\parallel}}{C}-OH$$

Mol. wt. 90.08
The end product of anaerobic metabolism. Used as the substrate for lactate dehydrogenase.

Lead nitrate
$Pb(NO_3)_2$
Mol. wt. 331.20
Used as a source of lead ions in the lead precipitation methods for demonstrating phosphatases.

Leucyl-4-methoxy-β-naphthylamide

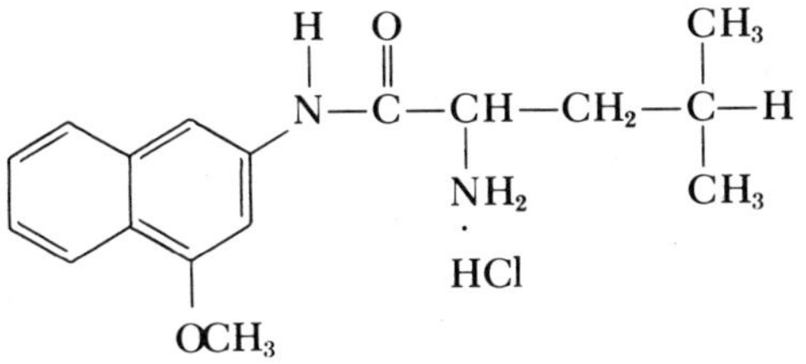

$$\text{Naphthyl—N—C—CH—CH}_2\text{—C—H}$$

with H, O above N and C; CH₃ above and CH₃ below the terminal C; NH₂ below CH; HCl; OCH₃ on the naphthyl ring.

Mol. wt. 323.83

Used as a substrate for aminopeptidase.

Lugol's iodine

Consists of iodine crystals and potassium iodide dissolved in water. Used to stain the substrate film in the method for demonstrating amylase.

Magnesium chloride

$MgCl \cdot 6H_2O$

Mol. wt. 203.33

Hygroscopic crystals. Used to increase the electrolyte concentration in the alcian blue–critical electrolyte concentration method.

Magnesium sulfate

$MgSO_4 \cdot 7H_2O$

Mol. wt. 246.50

Used as a source of magnesium ions to activate certain phosphatases.

Maleic acid

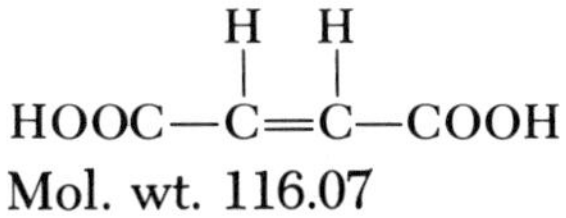

$$\text{HOOC—C}=\text{C—COOH}$$

with H, H above the two central carbons.

Mol. wt. 116.07

Used in the preparation of tris maleate buffer.

Malic acid

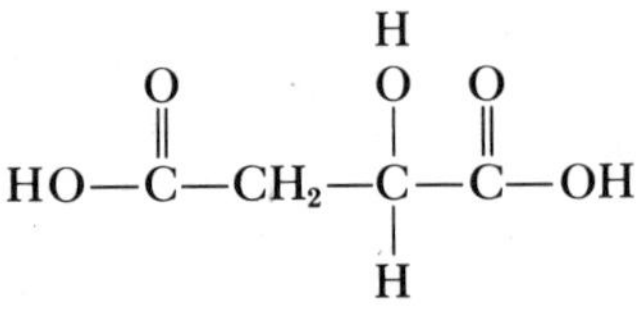

$$HO-\underset{\underset{O}{\parallel}}{C}-CH_2-\underset{\underset{H}{\mid}}{\overset{\overset{H}{\mid}}{C}}-\underset{\underset{}{\overset{\overset{O}{\parallel}}{C}}}{}-OH$$

Mol. wt. 134.09

A normal intermediate metabolite of the citric acid cycle. Used as the substrate for malic enzyme and for malic dehydrogenase.

Menadione (2-methyl-1,4-naphthoquinone)

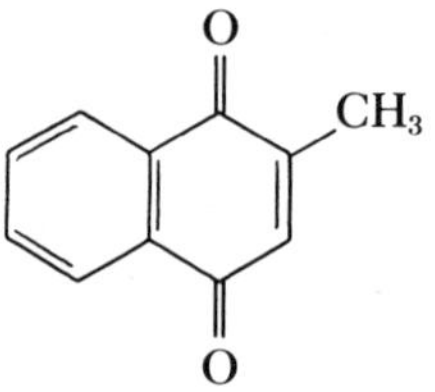

Mol. wt. 172.19

Has the physiological action of vitamin K. It is used as an electron carrier in the methods for demonstrating FAD-linked dehydrogenases.

Mercuric chloride

$HgCl_2$

Mol. wt. 271.50

A poisonous white crystalline compound. Used in certain fixatives.

Mercury orange (1-[4-chloromercuriphenylazo]-2-napthol)

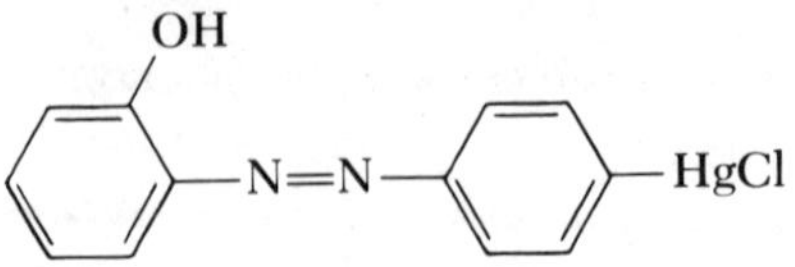

Mol. wt. 483.32

Used in a method for demonstrating sulfhydryl groups.

Metachromasia

The phenomenon in which certain basic dyes stain tissue components a color different from the color of a dilute solution of the dye.

Methenamine (hexamine)

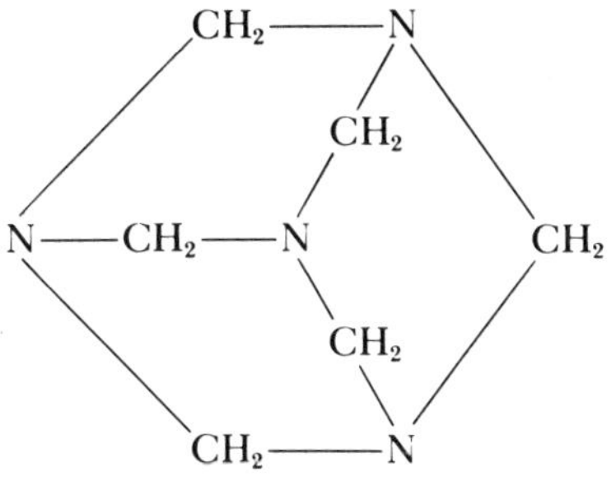

Mol. wt. 140.19
Used in preparing methenamine silver solution, which is used in certain silver impregnation stains.

Methanol (methyl alcohol)
CH_3OH
Mol. wt. 32.04
Methanol is very toxic and causes blindness. Used as a solvent.

_p_-Methoxy-_p_'-aminodiphenylamine

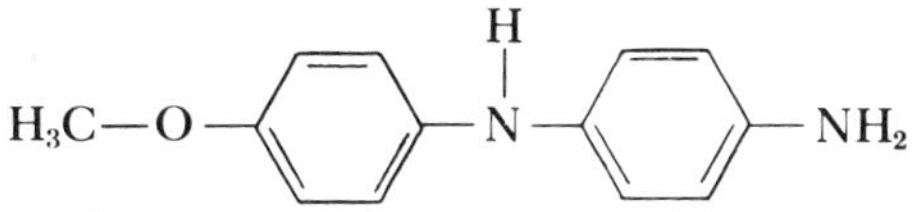

Mol. wt. 214.27
A substrate for the Nadi-type reaction for cytochrome oxidase.

Methoxyethanol
$H_3C—O—CH_2—CH_2OH$
Mol. wt. 76.10
Used as a solvent.

Methylene blue

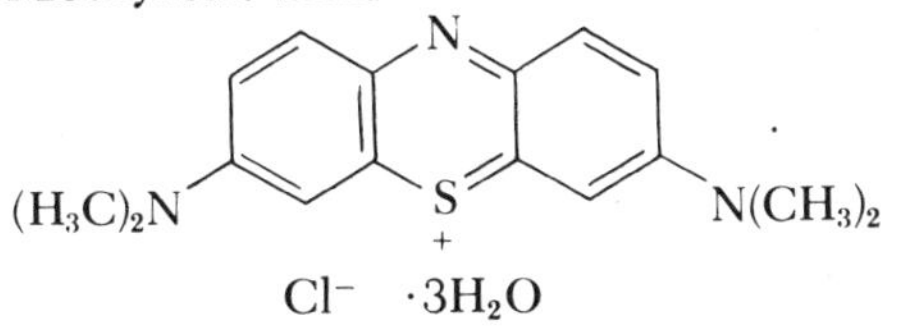

Cl^- $·3H_2O$
Mol. wt. 373.92   C.I. No. 52015
A thiazine basic dye.

Methyl green

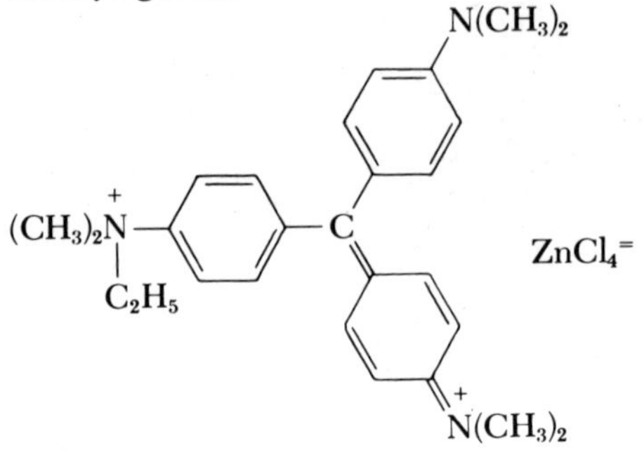

Mol. wt. 401.60 C.I. No. 42590
A triphenylmethane dye used in the methyl green–pyronin method for demonstrating DNA. May also be used as a nuclear counterstain.

Methyl salicylate (wintergreen oil)
$HOC_6H_4COOCH_3$
Mol. wt. 152.15
Used as a solvent and occasionally as a clearing agent.

Microperoxidase
A small polypeptide fragment of cytochrome c with very high peroxidase activity.

Mordant
A nondye intermediate, usually a metallic cation, that serves to hold the dye molecule firmly in place.

MTT 3-(4,5-dimethylthiazol-2-yl)-2,5-diphenyl-2H-tetrazolium bromide

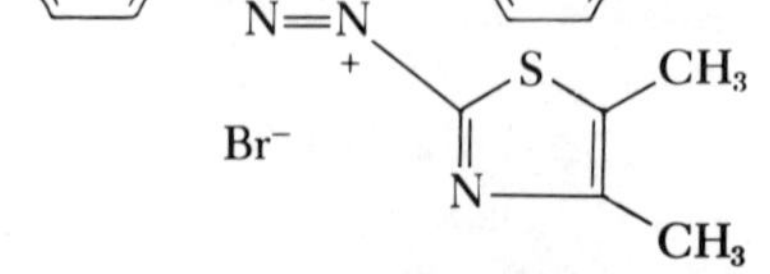

Mol. wt. of the bromide salt 414.33
A monotetrazolium compound useful for demonstrating dehydrogenases. It has been largely replaced with nitro blue tetrazolium.

D-Mucic acid (galactic acid)

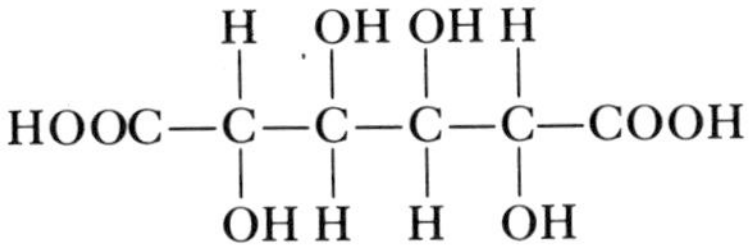

Mol. wt. 210.15
An inhibitor of β-glucuronidase.

α-Naphthol (1-naphthol)

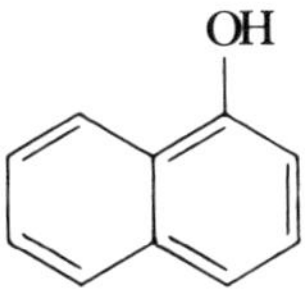

Mol. wt. 144.18
Used as a reagent in the Sakaguchi method for demonstrating arginine.

Naphthol-AS-BI-N-acetyl-β-glucosaminide

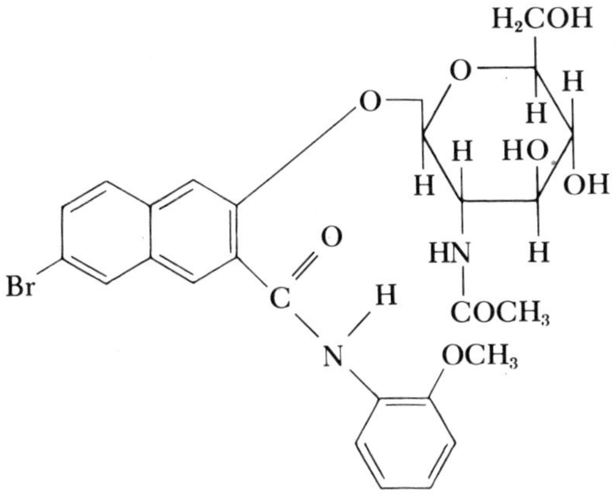

Mol. wt. 592.43
Used as a substrate for N-acetyl-β-glucosaminidase.

Naphthol-AS-BI-β-glucuronide

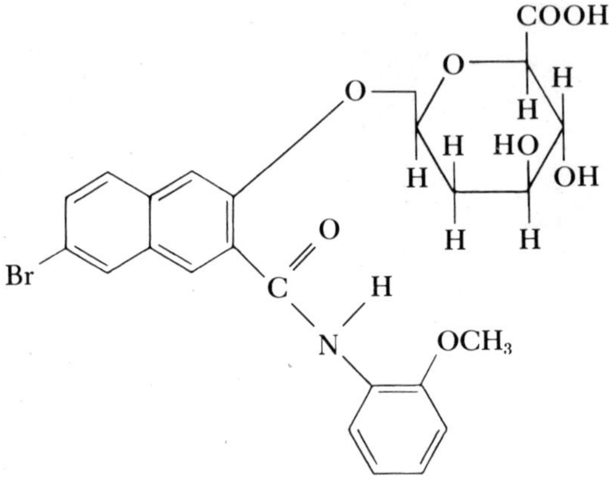

Mol. wt. 548.35

Used as a substrate for β-glucuronidase.

Naphthol AS-BI-phosphate, sodium salt

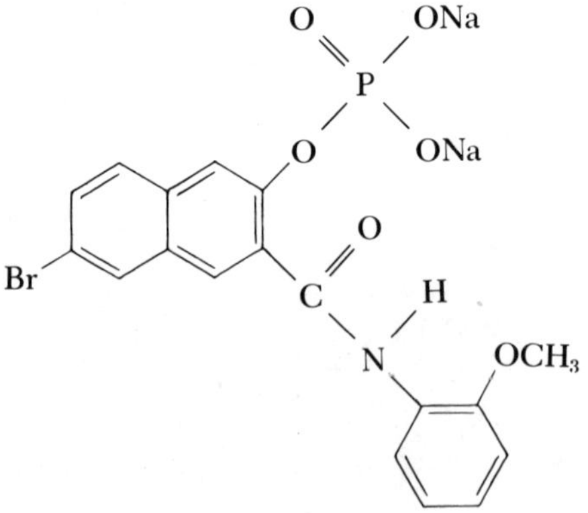

Mol. wt. 473.18

Used as a substrate for alkaline and acid phosphatase.

Naphthol AS-D-acetate

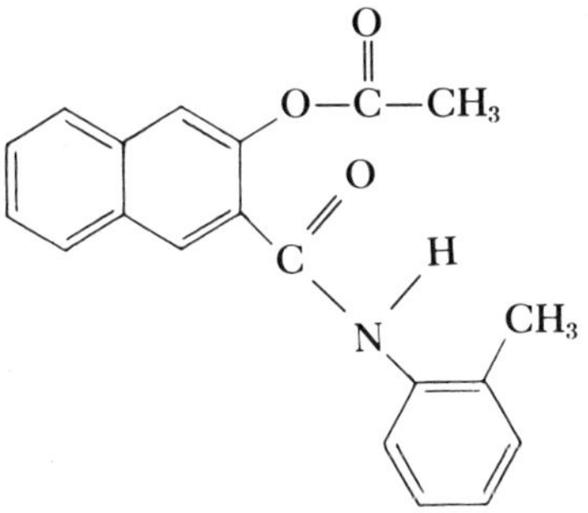

Mol. wt. 319.36
Used as a substrate for nonspecific esterase.

Naphthol AS-MX-phosphate, sodium salt

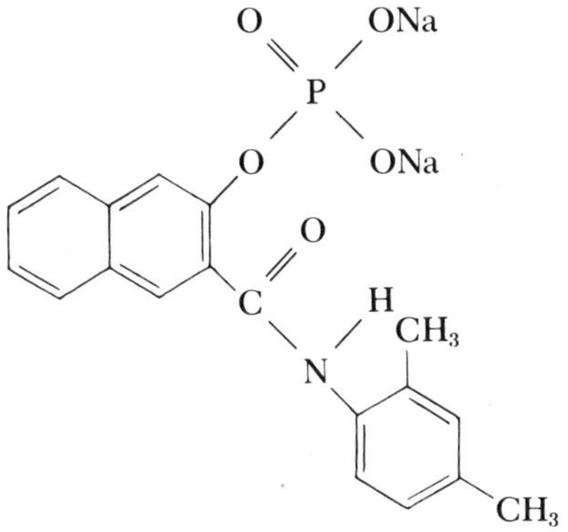

Mol. wt. 392.31
Used as a substrate for acid and alkaline phosphatase.

Naphthol AS-sulfate, sodium salt

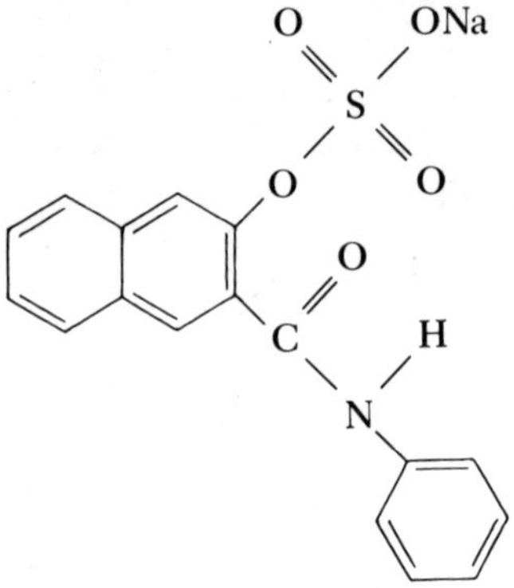

Mol. wt. 365.34
Used as a substrate for arylsulfatase.

Naphthol AS-TR-phosphate, sodium salt

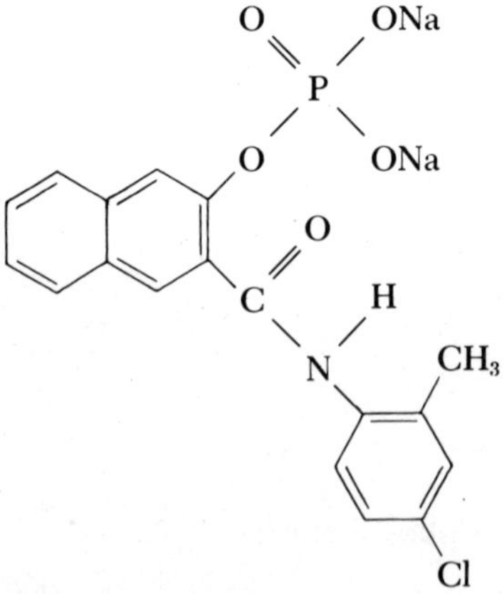

Mol. wt. 377.27
Used as a substrate for alkaline and acid phosphatase.

1,2-Naphthoquinone-4-sulfonic acid, sodium salt

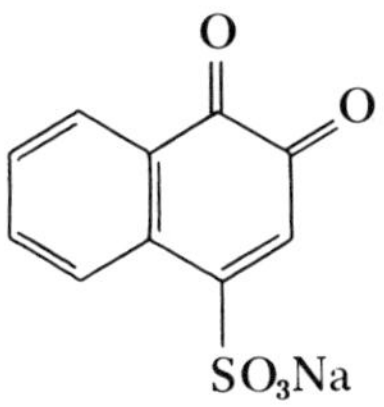

Mol. wt. 260.21
Used in a method for demonstrating cholesterol.

Naphthoresorcinol

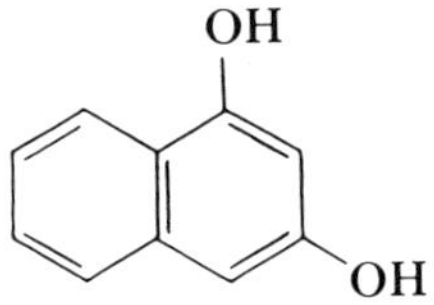

Mol. wt. 160.17
May be used as a coupling agent.

α-Naphthyl acetate

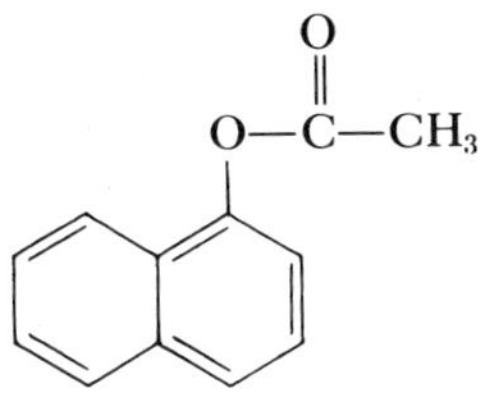

Mol. wt. 186.21
May be used as a substrate for nonspecific esterase.

β-Naphthyl acetate

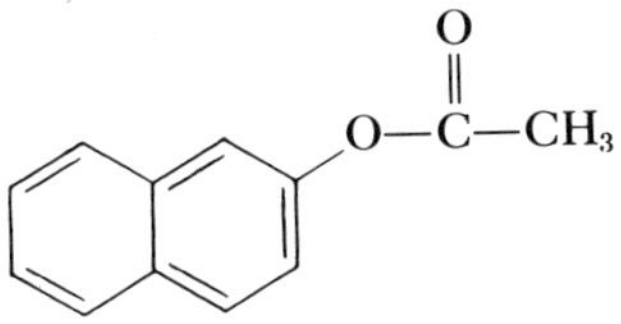

Mol. wt. 186.21
Can be used as a substrate for nonspecific esterase, but it is not recommended.

α-Naphthyl-N-acetyl-β-glucosamide

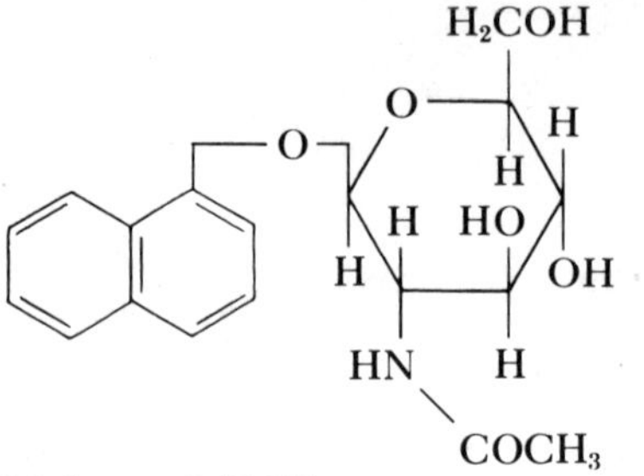

$$H_2COH$$

Mol. wt. 347.37

Used as a substrate for N-acetyl-β-glucosaminidase.

α-Naphthylamine

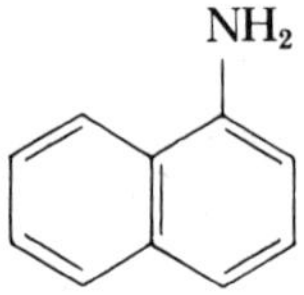

$$NH_2$$

Mol. wt. 143.19

Carcinogen. Used in the OTAN method for demonstrating lipids.

β-Naphthyl phosphate

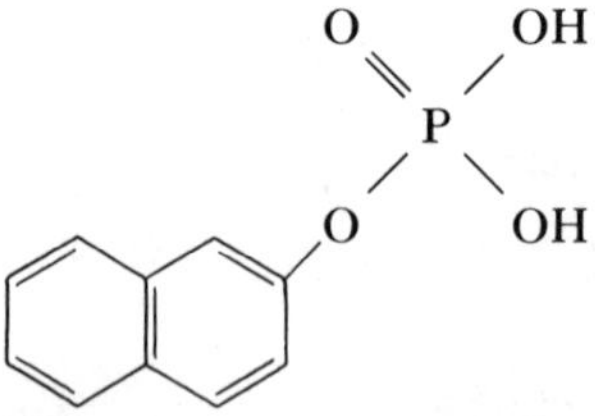

Mol. wt. 224.15

Formerly used as a histochemical substrate for alkaline phosphatase.

Neutral red

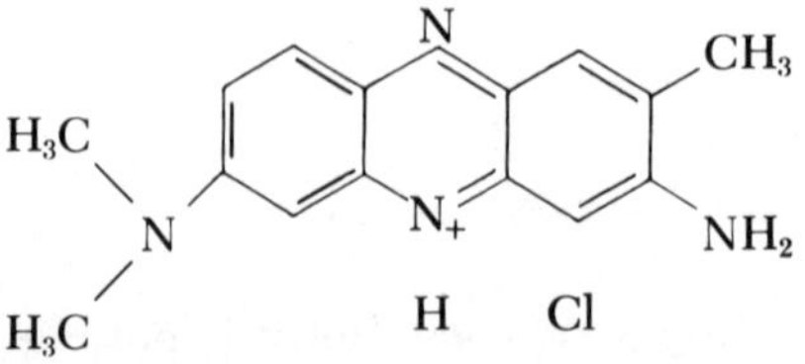

Mol. wt. 288.78 C.I. No. 50040

Used as a nuclear counterstain after certain histochemical reactions.

New fuchsin (magenta III)

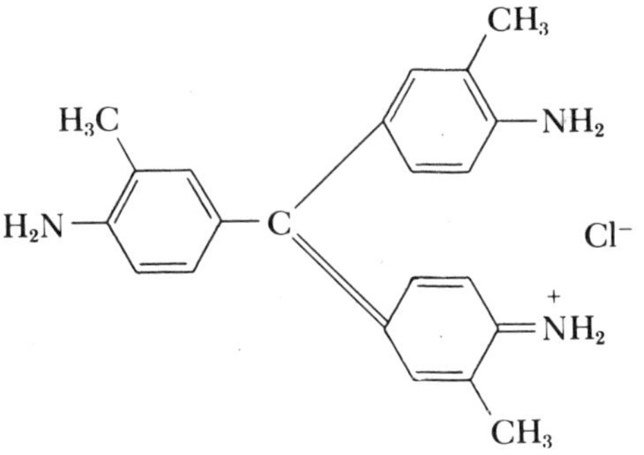

Mol. wt. 365.91 C.I. No. 42520
A minor component of basic fuchsin.

Nicotinamide

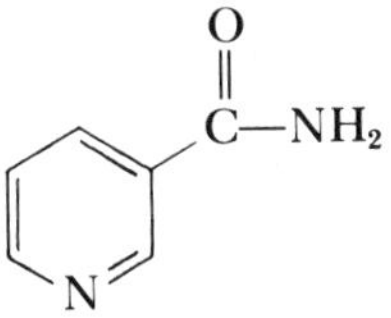

Mol. wt. 122.13
One of the B vitamins. Used as a reagent in the method for demonstrating 3-β-hydroxysteroid dehydrogenase.

Nile blue sulfate (nile blue A)

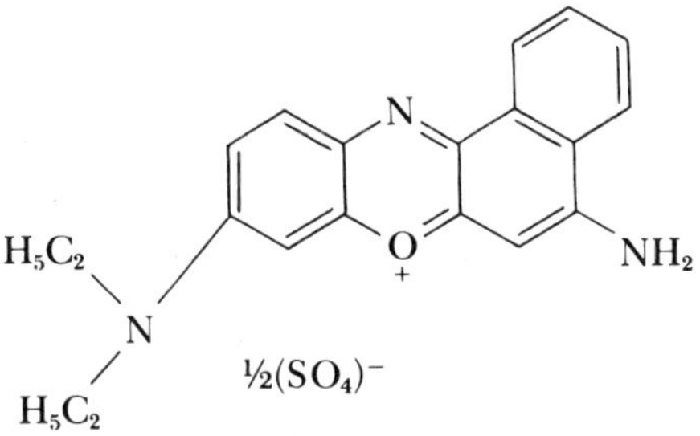

Mol. wt. of organic cation 318.40 C.I. No. 51180
An oxazine dye, used as a differential stain for lipids.

Nitric acid

HNO_3

Mol. wt. 63.01

A corrosive acid that causes yellow burns of the skin. Concentrated nitric acid is 70 to 71 percent. To make approximately a 1.0M solution, add 65 ml concentrated acid to water and bring to 1 liter with water.

Nitro blue tetrazolium (nitro BT)

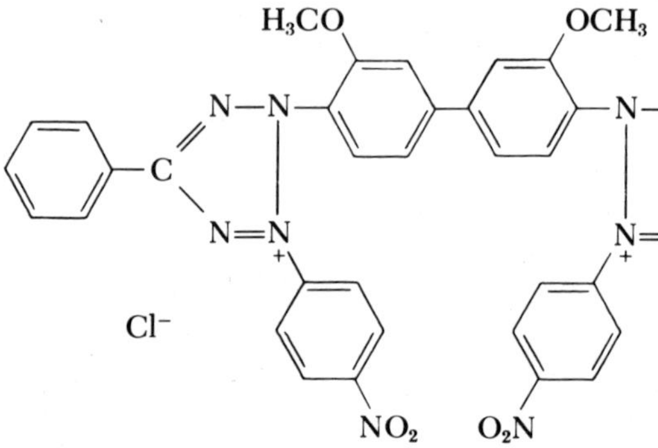

Mol. wt. 817.65

Used as an electron acceptor in procedures for demonstrating dehydrogenase enzymes.

p-Nitrocatechol sulfate, dipotassium salt (4-nitro-1,2-benzenediol mono [hydrogen sulfate])

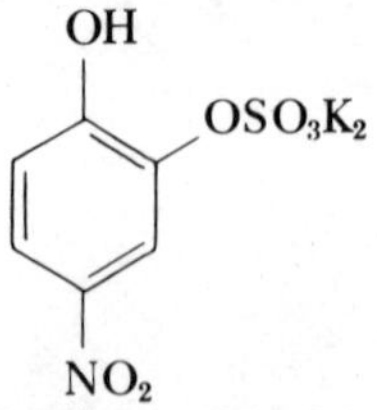

Mol. wt. 312.36

Used as a substrate for arylsulfatase.

p-Nitrophenyl phosphate, disodium salt

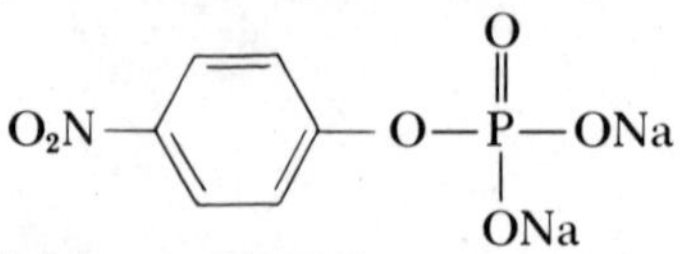

Mol. wt. 263.05

Used as a substrate for demonstrating Na^+/K^+-activated ATPase.

Nitro tetrazolium

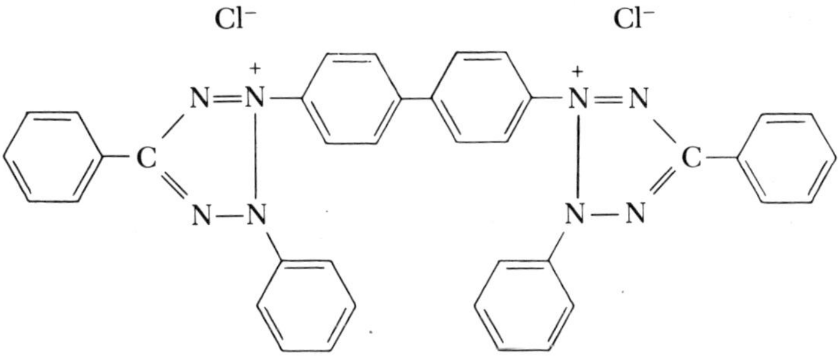

Mol. wt. 667.61
Formerly used as an electron acceptor in procedures for demonstrating dehydrogenase enzymes.

5-Norbornene-2,3-dicarboxylic acid

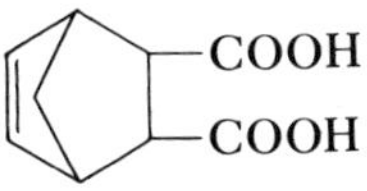

Mol. wt. 180.15
Used as a buffer for isolating enzymes from homogenates while preserving maximal activity.

Nuclear fast red

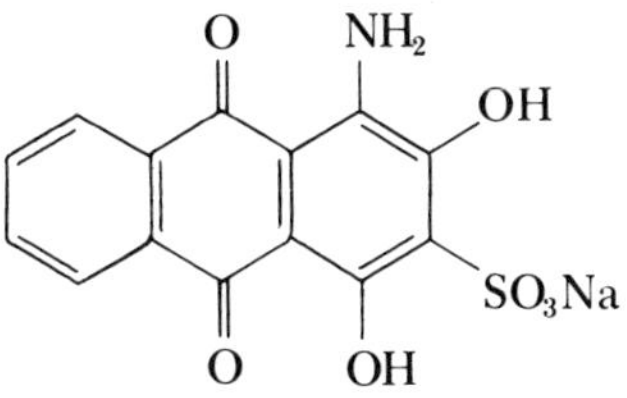

Mol. wt. 357.28
C.I. No. 60760
Used in a method for demonstrating calcium. Also used as a red nuclear counterstain following certain histochemical procedures.

Oil red 4B

(Structure uncertain)
Used as a stain for lipids.

Oil red O

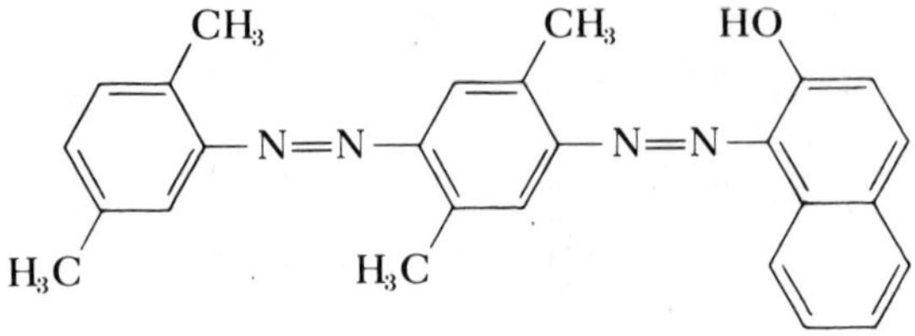

Mol. wt. 408.48
C.I. No. 26125
Used as a stain for lipids.

Orcinol monohydrate (methylresorcinol monohydrate)

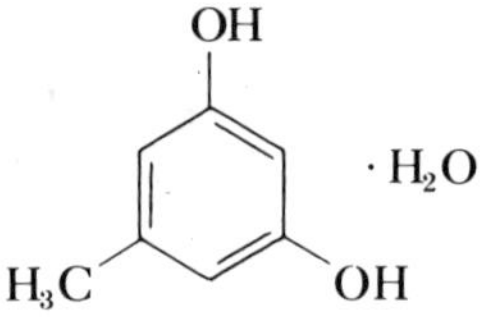

Mol. wt. 142.16
Used as one of the reagents in the Bial method for demonstrating sialic acid.

Orthochromasia

A term used in counterdistinction to *metachromasia* when certain basic dyes that are capable of staining metachromatically stain tissue components the same color as that of a dilute solution of the dye.

Osmium tetroxide

OsO_4
Mol. wt. 254.20
Used as a reagent for demonstrating unsaturated lipids and for enhancing the visibility of the polymeric reaction product of diaminobenzidine. Fumes are highly toxic.

Ouabain

$C_{29}H_{44}O_{12} \cdot 8H_2O$
Mol. wt. 728.79
A specific inhibitor of Na^+/K^+-activated ATPase.

Oxaloacetic acid

$$HOOC-CH_2-\overset{\displaystyle\|}{\underset{\displaystyle O}{C}}-COOH$$

Mol. wt. 132.07

One of the reaction products of glutamate oxaloacetate transaminase.

Paraffin (paraffins)

Consists of a series of long straight-chain molecules of variable lengths. The average length of the chains is probably about 30 carbons.

Paraformaldehyde

$(CH_2O)_x$

A dry, polymerized form of formaldehyde. Comes as a white powder.

Pararosanilin

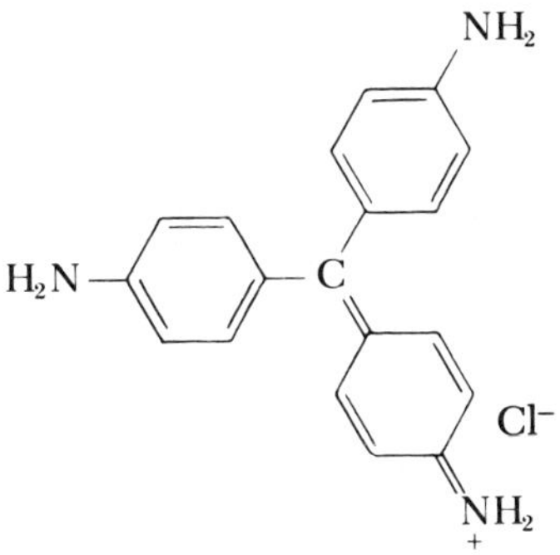

Mol. wt. 323.83

C.I. No. 42500

A triphenylmethane dye; the chief component of basic fuchsin.

Peracetic acid

$$CH_3-\overset{\displaystyle\overset{O}{\|}}{C}-O-OH$$

Mol. wt. 76.05

A toxic liquid with a disagreeable odor. Used as an oxidant.

Perchloric acid

$HClO_4$

Mol. wt. 100.46

Corrosive to skin and mucous membranes. Available as a 70% aqueous solution. Used in a method for demonstrating cholesterol.

Performic acid

$$H—C—O—OH$$
$$\overset{\|}{O}$$

Mol. wt. 62.03

May explode when brought in contact with metals, metallic oxides, or reducing agents. It is unstable and must be prepared fresh before each use. Used as an oxidant in a method for demonstrating unsaturated lipids.

Periodic acid

H_5IO_6

Mol. wt. 227.96

Used as the oxidizing agent in the periodic acid–Schiff reaction.

Phenazine methosulfate (5-methylphenazinium methyl sulfate)

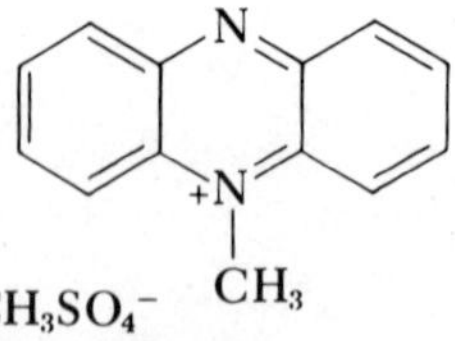

$CH_3SO_4^-$ CH_3

Mol. wt. 306.34

Used as an electron carrier in histochemical methods for certain dehydrogenases.

β-Phenylethylamine (phenethylamine)

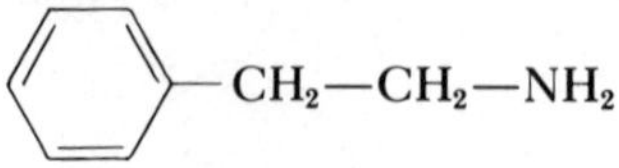

$CH_2—CH_2—NH_2$

Mol. wt. 121.18

A substrate that is specific for monoamine oxidase, A-form.

Phenylhydrazine hydrochloride

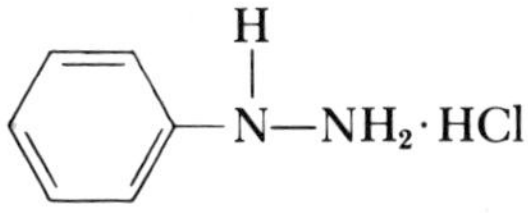

$$\langle\bigcirc\rangle - \overset{\overset{\textstyle H}{|}}{N} - NH_2 \cdot HCl$$

Mol. wt. 144.61

A reagent used in testing for carbonyl groups and also for blocking aldehyde groups in connection with the PAS reaction.

6-Phosphogluconic acid, trisodium salt

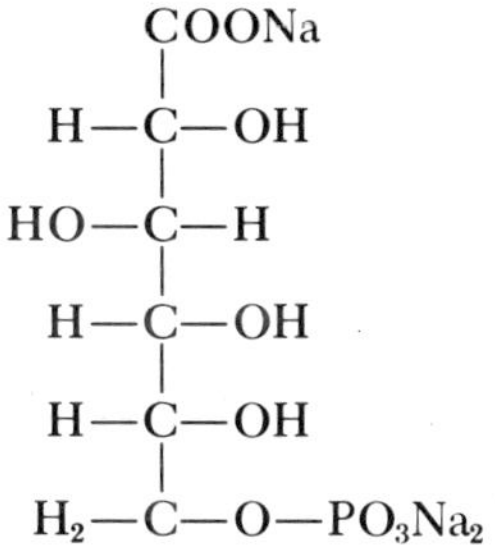

$$
\begin{array}{l}
\quad\ \text{COONa} \\
\quad\ \ | \\
\text{H}-\text{C}-\text{OH} \\
\quad\ \ | \\
\text{HO}-\text{C}-\text{H} \\
\quad\ \ | \\
\text{H}-\text{C}-\text{OH} \\
\quad\ \ | \\
\text{H}-\text{C}-\text{OH} \\
\quad\ \ | \\
\text{H}_2-\text{C}-\text{O}-\text{PO}_3\text{Na}_2
\end{array}
$$

Mol. wt. 311.11

An inhibitor of glucose phosphate isomerase; used as a substrate for phosphogluconate dehydrogenase.

Phosphoric acid, ortho

H_3PO_4

Mol. wt. 98.00

Used as one of the reagents in a modified Schultz method for cholesterol. To make approximately a 1.0M solution, dissolve 23.0 ml in water and add water to make 1 liter.

Picric acid (trinitrophenol)

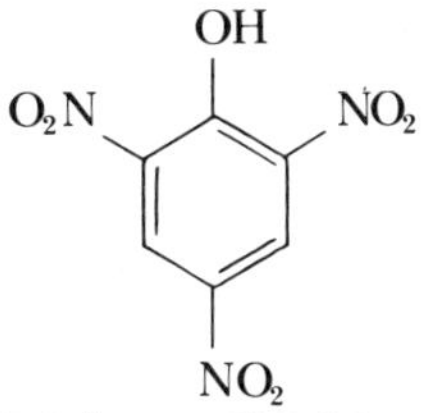

$$
\begin{array}{c}
\text{OH} \\
O_2N \diagdown \!\!\!\bigwedge\!\!\! \diagup NO_2 \\
| \\
NO_2
\end{array}
$$

Mol. wt. 229.11

Explosive when dry. Used as a component of certain fixatives.

Polyvinyl alcohol

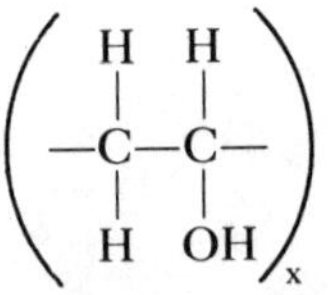

$$\left(\begin{array}{c} \mathrm{H} \quad \mathrm{H} \\ | \quad | \\ -\mathrm{C}-\mathrm{C}- \\ | \quad | \\ \mathrm{H} \quad \mathrm{OH} \end{array}\right)_x$$

A water-soluble synthetic polymer. Used in incubation media of soluble enzymes to decrease enzyme diffusion.

Polyvinyl pyrrolidone

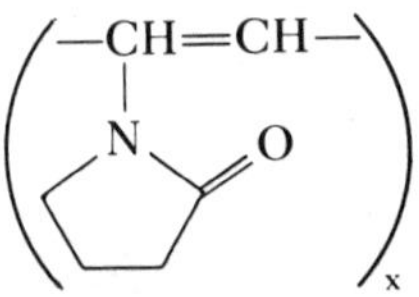

$$\left(\begin{array}{c} -\mathrm{CH}\!=\!\mathrm{CH}- \\ | \\ \mathrm{N} \qquad \mathrm{O} \end{array}\right)_x$$

Used to control diffusion in demonstrating certain soluble enzymes, particularly soluble dehydrogenases.

Potassium bromide

KBr
Mol. wt. 119.02
Used as one of the reagents in the bromine–silver nitrate method for demonstrating unsaturated lipids.

Potassium chlorate

$KClO_3$
Mol. wt. 122.56
Used as one of the reagents in the OTAN method for demonstrating phospholipids.

Potassium chloride

KCl
Mol. wt. 74.56
Used as one of the reagents in the method for demonstrating calcium-activated ATPase and sodium/potassium-activated ATPase.

Potassium cyanide

KCN

Mol. wt. 65.12

Used as an inhibitor of certain enzymes. In demonstrating dehydro-genases, it is sometimes used to block the terminal end of the electron transport system in order to divert the electrons toward the tetrazolium compound.

Potassium ferricyanide

$K_3Fe(CN)_6$

Mol. wt. 329.26

Used with ferric sulfate in the formation of ferric ferricyanide for dem-onstrating sulfhydryl groups.

Potassium ferrocyanide, trihydrate

$K_4Fe(CN)_6 \cdot 3H_2O$

Mol. wt. 422.41

Used in making a reagent for demonstrating iron by the prussian blue reaction. Ferrocyanide reacts with ferric ions to form a blue reaction product.

Potassium iodide

KI

Mol. wt. 166.01

Used in making an iodine stain for glycogen.

Potassium permanganate

$KMnO_4$

Mol. wt. 158.04

Used as an oxidant.

Potassium phosphate, dibasic

K_2HPO_4

Mol. wt. 174.18

Used in preparing phosphate buffers.

Potassium phosphate, monobasic

KH_2PO_4

Mol. wt. 136.09

Used in preparing phosphate buffers.

Propylene glycol (1,2-propanediol)

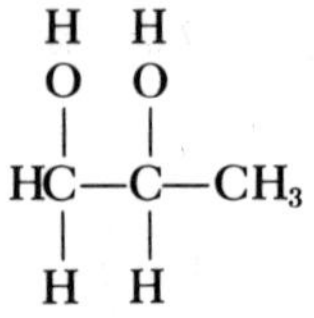

$$\begin{array}{c} \text{H} \quad \text{H} \\ \text{O} \quad \text{O} \\ | \quad\; | \\ \text{HC--C--CH}_3 \\ | \quad\; | \\ \text{H} \quad \text{H} \end{array}$$

Mol. wt. 76.10

A viscous liquid that is used as a solvent for fat stains.

Pyridine

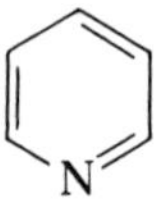

Mol. wt. 79.10

A flammable liquid with a disagreeable odor. Used as an organic solvent.

Pyridoxal phosphate

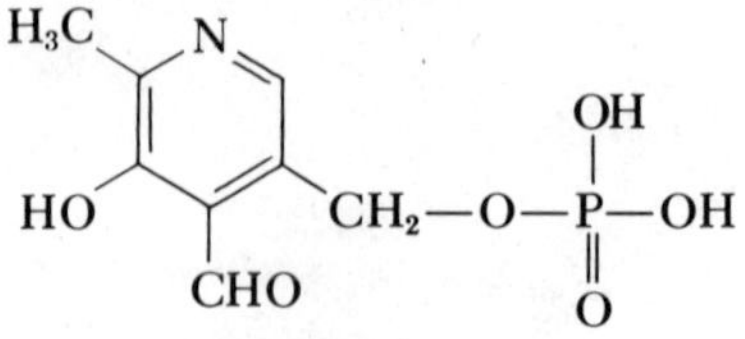

Mol. wt. 247.15

The biologically active form of vitamin B_6. It serves as the coenzyme for a group of enzymes that are involved in removing or transferring amino groups, removing carboxyl groups, and removing sulfhydryl groups.

Pyrogallol (pyrogallic acid; 1,2,3-trihydroxybenzene)

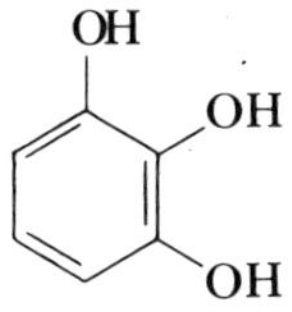

Mol. wt. 126.11
May be used as a coupling agent.

Pyronin Y

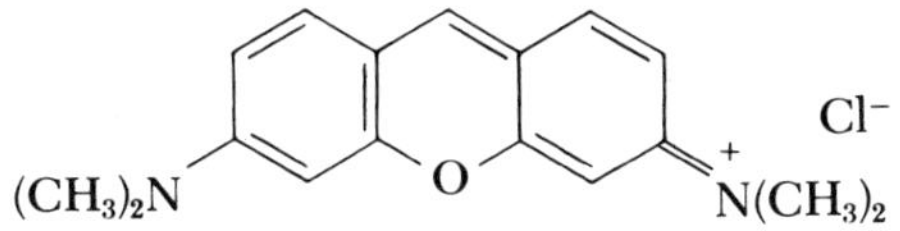

Cl⁻

$(CH_3)_2N$ O $N(CH_3)_2$

Mol. wt. 302.80
C.I. No. 45005
Used in the methyl green–pyronin method for staining nucleic acids.

Pyruvic acid

$$H_3C-\overset{\overset{\textstyle O}{\|}}{C}-COOH$$

Mol. wt. 88.07
A normal metabolic intermediate of aerobic metabolism.

Resorcinol

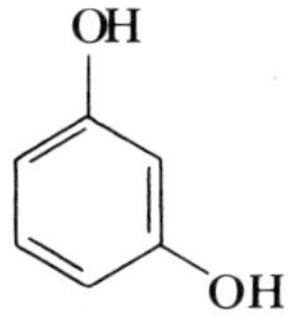

Mol. wt. 110.11
May be used as a coupling agent.

Rosanilin

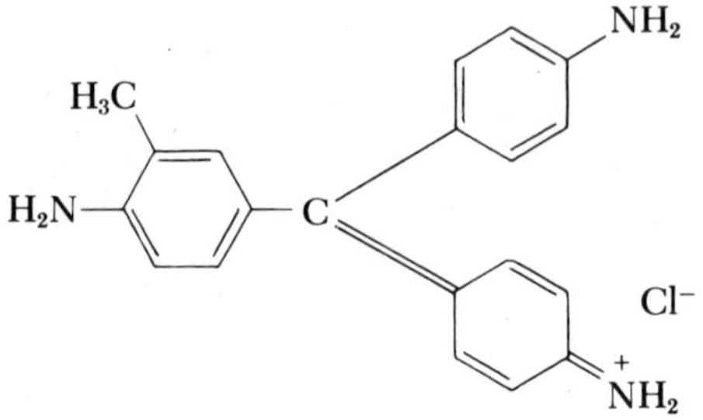

Mol. wt. of chloride salt 337.86
C.I. No. 42500
One of the dyes of basic fuchsin.

Rubeanic acid (dithiooxamide)

H_2N—C—C—NH_2
          ‖   ‖
          S   S

Mol. wt. 120.20
Used as the principal reagent in the method for demonstrating copper.

D-Saccharic acid (D-glucaric acid)

```
        COOH
         |
   H—C—OH
         |
 HO—C—H
         |
   H—C—OH
         |
   H—C—OH
         |
        COOH
```

Mol. wt. 210.14
A specific inhibitor of β-glucuronidase.

S-acid (8-amino-1-naphthol-5-sulfonic acid)

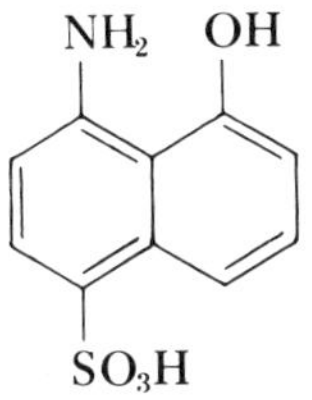

NH₂ OH

SO₃H

Mol. wt. 239.29
Used as a chromogenic coupling agent.

Safranin O

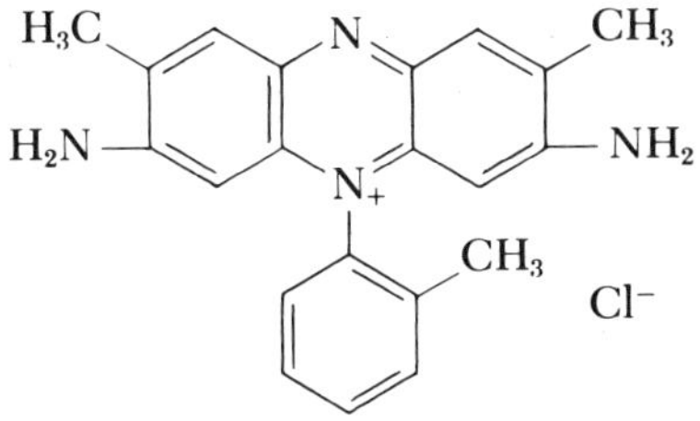

H₃C   N   CH₃

H₂N   NH₂
          N+
              CH₃
                    Cl⁻

Mol. wt. of chloride salt 364.89
C.I. No. 50240
A metachromatic stain of the azine group.

Saponification
Treating a tissue section with 0.5 to 1% potassium hydroxide in 70% ethanol in order to deesterify carboxyl groups or sialic acid residues.

Schiff's reagent (leucofuchsin)
Originally made by passing sulfur dioxide gas through a solution of basic fuchsin or rosanilin until the solution became clear. The solution is readily recolorized by aldehydes (and more slowly by ketones). It is widely used in histochemistry for demonstrating aldehydes in tissues.

Silver bromide
AgBr
Mol. wt. 187.80
A by-product in the bromination method for demonstrating unsaturated lipids.

Silver nitrate

$AgNO_3$

Mol. wt. 169.87

Used in many histological methods for demonstrating nervous tissue as well as in the bromination method for demonstrating unsaturated lipids.

Sodium acetate

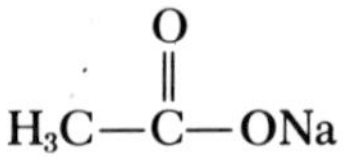

$$H_3C-\overset{\overset{\displaystyle O}{\displaystyle \|}}{C}-ONa$$

Mol. wt. 82.03

Used in preparing acetate buffers.

Sodium azide

NaN_3

Mol. wt. 65.02

Used as an inhibitor of certain enzymes.

Sodium borohydride

$NaBH_4$

Mol. wt. 37.84

Reacts with water to form sodium hydroxide and evolve hydrogen gas (and is therefore a dangerous fire risk). Used to block carbonyl groups.

Sodium carbonate

$NaHCO_3$

Mol. wt. 84.01

Used in preparing carbonate buffers and as one of the reagents in the Tween method for demonstrating lipase.

Sodium carbonate, anhydrous

Na_2CO_3

Mol. wt. 105.99

Used in preparing a fixative for tissues in which calcium is to be demonstrated.

Sodium chloride
NaCl
Mol. wt. 58.44
Ordinary table salt. Frequently used to adjust osmotic pressures of solutions.

Sodium cyanide
NaCN
Mol. wt. 49.01
Poison. Used as one of the reagents in the dithiozone method for demonstrating zinc.

Sodium fluoride
NaF
Mol. wt. 41.99
Used as one of the reagents in the preincubation medium in the ferric ion precipitation method for demonstrating glycogen phosphorylase.

Sodium hydrosulfite (sodium dithionite)
$Na_2S_2O_4$
Mol. wt. 174.11
A reducing agent. Used in the DNFB method for proteins to reduce nitro groups to amine groups.

Sodium hypochlorite
NaOCl
Mol. wt. 74.45
Used as one of the reagents in the modified Sakaguchi method for demonstrating arginine.

Sodium metabisulfite
$Na_2S_2O_5$
Mol. wt. 190.11
Frequently used as the reducing agent in making Schiff's reagent.

Sodium nitrite
$NaNO_2$
Mol. wt. 69.00
Used as a reducing reagent in the diazotization reaction.

Sodium nitroferricyanide, dihydrate
$Na_2Fe(CN)_5NO \cdot 2H_2O$
Mol. wt. 297.97
Used as an inhibitor of certain enzymes.

Sodium sulfate
Na_2SO_4
Mol. wt. 142.04
Used as one of the reagents in the tetrazolium method for demonstrating monoamine oxidase.

Sodium sulfite
Na_2SO_3
Mol. wt. 126.05
Used in preparing Schiff's reagent in the Feulgen–methylene blue method for demonstrating nucleic acids.

Sodium thiosulfate, pentahydrate
$Na_2S_2O_3 \cdot 5H_2O$
Mol. wt. 248.19
A reducing reagent used in the methenamine silver method for demonstrating melanin and in the dithizone method for demonstrating zinc.

Sorbitan 6-phosphate

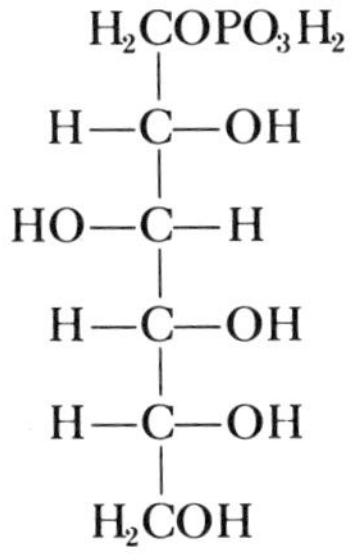

$$
\begin{array}{l}
\text{H}_2\text{COPO}_3\text{H}_2 \\
\quad | \\
\text{H}-\text{C}-\text{OH} \\
\quad | \\
\text{HO}-\text{C}-\text{H} \\
\quad | \\
\text{H}-\text{C}-\text{OH} \\
\quad | \\
\text{H}-\text{C}-\text{OH} \\
\quad | \\
\text{H}_2\text{COH}
\end{array}
$$

Mol. wt. 262.15
A specific inhibitor of glucose 6-phosphatase.

Sorbitol

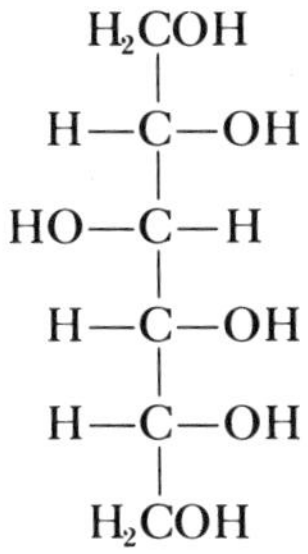

$$
\begin{array}{l}
\text{H}_2\text{COH} \\
\quad | \\
\text{H}-\text{C}-\text{OH} \\
\quad | \\
\text{HO}-\text{C}-\text{H} \\
\quad | \\
\text{H}-\text{C}-\text{OH} \\
\quad | \\
\text{H}-\text{C}-\text{OH} \\
\quad | \\
\text{H}_2\text{COH}
\end{array}
$$

Mol. wt. 182.18
A polyalcohol forming a subunit of the Tweens.

Stains-all

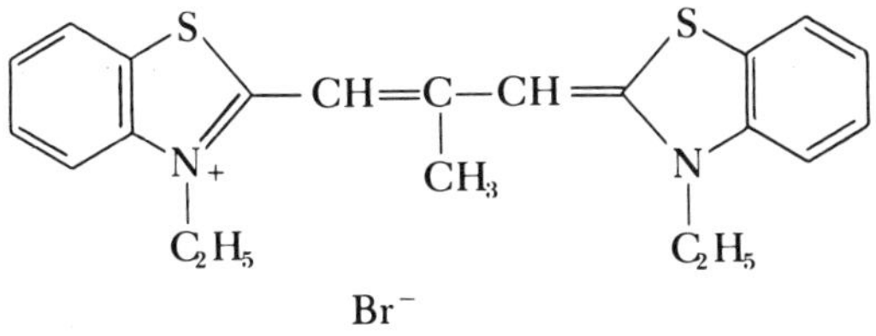

Mol. wt. 559.60
Used in a procedure for demonstrating various proteins and carbohydrates.

Substantivity
The affinity of a chemical substance for tissue proteins. Usually applied to the final reaction product of enzyme histochemical reactions.

Succinic acid

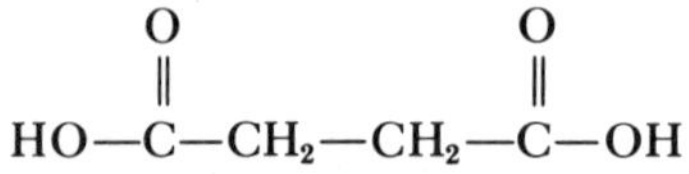

$$\underset{\text{HO}}{}-\overset{\overset{\text{O}}{\|}}{\text{C}}-\text{CH}_2-\text{CH}_2-\overset{\overset{\text{O}}{\|}}{\text{C}}-\text{OH}$$

Mol. wt. 118.09
Used as the substrate for succinate dehydrogenase.

Sudan IV

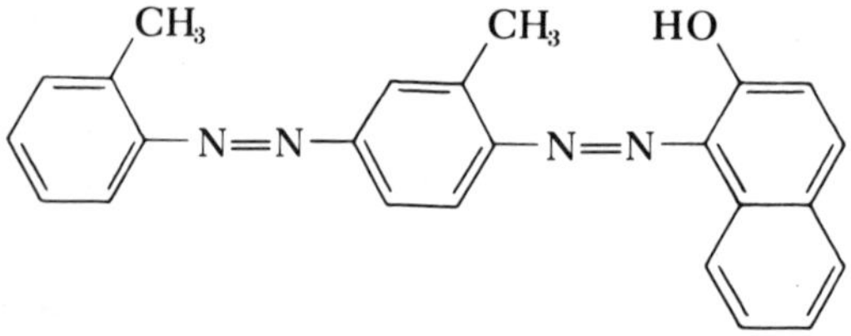

Mol. wt. 380.45
C.I. No. 26105
Used as a stain for lipids.

Sudan black B

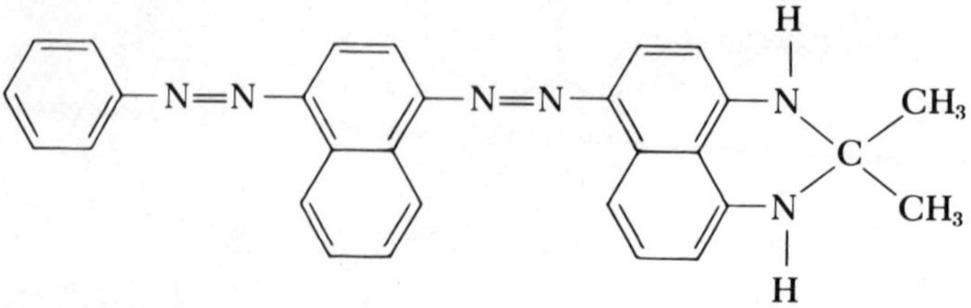

Mol. wt. 456.55
C.I. No. 26150
Used as a stain for lipids.

Sulfanilic acid

$$H_2N-\langle\rangle-SO_3H$$

Mol. wt. 173.19
Used as one of the reagents in a method for demonstrating histidine.

Sulfuric acid

H_2SO_4
Mol. wt. 98.08   Sp. gr. 1.84
A strong mineral acid; a strong oxidant. Vapors are highly irritating to mucous membranes. To make approximately a 1.0M solution, add 28.0 ml of the concentrated acid to water and add more water to make 1 liter.

Taurocholic acid

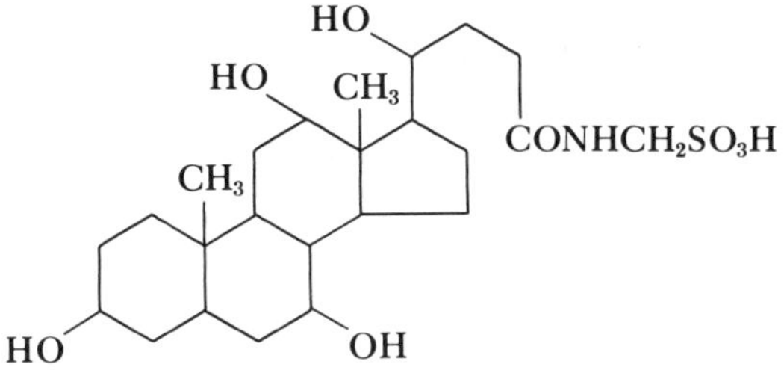

$CONHCH_2SO_3H$

Mol. wt. 515.72
A reducing agent. Can be used to block sulfhydryl groups on proteins.

Tetraisopropylpyrophosphoramide (iso OMPA)

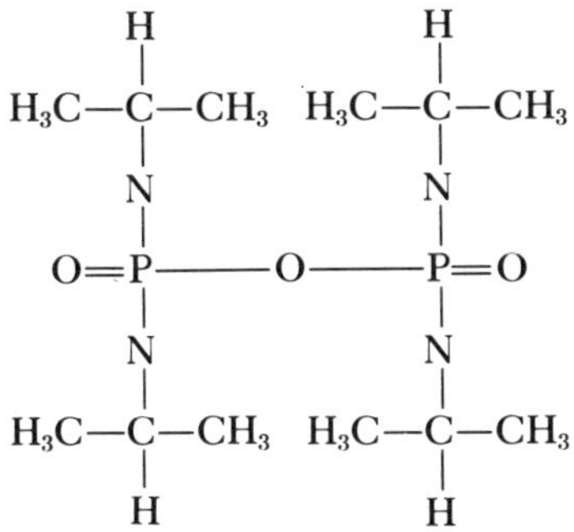

Mol. wt. 338.33
Used as an inhibitor of pseudocholinesterase.

L-Tetramisole Hydrochloride

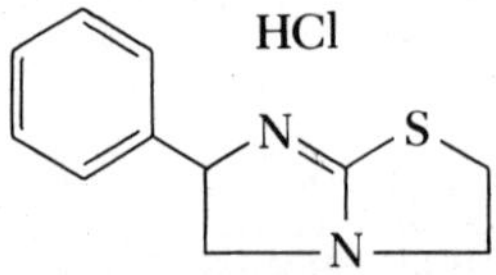

Mol. wt. 240.8

A specific inhibitor of alkaline phosphatase (found in bone, kidney, and liver) but not of intestinal alkaline phosphatase.

Tetranitro blue tetrazolium (TNBT)

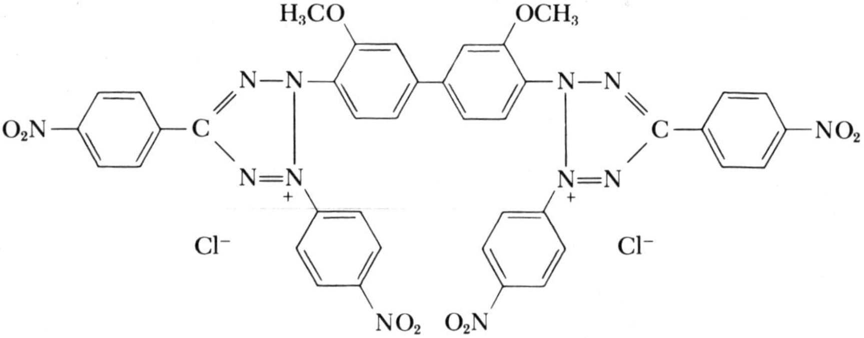

Mol. wt. 907.64

Used as an electron acceptor in histochemical reactions for dehydrogenase enzymes.

Thiocarbohydrazide (3-thiocarbozide)

```
      H   S   H
      |   ‖   |
H2N—N—C—N—NH2
```

Mol. wt. 106.15

Used as an osmophilic coupling agent.

Thioglycolic acid (mercaptoacetic acid)

$HS—CH_2—COOH$

Mol. wt. 92.12

May be used to block sulfhydryl groups. May cause burns and blistering of skin.

2-Thiolacetoxybenzanilide

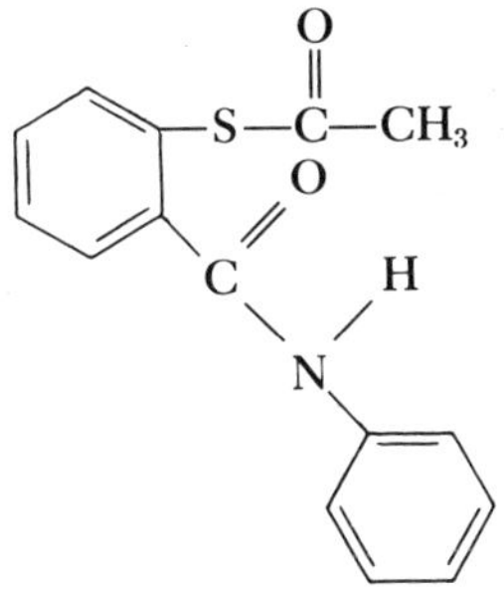

Mol. wt. 271.34
Used as a substrate in the Hatchett's brown method for demonstrating nonspecific esterase.

Thionin

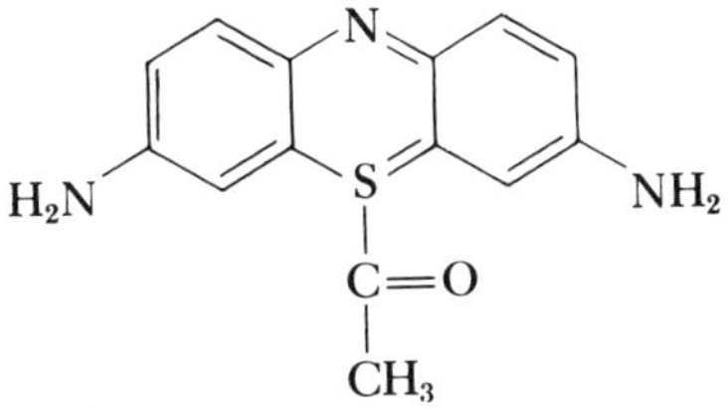

Mol. wt. 271.34   C.I. No. 52000
Used as a histological stain; may be reduced to a Schiff-type reagent.

Thionyl chloride

$SOCl_2$
Mol. wt. 118.98
A colorless liquid compound that hydrolyzes spontaneously in water to form SO_2 and HCl. It is sometimes used as the source of SO_2 in the preparation of Schiff's reagent.

Toluidin blue

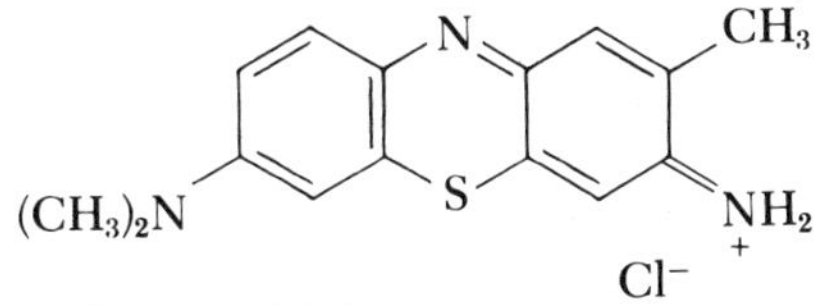

Mol. wt. 305.82   C.I. No. 52040
A thiazan dye with metachromatic properties.

Trichloroacetic acid

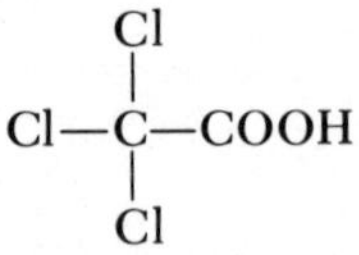

$$\begin{array}{c} Cl \\ | \\ Cl-C-COOH \\ | \\ Cl \end{array}$$

Mol. wt. 163.39

Corrosive crystals. A 1% solution in 80% ethanol has been recommended as a fixative when the tissues are intended for demonstrating tryptophan by the DMAB method.

Triethyl phosphate

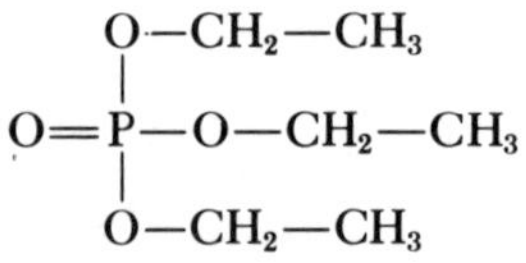

$$\begin{array}{c} O-CH_2-CH_3 \\ | \\ O=P-O-CH_2-CH_3 \\ | \\ O-CH_2-CH_3 \end{array}$$

Mol. wt. 182.16

Used as a solvent for fat stains.

Triglycerides

Three molecules of long-chain fatty acids are esterified with one molecule of glycerol to form a triglyceride molecule. Triglycerides are the natural substrates for lipase but cannot serve as its histochemical substrate.

Triphenyl tetrazolium chloride

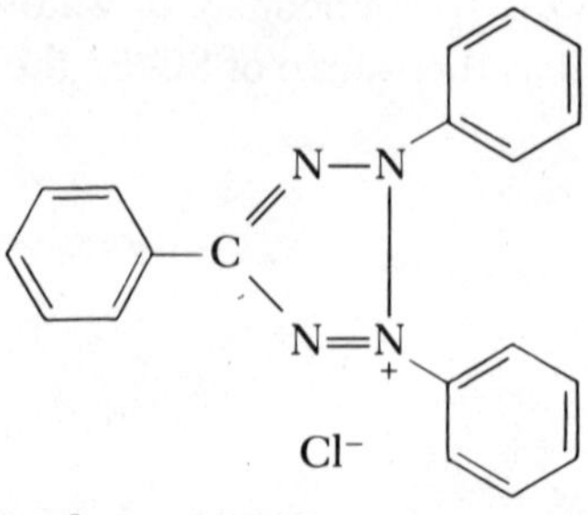

Mol. wt. 334.81

One of the first tetrazolium salts to be used for demonstrating dehydrogenase enzymes.

Tryptamine hydrochloride

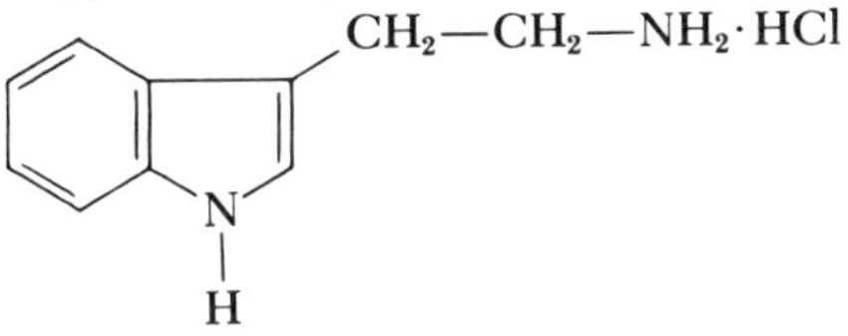

$CH_2-CH_2-NH_2 \cdot HCl$

Mol. wt. 196.68

Used as a substrate for monoamine oxidase.

Tween

Tweens are polyalcohols (such as sorbitol or mannitol) esterified with a long-chain fatty acid. They are classified according to length of the fatty acids. Tween 20 is a polyalcohol monolaurate; Tween 40 is a monopalmitate; Tween 60 is a monostearate; and Tween 80 is a monooleate. The remaining hydroxyl groups are esterified with polyethylene oxide. (Tween® is a trademark of Atlas Chemical Industries, Inc.)

Tyramine

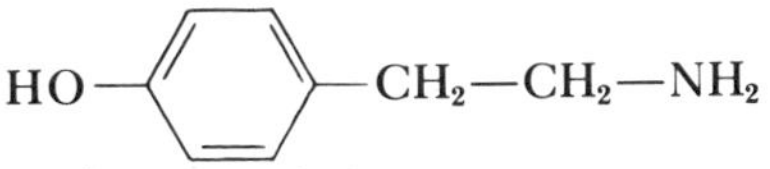

$HO-\langle\ \rangle-CH_2-CH_2-NH_2$

Mol. wt. 137.18

Used as a substrate for monoamine oxidase.

Uridine diphosphoglucose, disodium salt

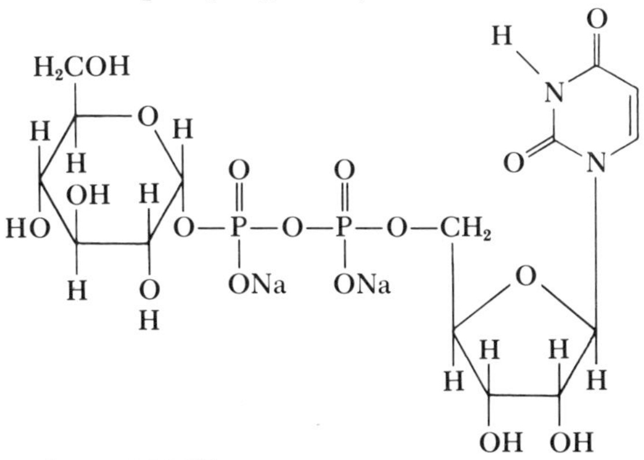

Mol. wt. 610.27

Used as the substrate for UDP-glucose dehydrogenase.

Uridine diphosphoglucuronic acid, disodium salt

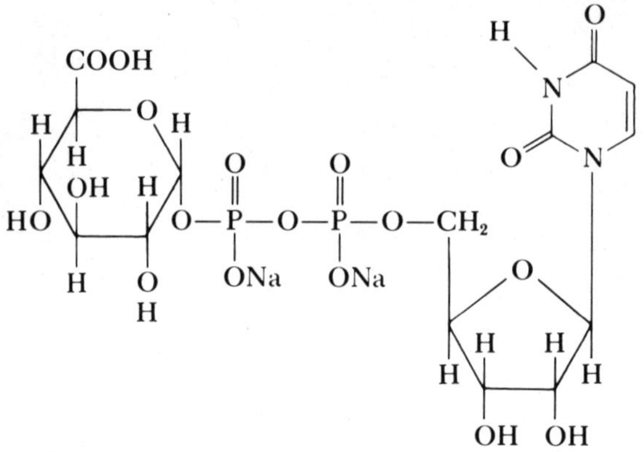

Mol. wt. 624.26

Reaction product of UDP-glucose dehydrogenase. Also used as the substrate for UDP-glucuronate-5-epimerase.

Uridine diphosphoiduronic acid, disodium salt

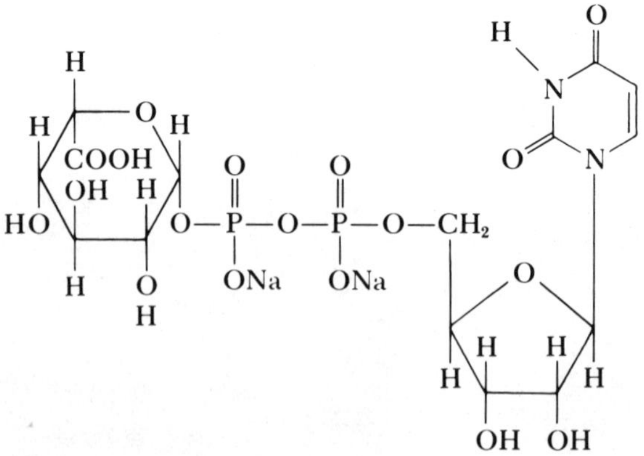

Mol. wt. 624.26

The reaction product of UDP-glucuronate-5-epimerase.

Veronal (barbitone; barbital)

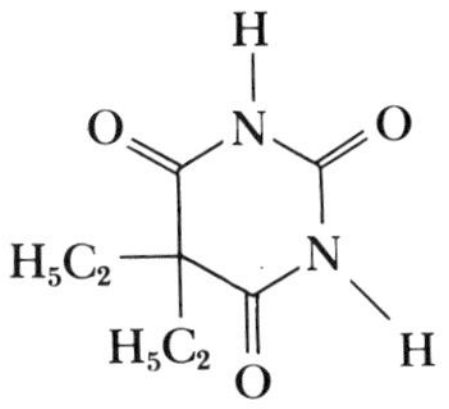

Mol. wt. 184.20

Used in preparing veronal (barbital) buffer.

Xylcne (xylol)

$C_6H_4(CH_3)_2$

(A mixture of ortho meta and para compounds)

Mol. wt. 106.07

Used as a solvent for paraffin and as a clearing agent for tissues.

INDEX

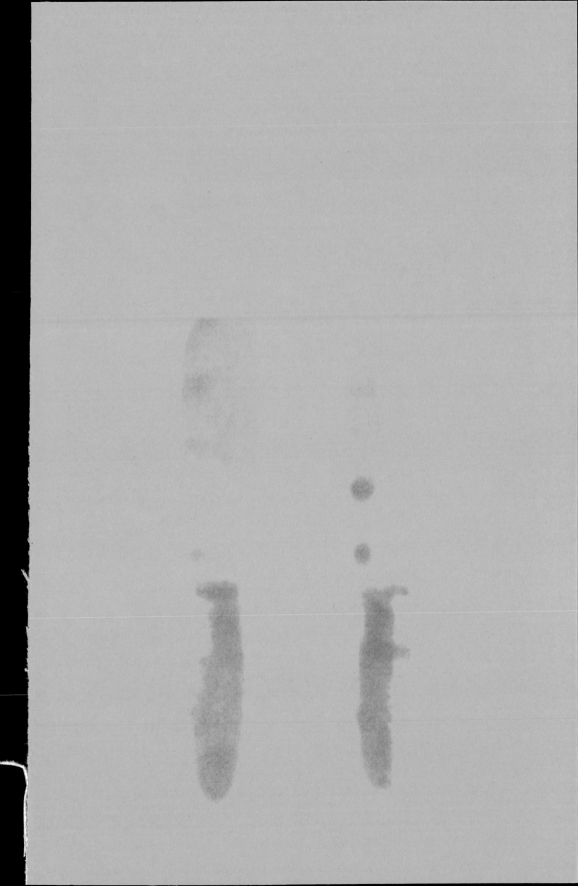